TOTAL LEAST SQUARES AND ERRORS-IN-VARIABLES MODELING

Total Least Squares and Errors-in-Variables Modeling

Analysis, Algorithms and Applications

edited by

Sabine Van Huffel

Department of Electrical Engineering,
Katholieke Universiteit, Leuven-Heverlee, Belgium

and

Philippe Lemmerling

Department of Electrical Engineering,
Katholieke Universiteit, Leuven-Heverlee, Belgium

SPRINGER-SCIENCE+BUSINESS MEDIA, B.V.

A C.I.P. Catalogue record for this book is available from the Library of Congress.

ISBN 978-90-481-5957-4 ISBN 978-94-017-3552-0 (eBook)
DOI 10.1007/978-94-017-3552-0

Printed on acid-free paper

Contents

Preface ix

Introduction to Total Least Squares and Errors-in-Variables Modeling 1
Sabine Van Huffel and Philippe Lemmerling

Part I Basic Concepts and Analysis in EIV Modeling

TLS and its Improvements by Semiparametric Approach 15
Shun-ichi Amari and Motoaki Kawanabe

Unifying Least Squares, Total Least Squares and Data Least Squares 25
Christopher C. Paige and Zdeněk Strakoš

Bounds for the Least Squares Residual Using Scaled TLS 35
Christopher C. Paige and Zdeněk Strakoš

Part II Total Least Squares Algorithms

Recent Developments in Rank Revealing and Lanczos Methods for Total Least Squares Related Problems 47
Ricardo D. Fierro and Per Christian Hansen

A Regularized Total Least Squares Algorithm 57
Hongbin Guo and Rosemary A. Renaut

The Parametric Quadratic Form Method for Solving TLS Problems with Elementwise Weighting 67
Amedeo Premoli and Maria Luisa Rastello

Part III Structured Total Least Squares Problems

Structured Total Least Squares: Analysis, Algorithms and Applications 79
Philippe Lemmerling and Sabine Van Huffel

Fast Structured Total Least Squares Algorithms via Exploitation of the Displacement Structure 93
Nicola Mastronardi, Philippe Lemmerling and Sabine Van Huffel

The Extended STLS Algorithm for Minimizing the Extended LS Criterion 107
Arie Yeredor

Part IV Nonlinear EIV Models and Statistical Estimators

Bayesian Smoothing for Measurement Error Problems 121
Scott M. Berry, Raymond J. Carroll and David Ruppert

On the Polynomial Measurement Error Model 131
Chi-Lun Cheng and Hans Schneeweiss

On Consistent Estimators in Nonlinear Functional EIV Models 145
Alexander Kukush and Silvelyn Zwanzig

On Consistent Estimators in Linear and Bilinear Multivariate Errors-in-Variables Models 155
Alexander Kukush, Ivan Markovsky and Sabine Van Huffel

Identification of Semi-Linear Models Within an Errors-in-Variables Framework 165
Rik Pintelon and Johan Schoukens

Cox's Proportional Hazards Model under Covariate Measurement Error 179
Thomas Augustin and Regina Schwarz

Part V EIV Modeling with Bounded Uncertainties

State-Space Estimation with Uncertain Models 191
Ali H. Sayed and Ananth Subramanian

Models for Robust Estimation and Identification 203
Shivkumar Chandrasekaran and Keith Schubert

Robust Solutions to Linear Approximation Problems Under Ellipsoidal Uncertainty 213
G. Alistair Watson

Part VI Curve Fitting

QR Factorization of the Jacobian in Some Structured Nonlinear Least Squares Problems 225
Åke Björck

Neural Minor Component Analysis and TLS 235
Giansalvo Cirrincione and Maurizio Cirrincione

On the Structural Line Segment Model 251
Ori Davidov, Alexander Goldenshluger and Roman Reidman

Model Fitting for Multiple Variables by Minimising the Geometric Mean Deviation 261
Chris Tofallis

Part VII EIV Estimation in System Identification

Perspectives on Errors-in-Variables Estimation for Dynamic Systems 271
Torsten Söderström, Umberto Soverini and Kaushik Mahata

Errors-in-Variables Filtering in Behavioural and State-Space Contexts 281
Roberto Guidorzi, Roberto Diversi and Umberto Soverini

Weighted Total Least Squares, Rank Deficiency
and Linear Matrix Structures 293
Bart De Moor

Frequency-Domain TLS and GTLS Algorithms
for Modal Analysis Applications 305
Peter Verboven, Patrick Guillaume and Eli Parloo

Part VIII EIV Estimation in Signal Processing

A New Decimative Spectral Estimation Method with
Unconstrained Model Order and Decimation Factor 321
Stavroula-Evita Fotinea, Ioannis Dologlou and George Carayannis

Modeling Audio with Damped Sinusoids Using
Total Least Squares Algorithms 331
*Werner Verhelst, K. Hermus, P. Lemmerling,
P. Wambacq and S. Van Huffel*

Real-Time TLS Algorithms in Gaussian
and Impulse Noise Environments 341
Da-Zheng Feng, Zheng Bao and Xian-Da Zhang

Part IX EIV Applications in Other Fields

Efficient Computation of the Riemannian SVD in
Total Least Squares Problems in Information Retrieval 353
Ricardo D. Fierro and Michael W. Berry

Constrained Total Least Squares for Color Image Reconstruction 365
Michael K. Ng, Nirmal K. Bose and Jaehoon Koo

Total Least Squares in Astronomy 375
Richard L. Branham, Jr.

TLS and Constrained TLS Neural Networks for Computer Vision 385
Giansalvo Cirrincione

Index 397

The **goal** of this book is to present an overview of the progress in Total Least Squares (TLS) and Errors-In-Variables (EIV) techniques in the last five years. This third book nicely *completes* my first two books, entitled "*The total least squares problem: computational aspects and analysis*" (authors: S. Van Huffel and J. Vandewalle, SIAM, 1991) and "*Recent advances in total least squares techniques and errors-in-variables modeling*" (editor: S. Van Huffel, SIAM, 1997). While the first book is entirely devoted to TLS, the second book presents the progress in TLS and in the broader field of errors-in-variables modeling from 1990 till 1996.

Since the appearance of these two books, many new TLS algorithms have been developed, such as rank-revealing, neural based and regularized TLS algorithms, as well as fast algorithms for solving TLS problems with elementwise weighting. A lot of attention was focused to structured TLS problems: fast and robust algorithms have been described. In particular for Toeplitz- or Hankel-structured problems, arising in many signal processing and system identification applications, a significant reduction in computation time is achieved by exploiting the displacement structure. New properties of EIV models in the presence of colored and nongaussian noise distributions, outliers and latency errors have been investigated, and new generalizations have been presented, in particular extensions to nonlinear and polynomial EIV models and to Cox's proportional Hazards model. The statistical properties of the estimators of the unknown model parameters are discussed and algorithms are derived for computing these estimators in a reliable way. In addition, various types of bounded uncertainties have been proposed in order to improve robustness of the estimators under various noise conditions and algorithms are outlined. Another hot topic that uses the TLS idea is orthogonal curve fitting, where one aims to minimize the orthogonal distances from the measurement points to a prescribed geometry: line, hyperplane, etc. New and faster algorithms for computing these geometries have been developed and analyzed. And last but not least, the number of applications in TLS and EIV modeling has increased exponentially, because of its emergence in new fields such as computer vision, image reconstruction, speech and audio processing, and its gain in popularity in fields as signal processing, modal and spectral analysis, system identification and astronomy. In the book, the use of TLS and errors-in-variables models in the most important application fields, such as signal processing and system identification, are overviewed and new algorithms that apply the TLS concept to the model characteristics used in that field are described. For example, in system identification the well-known Kalman filtering is extended to the errors-in-variables context in which noise on the inputs as well as on the outputs is taken into account thereby improving the filtering performance.

In response to this growing interest in TLS and EIV modeling by researchers and practitioners, I invited well-known experts from several disciplines to prepare a survey paper and present it at the third international workshop on TLS and EIV modeling held in Leuven, Belgium, August 27–29, 2001. These invited papers, representing two-thirds of the book, together with a selection of other presented contributions, are included in this book and yield a complete overview of the main scientific achievements since 1996 in TLS and errors-in-variables modeling. Not only computational issues, but also statistical, numerical, algebraic properties are described, as well as many new generalizations and applications. In this way, the topic is studied from different points of view as was also the case in the first two books. This *interdisciplinary* character points out the strongness of this book and enables the reader to get a complete overview of the state of the art of the research in TLS and errors-in-variables modeling as a whole.

All these methods and applications haven't been described in the first two books on TLS (SIAM, 1991 and 1997). Therefore, the new book nicely completes the earlier ones by updating the description of the state of the art on TLS since 1996. Being aware of the growing interest in these techniques by researchers and applied engineers, I strongly believe that this book will aid and stimulate users to apply the new techniques and models correctly to their own practical problems.

SABINE VAN HUFFEL

INTRODUCTION TO TOTAL LEAST SQUARES AND ERRORS-IN-VARIABLES MODELING

Bridging the Gap between Statistics, Computational Mathematics and Engineering

Sabine Van Huffel and Philippe Lemmerling
ESAT-SISTA, K.U.Leuven
Kasteelpark Arenberg 10
B-3001 Leuven-Heverlee, Belgium
{Sabine.Vanhuffel, Philippe.Lemmerling}@esat.kuleuven.ac.be

1. Introduction, formulation and state of the art until 1996

1.1 Introduction and problem formulation

The main purpose of this book is to present an overview of the progress in the last five years of a modeling technique which is known as Total Least Squares (TLS) in computational mathematics and engineering, and as Errors-In-Variables (EIV) modeling or orthogonal regression in the statistical community.

The basic motivation for TLS is the following. Let a set of multidimensional data points (vectors) be given. How can we obtain a linear model that explains the data? The idea is to modify all data points in such a way that some norm of the modification is minimized subject to the constraint that the modified vectors satisfy a linear relation. The origin of this basic insight can be traced back to the beginning of the previous century. It was rediscovered many times, often independently, mainly in the statistical and psychometric literature (where it is known as errors-in-variables or orthogonal regression and where its relation to principal component analysis and factor analysis was thoroughly investigated in the past half century). However, it is only in the last 20 years that the technique penetrated in scientific and engineering applications. One of the main reasons for its sudden popularity is the availability of efficient and numerically robust algorithms, in which the Singular Value

S. Van Huffel and P. Lemmerling (eds.), Total Least Squares and Errors-in-Variables Modeling, 1–11.

Decomposition (SVD) plays a prominent role. Another reason is the fact that TLS is really an applications oriented procedure. It is ideally suited for situations in which all data are corrupted by noise, which is almost always the case in engineering applications. In this sense it is a powerful extension of the classical Least Squares (LS) idea, which corresponds only to a partial modification of the data.

A comprehensive description of the state of the art on TLS from its conception up to the summer of 1990 and its use in parameter estimation has been presented in the monograph "*The total least squares problem: computational aspects and analysis*" by S. Van Huffel and J. Vandewalle (SIAM, 1991). While the latter book is entirely devoted to TLS, a second book, entitled "*Recent advances in total least squares techniques and errors-in-variables modeling*" (editor: S. Van Huffel, SIAM, 1997), presents the progress in TLS and in the broader field of errors-in-variables modeling from 1990 till 1996. Hence, this new volume nicely *completes* these first two books.

The problem of *linear parameter estimation* arises in a broad class of scientific disciplines such as signal processing, automatic control, system theory and in general engineering, statistics, physics, economics, biology, medicine, etc. It starts from a model described by a linear equation:

$$\alpha_1 x_1 + \cdots + \alpha_n x_n = \beta \tag{1}$$

where $\alpha_1, \ldots, \alpha_n$ and β denote the variables and $x = [x_1, \ldots, x_n]^T \in \mathbb{R}^n$ plays the role of a parameter vector that characterizes the specific system. A basic problem of applied mathematics is to determine an estimate of the *true* but *unknown* parameters from certain measurements of the variables. This gives rise to an *overdetermined* set of m linear equations $(m > n)$:

$$Ax \approx b \tag{2}$$

where the i-th row of the data matrix $A \in \mathbb{R}^{m\times n}$ and the vector of observations $b \in \mathbb{R}^m$ contain the measurements of the variables $\alpha_1, \ldots, \alpha_n$ and β respectively.

In the classical least squares approach the measurements A of the variables α_i (the lefthand side of (2)) are assumed to be free of error and hence, all errors are confined to the observation vector b (the righthand side of (2)). However, this assumption is frequently unrealistic: sampling errors, human errors, modeling errors and instrument errors may imply inaccuracies of the data matrix A as well. One way to take errors in A into account is to introduce *perturbations also in* A and consider the following TLS problem ($R(A)$ denotes the range of A):

Definition 1 (Total Least Squares problem). *Given an overdetermined set of m linear equations $Ax \approx b$ in n unknowns x. The total least squares problem seeks to*

$$\underset{[\widehat{A}\,\widehat{b}] \in \mathbb{R}^{m\times(n+1)}}{\text{minimize}} \quad \| [A\,b] - [\widehat{A}\,\widehat{b}] \|_F \tag{3}$$

$$\text{subject to} \quad \widehat{b} \in R(\widehat{A}) \tag{4}$$

Once a minimizing $[\widehat{A}\,\widehat{b}]$ is found, then any x satisfying

$$\widehat{A}x = \widehat{b} \tag{5}$$

is called a TLS solution.

1.2 Computation

In most problems, the TLS problem (3-5) has a unique solution which can be obtained from a simple scaling of the right singular vector of $[A\ b]$ corresponding to its smallest singular value. Extensions of this basic TLS problem to multidimensional TLS problems $AX \approx B$ having more than one righthand side vector, to problems in which the TLS solution is no longer unique or fails to have a solution altogether and to mixed LS-TLS problems that assume some of the columns of A to be error-free, are also considered in the first book on TLS (SIAM, 1991). In addition, it is shown how to speed up the TLS computations directly by computing the SVD only partially or iteratively if a good starting vector is available. More recent advances, e.g. recursive TLS algorithms, rank-revealing TLS algorithms, regularized TLS algorithms, TLS algorithms for large scale problems, etc., were presented during the second workshop on TLS and Errors-in-Variables Modeling (Leuven, August 21–24, 1996) and reviewed in the book, entitled "*Recent advances in total least squares techniques and errors-in-variables modeling*" (editor: S. Van Huffel, SIAM, 1997).

1.3 Properties

The TLS principle, as introduced in numerical analysis, is strongly related to the concept of EIV or measurement modeling in the field of statistics, studied by Adcock (1878) and Pearson (1902) already more than one century ago. These models with *only measurement error*, referred to as classical *errors-in-variables* (EIV) models, are characterized by the fact that the true values of the observed variables satisfy one or more unknown but exact linear relations of the form (1). In particular, in case of one underlying linear relation, we assume that the m

measurements in A, b are related to n unknown parameters x by :

$$A_0 x = b_0 \qquad A = A_0 + \Delta A \text{ and } b = b_0 + \Delta b \tag{6}$$

where $\Delta A, \Delta b$ represent the measurement errors and all rows of $[\Delta A; \Delta b]$ are independently and identically distributed with zero mean and covariance matrix $\mathcal{C}$, known up to a scalar multiple σ_ν^2. If additionally $\mathcal{C} = \sigma_\nu^2 I$ is assumed with I the identity matrix (i.e. Δa_{ij} and Δb_i are uncorrelated random variables with equal variance) and A_0 has full rank, then it can be proved that the TLS solution x_{TLS} of $Ax \approx b$ estimates the *true* parameter values x_0, given by $A_0^\dagger b_0$, *consistently*, i.e. x_{TLS} converges to x_0 as m tends to infinity. This property of TLS estimates does not depend on any assumed distributional form of the errors. Note also that the LS estimates are inconsistent in this case. In these cases, TLS gives better estimates than does LS, as confirmed by simulations. This situation may occur far more often in practice than is recognized. It is very common in agricultural, medical and economic science, in humanities, business and many other data analysis situations. Hence TLS should be a quite useful tool to data analysts. In fact, the keyrole and importance of LS in regression analysis is the same as that of TLS in EIV regression. Nevertheless, a lot of confusion exists in the fields of numerical analysis and statistics about the principle of TLS and its relation to EIV modeling. In particular, the name "Total Least Squares" is still largely unknown in the statistical community, while inversely the concept of EIV modeling did not penetrate sufficiently well in the field of computational mathematics and engineering. Roughly speaking, we can say that TLS is a special case of EIV variables estimation and, as such, TLS is reduced to a method in statistics but, on the other hand, TLS is used in other fields like signal processing, system identification, modal analysis, etc., where mainly the data modification idea is used and explained from a geometric point of view, independently from its statistical interpretation.

Let us now discuss some of the main properties of the TLS method. The properties of TLS are best understood by comparing them with those of LS. First of all, a lot of insight can be gained by comparing their analytical expressions, given by:

$$\text{LS:} \quad x_{LS} = (A^T A)^{-1} A^T b \tag{7}$$

$$\text{TLS:} \quad x_{TLS} = (A^T A - \sigma_{n+1}^2 I)^{-1} A^T b \tag{8}$$

with A of full rank and σ_{n+1} the smallest singular value of the augmented data matrix $[A\ b]$.

From a numerical analyst's point of view, these formulas tell us that the TLS solution is more ill-conditioned than the LS solution since it has a higher condition number. This implies that errors in the data more likely affect the TLS solution than the LS solution. This is particularly true under worst case perturbations. Hence, TLS can be considered as a kind of *de*regularizing procedure. However, from a statistical point of view, these formulas tell us that TLS is doing the right thing in the presence of independently and identically distributed (i.i.d.) equally sized errors since it removes (asymptotically) the bias by subtracting the error covariance matrix (estimated by $\sigma_{n+1}^2 I$) from the data covariance matrix $A^T A$.

Secondly, while LS minimizes a sum of squared residuals, TLS minimizes a sum of squared *normalized* residuals, expressed as follows:

$$\text{LS:} \qquad \min_x \|Ax - b\|^2 \tag{9}$$

$$\text{TLS:} \qquad \min_{z=\left[\begin{smallmatrix} x \\ -1 \end{smallmatrix}\right]} \frac{\|[A\ b]z\|^2}{\|z\|^2} = \min_x \frac{\|Ax - b\|^2}{\|x\|^2 + 1} \tag{10}$$

From a numerical analyst's point of view, we say that TLS minimizes the Rayleigh quotient. From a statistical point of view, we say that we normalize the residuals by multiplying them with the inverse of their covariance matrix (up to a scaling factor) in order to derive consistent estimates.

Other properties of TLS, which were studied in the field of numerical analysis, are its sensitivity in the presence of errors on all data. Differences between the LS and TLS solution are shown to increase when the ratio $\sigma_n([A\ b])/\sigma_n(A)$ (with $\sigma_n(X)$ the nth largest singular value of a matrix X) is growing. This is the case when the set of equations $Ax \approx b$ becomes less compatible, when the vector b is growing in length and when A tends to be rank-deficient. Assuming i.i.d. equally sized errors, the improved accuracy of the TLS solution compared to that of TLS is maximal when the orthogonal projection of b is parallel with the nth singular vector of A, corresponding to its smallest singular value $\sigma_n(A)$. Additional algebraic connections and sensitivity properties of the TLS and LS problem, as well as many more statistical properties of the TLS estimators, based on knowledge of the distribution of the errors in the data, are analyzed in the first two books on TLS (SIAM, 1991 and 1997).

1.4 Extensions

Furthermore, several extensions of the TLS problem have been investigated. In particular, the *mixed LS-TLS* problem formulation allows to extend consistency of the TLS estimator in EIV models, where some of the variables α_i are measured without error. The *data least squares* problem refers to the special case in which all variables except β are measured with error and was introduced in the field of signal processing by Dowling and DeGroat in the mid nineties. Whenever the errors are independent but unequally sized, *weighted TLS* problems should be considered using appropriate diagonal scaling matrices in order to maintain consistency. If additionally, the errors are also correlated then the *generalized TLS* problem formulation allows to extend consistency of the TLS estimator in EIV models, provided the corresponding error covariance matrix is known up to a factor of proportionality. More general problem formulations, such as *restricted TLS*, which also allow the incorporation of equality constraints, have been proposed, as well as equivalent problem formulations using other L_p norms and resulting in the so-called *Total L_p approximations*. The latter problems proved to be useful in the presence of outliers. Robustness of the TLS solution is also improved by adding regularization, resulting in the *regularized TLS* methods, or by exploiting the matrix structure (e.g. Toeplitz, Hankel) in the problem formulation, leading to the so-called *structured TLS* problems. The latter problems are also studied in more detail in this volume and fast algorithms are presented, by exploiting the matrix structure in the computations.

Finally, we mention the important extension to *nonlinear EIV* models, nicely studied in the book of Caroll, Ruppert and Stefanski (1995). In these models, the relationship between the variables α_i and β is assumed to be nonlinear. It is important to notice here that the close relationship between nonlinear TLS and EIV stops to exist. Indeed, consider the bilinear EIV model $AXB \approx C$, in which A, B, and C are noisy. Applying the TLS principle to this model leads to the following bilinear TLS problem:

$$\min_{\Delta A, \Delta B, \Delta C, X} \quad \|\Delta A\|_F^2 + \|\Delta B\|_F^2 + \|\Delta C\|_F^2 \tag{11}$$

$$\text{subject to} \quad (A - \Delta A)\, X\, (B - \Delta B) = C - \Delta C \tag{12}$$

However, solving this problem yields inconsistent estimates of X. As recently proved by Kukush, a consistent estimate can be obtained using the adjusted LS estimator (the full rank case is considered here for reasons of simplicity):

$$X_{ALS} = (A^T A - V_A)^{-1}(A^T C B^T)(B B^T - V_B)^{-1} \tag{13}$$

with $V_A = \mathbf{E}\widetilde{A}^T\widetilde{A}$ and $V_B = \mathbf{E}\widetilde{B}^T\widetilde{B}$ where $\mathbf{E}$ denotes the expected value and $\widetilde{A}$ and $\widetilde{B}$ represent the errors on A and B respectively. Corrections for small samples are derived in this volume and shown to give superior performance for small sized problems.

2. New evolutions since 1996: outline of the book

Since the appearance of the first two books on TLS (SIAM, 1991 and 1997), many new TLS algorithms have been developed, such as rank-revealing, neural based and regularized TLS algorithms, as well as new algorithms for solving TLS problems with elementwise weighting. A lot of attention was focused to structured TLS problems: fast and robust algorithms have been described. In particular for Toeplitz- or Hankel-structured problems, arising in many signal processing and system identification applications, a significant reduction in computation time is achieved by exploiting the displacement structure. New properties of EIV models in the presence of colored and nongaussian noise distributions, outliers and latency errors have been investigated, and new generalizations have been presented, in particular extensions to nonlinear and polynomial EIV models and to Cox's proportional Hazards model. The statistical properties of the estimators of the unknown model parameters are discussed and algorithms are derived for computing these estimators in a reliable way. In addition, various types of bounded uncertainties have been proposed in order to improve robustness of the estimators under various noise conditions and algorithms are outlined. Another hot topic that uses the TLS idea is orthogonal curve fitting, where one aims to minimize the orthogonal distances from the measurement points to a prescribed geometry: line, hyperplane, etc. New and faster algorithms for computing these geometries have been developed and analyzed. And last but not least, the number of applications in TLS and EIV modeling has increased exponentially, because of its emergence in new fields such as computer vision, image reconstruction, speech and audio processing, and its gain in popularity in fields as signal processing, modal and spectral analysis, system identification and astronomy.

In this volume, the use of TLS and errors-in-variables models in the most important application fields, such as signal processing and system identification, are surveyed and new algorithms that apply the TLS concept to the model characteristics used in those fields are described. For example, in system identification the well-known Kalman filtering is extended to the errors-in-variables context in which noise on the inputs as well as on the outputs is taken into account thereby improving the filtering performance.

All these methods and applications haven't been described in the first two books on TLS (SIAM, 1991 and 1997). Therefore, the new volume nicely completes the earlier ones by updating the description of the state of the art on TLS since 1996. Moreover, the new volume broadens the view on errors-in-variables modeling by also discussing other useful types of errors-in-variables models with bounded uncertainties, together with the algorithms to compute the corresponding model parameters, and by discussing new application fields such as modal analysis, speech processing, information retrieval, computer vision and galactic kinematics.

All contributions evolve around the following nine themes, corresponding to the nine parts of this book:

Part I:	Basic concepts and analysis in errors-in-variables modeling
Part II:	Total least squares algorithms
Part III:	Structured total least squares problems
Part IV:	Nonlinear errors-in-variables models and statistical estimators
Part V:	Errors-in-variables modeling with bounded uncertainties
Part VI:	Curve fitting
Part VII:	Errors-in-variables estimation in system identification
Part VIII:	Errors-in-variables estimation in signal processing
Part IX:	Errors-in-variables applications in other fields

In **Part I**, the TLS estimator is analyzed in the framework of information geometry of semiparametric statistical models and in a scaled TLS framework, unifying LS, TLS and data LS problems. **Part II** overviews recent progress in TLS algorithms, based on the implicitly restarted Lanczos and using the rank-revealing orthogonal decomposition. In addition, new algorithms are presented for solving ill-conditioned TLS problem with Tikhonov regularization and TLS problems with independent but elementwise differently sized errors. **Part III** focuses on structured TLS problems: different problem formulations are analyzed, various computational algorithms are outlined and shown to reduce the computation time by exploiting the displacement structure. In addition, it is shown how to extend the problem and solve it, if latency or modeling errors are included. In **Part IV**, various types of nonlinear EIV models, including bilinear, polynomial, nonlinear functional, semi-linear and Cox's proportional Hazards models, are considered and consistent estimators are derived. In addition, the advantages of using Bayesian approaches are investigated. **Part V** presents new EIV models with various types of bounded uncertainties, including ellipsoidal and parametric uncertainties, based on robust regularized LS formulations. It is shown how these models improve the robustness of the TLS estimator and how estimators can be computed efficiently. **Part VI** focuses on curve fitting, including orthogonal distance regression or geometric fitting, where we aim to minimize the sum of squared orthogonal distances from the data points to a prescribed geometry. Fast algorithms for orthogonal distance regression, making use of the sparsety structure in the QR factorization of the associated Jacobian, as well as new algorithms, based on neural computing, for minor component analysis (which leads directly to the

TLS solution) and for structural line segment fitting, using the method of moments estimator, are investigated. In addition, the properties of a new approach, called least volume model fitting, are investigated.

Finally, Parts VII, VIII and IX review TLS applications in various applied fields to illustrate the improvements obtained so far with TLS based estimators. First of all in **Part VII**, applications in the field of system identification and input-output system modeling are discussed. Various methods for identifying dynamic EIV systems are reviewed and it is shown how to extend the Kalman filter to an EIV context in case the inputs as well as the outputs are contaminated by noise. **Part VIII** focuses on applications in signal processing, in particular in-vivo magnetic resonance spectroscopy and audio coding, and presents new state-space based methods for spectral estimation, using decimation and a multichannel approach. In addition, it is shown how to extend the least mean squares (LMS) algorithm to the EIV context for use in adaptive signal processing and various noise environments. Finally, **part IX** discusses applications in other fields, including information retrieval, image reconstruction, astronomy, and computer vision. It is shown how the TLS methods and its generalizations, including structured, regularized and generalized TLS, can be successfully applied.

This list of applications of TLS and EIV modeling is certainly not exhaustive and clearly illustrates the increased interest over the past five years. It is hoped that this book will aid and stimulate the reader to apply TLS and EIV modeling in his or her own applications and problems.

3. Using the Book

This book presents a state of the art of the progress in TLS and EIV modeling over the last five years and consists of invited overview papers, as well as selected contributed papers, presented at the third international workshop on TLS and EIV modeling held in Leuven, Belgium, August 27–29, 2001. All contributions are self-contained and can be read separately. They are written for an interdisciplinary audience, thereby avoiding to be too technical or too specialized.

To understand the book, the reader is assumed to have basic knowledge of the TLS problem and EIV modeling. Therefore, the reader is encouraged to read first the book on TLS by S. Van Huffel and J. Vandewalle (SIAM, 1991), although this is not a strong requirement. In addition, the reader is assumed to have basic knowledge of linear algebra, matrix computations and elementary statistics.

All used notations are clearly defined in each contribution, except for the following very common notations in linear algebra which are globally defined here for the whole book:

- A matrix is always denoted by a capital letter, e.g. A. The corresponding lowercase letter with the subscript i and ij refers to the i-th column and (i,j)-th entry respectively, e.g. a_i, a_{ij}.
- A vector is represented by a lowercase letter, e.g. b. The individual components are denoted with single subscripts, e.g. b_i.
- The superscripts ^{T}, ^{H}, and † denote respectively the transpose, conjugate transpose, and Moore-Penrose pseudo-inverse of a vector or matrix.
- The 2-norm or Euclidean norm of a vector s is denoted by $\|s\|_2$ or, simply, by $\|s\|$, while $\|S\|_2$ or $\|S\|$ denotes the 2-norm of a matrix S. The Frobenius norm of a matrix S is denoted by $\|S\|_F$, while the L_p-norm with $1 \leq p \leq \infty$ is denoted by $\|S\|_p$.
- If S is an $m \times n$ matrix and diagonal, then write

  $$S = \text{diag}(\sigma_1, \ldots, \sigma_p), \quad \text{or} \quad S = \text{diag}(\sigma), \qquad p = \min\{m, n\},$$

  i.e. $s_{ij} = 0$ whenever $i \neq j$ and $s_{ii} = \sigma_i$ for $i = 1, \ldots, p$, where $\sigma = [\sigma_1, \ldots, \sigma_p]^T$.
- The $m \times m$ identity matrix is denoted by I_m or, simply, by I.
- $\mathbf{E}x$ denotes the expected value of x.
- The kth unit vector, with kth component equal to one and zeroes elsewhere, is denoted by e_k.
- Denote the singular value decomposition (SVD) of an $m \times n$ matrix S by

  $$S = U \Sigma V^T$$

  $$\begin{aligned} U &= [u_1, \ldots, u_m], \quad u_i \in \mathbb{R}^m, \ U^T U = I_m, \\ V &= [v_1, \ldots, v_n], \quad v_i \in \mathbb{R}^n, \ V^T V = I_n, \\ \Sigma &= \text{diag}(\sigma_1, \ldots, \sigma_p), \ p = \min\{m, n\} \ \sigma_1 \geq \cdots \geq \sigma_p \geq 0. \end{aligned}$$

  The σ_i are the singular values of S. The vectors u_i and v_i are respectively the i-th left and right singular vector of S. If differentiation is made between the SVD of several matrices, accents or primes are added to each symbol in the SVD.

Acknowledgments

It is our pleasure to acknowledge the assistance of many well-known contributors responsible for the success of this book and the third international workshop on TLS and EIV modeling. To all these authors and to the workshop participants we express our warmest thanks. Furthermore, we express our feelings of thanks to our research group ESAT-SISTA for the help during the workshop. In particular, we are indebted to Ida Tassens for her invaluable assistance in organizing the workshop and to Ivan Markovsky for his careful help in editing this volume. In addition, we would like to thank the sponsors of our workshop at which all contributions in this book have been presented: the Fund for Scientific Research — Flanders (FWO) and its Scientific Research Community on numerical methods for mathematical modeling, the European Association of Signal Processing (EURASIP), and the Katholieke Universiteit Leuven. We hope that this forum of interdisciplinary communication has encouraged the participants to broaden their perception on TLS and EIV modeling.

Finally, we wish to thank Kluwer Academic Publishers, in particular the editing staff, for their assistance in the publication of this book.

Sabine Van Huffel and Philippe Lemmerling

Part I

BASIC CONCEPTS AND ANALYSIS IN ERRORS-IN-VARIABLES MODELING

TLS AND ITS IMPROVEMENTS BY SEMIPARAMETRIC APPROACH

Shun-ichi Amari
RIKEN Brain Science Institute
Wako-shi, Hirosawa 2-1,
Saitama 351-0198 Japan
amari@brain.riken.go.jp

Motoaki Kawanabe
The IDA group, Fraunhofer FIRST
Kekuléstrasse 7,
12489 Berlin, Germany
nabe@first.gmd.de

Abstract The total least squares method seems to give a good consistent estimator in the linear error-in-variables model. However, this is not the optimal one. We give a simple adaptive method of improving the TLS estimator. The theory is based on information geometry of semiparametric statistical models, and is applicable to many other problems. Its intuitive introduction is also given.

Keywords: information geometry, semiparametric approach, linear errors-in-variables, estimating function, total least squares, structural relationship.

1. Introduction

Let us consider a very simple statistical problem of estimating the coefficient of a linear relation in two random variables x and y. We assume a linear relation

$$y = \theta x, \tag{1}$$

where θ is the parameter to be estimated. Let us use a number of observations $(x_1, y_1), (x_2, y_2), \ldots, (x_n, y_n)$, where all of these values are contaminated by noise. Let ξ_i be the true value for x_i, which we do not know. Then the true values of y_i are $\theta\xi_i$. By introducing noise, we can

S. Van Huffel and P. Lemmerling (eds.), Total Least Squares and Errors-in-Variables Modeling, 15–24.

write the following relations

$$x_i = \xi_i + n_i \tag{2}$$

$$y_i = \theta\xi_i + n_i' \tag{3}$$

Here, we assume that n_i and n_i' are independent random variables subject to the Gaussian distribution with mean 0 and variance σ^2.

This is a simple special problem, in which we assumed that the relation of y to x is linear, passing through the origin. We also assumed that the variances of the noise are the same for both x and y. In order to fit this assumption, we need to preprocess the original observations by changing scales of x and y depending on the variances of the noise in x and y.

Given data $(x_1, y_1), \ldots, (x_n, y_n)$, we can plot them in the x-y plane. The problem is to obtain the best linear fit. However, it is not easy to obtain the best estimator of θ. This is a typical example of the Neyman-Scott problem [8], where the number of unknown parameters $(\theta; \xi_1, \ldots, \xi_n)$ increases with the number n of observations. In such a situation, the maximum likelihood estimator is not necessarily consistent nor efficient. One may formulate the problem in the framework of a semiparametric statistical model [5, 6], which includes a nuisance parameter of infinite degrees of freedom or of functions.

The simplest idea is to use ordinary least squares, that minimizes the squared error from the regressor $y = \theta x$,

$$E_1(\theta) = \sum (y_i - \theta x_i)^2 . \tag{4}$$

The solution is easily given by

$$\hat{\theta}_1 = \frac{\sum x_i y_i}{\sum x_i^2}. \tag{5}$$

However, this does not take errors in variables x_i into account, and the estimator $\hat{\theta}_1$ is not even consistent, that is, when n tends to infinity, $\hat{\theta}_1$ does not converge to the true θ.

Total least squares (TLS) takes the noise in both variables into account, and minimizes the sum of the squares of the orthogonal projection errors of data points to the regressor,

$$E_2(\theta) = \sum \frac{(y_i - \theta x_i)^2}{1 + \theta^2}. \tag{6}$$

By differentiating this with respect to θ, the estimator $\hat{\theta}_2$ is obtained by solving

$$\sum (y_i - \theta x_i)(x_i + \theta y_i) = 0 \tag{7}$$

We can prove that this $\hat{\theta}_2$ is a consistent estimator, converging to the true θ in the order of $1/\sqrt{n}$ as n increases.

Another idea is to maximize the likelihood with respect to all the unknown variables $\theta, \xi_1, \ldots, \xi_n$. By eliminating $\xi_1, \ldots, \xi_n$, we have the same equation as (7), so that $\hat{\theta}_2$ is the maximum likelihood estimator (mle) in the case of Gaussian noise.

By adding all x_i and all y_i and taking their ratio, we have a simple primitive estimator

$$\hat{\theta}_3 = \frac{\sum y_i}{\sum x_i}. \tag{8}$$

This is also a consistent estimator. Then, an interesting question arises: which is better, $\hat{\theta}_2$ or $\hat{\theta}_3$? Or does there exist an estimator better than both? More generally, what is the best asymptotic estimator? The present paper addresses these questions in the linear regression case, which are derived from the point of view of information geometry [3, 4], by using the differential geometrical framework in statistics [1].

2. Semiparametric statistical model and estimating function

We formulate this problem as a semiparametric statistical model. The probability density of x and y, when ξ and θ are given, is

$$q(x, y; \theta, \xi) = \frac{1}{2\pi\sigma^2} \exp\left[-\frac{1}{2\sigma^2}\left\{(x-\xi)^2 + (y-\theta\xi)^2\right\}\right]. \tag{9}$$

We assume that $\xi_1, \ldots, \xi_n$ are also n independent realizations of a random variable ξ which is subject to an unknown probability distribution $s(\xi)$. Then, (x, y) are regarded as iid observations from the distribution

$$p(x, y; \theta, s) = \int q(x, y; \theta, \xi) s(\xi) d\xi. \tag{10}$$

The probability model (10) includes two unknown parameters, θ and s. Here, θ is the parameter to be estimated, called the parameter of interest or the structural parameter, whereas s is called the incidental or nuisance parameter. However, this $s(\xi)$ is a function having infinite degrees of freedom. Such a statistical model is called a semiparametric statistical model [5]. The Neyman-Scott problem belongs to this class.

We use unbiased estimating functions [7] to solve the problem. An unbiased estimating function $f(x, y; \theta)$ is a function of (x, y) and θ, but it does not include the nuisance parameter s, such that its expectation is 0 whatever s is,

$$\mathbf{E}_{\theta,s}\left[f(x, y; \theta)\right] = 0, \tag{11}$$

where $\mathbf{E}_{\theta,s}$ denotes expectation with respect to $p(x, y; \theta, s)$. In order to avoid trivial functions, we require one more condition, for $\theta \neq \theta'$,

$$\mathbf{E}_{\theta',s}\left[f(x, y; \theta)\right] \neq 0. \tag{12}$$

When an unbiased estimating function exists, by replacing the expectation (11) by the sample mean, we obtain the following estimating equation

$$\sum_{i=1}^{n} f\left(x_i, y_i; \theta\right) = 0. \tag{13}$$

Since f does not include s, we can easily solve the equation. Moreover, the estimator $\hat{\theta}$ is consistent whatever $s(\xi)$ is. The squared error of the estimator is evaluated asymptotically by

$$\mathbf{E}\left[\left(\hat{\theta} - \theta_0\right)^2\right] \approx \frac{1}{n} \frac{\mathbf{E}\left[\left\{f\left(x, y; \theta_0\right)\right\}^2\right]}{\left\{\mathbf{E}\left[\frac{\partial}{\partial\theta} f\left(x, y; \theta_0\right)\right]\right\}^2}, \tag{14}$$

where θ_0 is the true parameter and expectation is taken with respect to the true distribution $p\left(x, y; \theta_0, s\right)$ [7].

Are there unbiased estimating functions in the present problem? What is the optimal one, provided there exist many? To answer these problems, let us consider a function of the form

$$f(x, y; \theta) = (y - \theta x) h(x + \theta y), \tag{15}$$

where $h(t)$ is an arbitrary function. We can easily prove that this is an unbiased estimating function for any h, because of

$$\mathbf{E}_{\theta,s}\left[(y - \theta x) h(x + \theta y)\right] = 0. \tag{16}$$

This is proved by changing the variables from (x, y) to $z = y - \theta x$ and $t = x + \theta y$. They are independent random variables, where $\mathbf{E}[z] = 0$, $\mathbf{E}[t] = \xi\left(1 + \theta^2\right)$, and their variances are $\left(1 + \theta^2\right)\sigma^2$. The TLS estimator is given by putting

$$h(t) = t, \tag{17}$$

while the total average estimator (8) is derived from the estimating function of $h(t) = 1$. On the other hand, the ordinary least squares estimator is given from

$$f(x, y; \theta) = x(y - \theta x) \tag{18}$$

but this function f is not unbiased, that is, it does not satisfy (11). Hence, the derived ordinary least squares estimator is not consistent.

We can prove that the above class (15) of estimating functions are admissible, implying that, for any estimating functions, there exists one in the class that gives a better or at least equivalent estimator, for some s. Obviously, the best estimating function depends on the unknown s. When s is known, the problem reduces to the ordinary statistical estimation problem, and the mle is one of the asymptotically best estimators, that is, Fisher efficient.

3. Adaptive choice of estimating function

It is computationally not easy to find the optimal estimating function from the observed data, because we need to estimate the unknown density function $s(\xi)$. Instead, we consider the following one parameter family of estimating functions,

$$f_c(x, y; \theta) = (y - \theta x)(x + \theta y + c) \tag{19}$$

parameterized by c. We call the estimator derived from f_c the c-estimator. When $c = 0$, this gives the total least squares estimator, and when c goes to infinity, it gives the total average estimator.

We calculate the asymptotic error of the c-estimator from (14). We have, after tedious calculations by using z and t,

$$\mathbf{E}\left[\{f_c(x, y; \theta)\}^2\right] = \sigma^2 \left(1 + \theta^2\right) \left[\sigma^2 \left(1 + \theta^2\right) + c^2 + 2c \left(1 + \theta^2\right) \bar{\xi} + \left(1 + \theta^2\right)^2 \bar{\xi^2}\right] \tag{20}$$

$$\mathbf{E}\left[\partial_\theta f_c(x, y; \theta)\right] = -\left[c\bar{\xi} + \left(1 + \theta^2\right) \bar{\xi^2}\right], \tag{21}$$

where $\bar{\xi}$ and $\bar{\xi^2}$ are the expectations of ξ and ξ^2 with respect to $s(\xi)$. Hence the asymptotic error (14) of the c-estimator depends on the unknown $s(\xi)$ only through the expectation $\bar{\xi}$ and the variance V_ξ of ξ. Moreover, the optimal c, which gives the optimal estimator in the class of c-estimators, is

$$c^* = \frac{\sigma^2 \bar{\xi}}{V_\xi}, \tag{22}$$

It is easy to obtain the estimators of $\bar{\xi}$ and $V_\xi = \bar{\xi^2} - \left(\bar{\xi}\right)^2$ as

$$\hat{\xi} = \frac{1}{n} \sum x_i, \tag{23}$$

$$\hat{\xi^2} = \frac{1}{n} \sum x_i^2 - \sigma^2. \tag{24}$$

When the noise variance σ^2 is unknown, it can easily be estimated from the data by using any consistent estimators and evaluating the error of the data points.

The present analysis shows that the TLS estimator ($c = 0$) is not optimal. It is a good one, when the noise variance is small, or when the average of $\bar{\xi}$ is small. However, when they are large, or when V_ξ is small, the simple total average estimator ($c = \infty$) is better. The c^*-estimator is better than both of them.

The c^*-estimator is not the optimal one. When ξ is subject to the Gaussian distribution, we can prove that this is the optimal estimator. Otherwise, it is not. In such a case, we may consider a more complicated family of estimating functions, for example

$$h(t) = t^2 + c_1 t + c_2 \tag{25}$$

which uses higher order moments of ξ. However, improvement by (25) is usually not remarkable, so that the simple c^*-estimator is sufficient for practical applications.

4. Extension to multivariate case

In the multivariate case of $y = \boldsymbol{\theta} \cdot \boldsymbol{x}$,

$$\boldsymbol{x}_i = \boldsymbol{\xi}_i + \boldsymbol{n}_i, \quad y_i = \boldsymbol{\theta} \cdot \boldsymbol{x}_i + n_i', \tag{26}$$

we have the vector $\mathbf{c}$-estimating function

$$\boldsymbol{f}_c(\boldsymbol{x}, y; \boldsymbol{\theta}) = (y - \boldsymbol{\theta} \cdot \boldsymbol{x})(\boldsymbol{x} + y\boldsymbol{\theta} + \mathbf{c}). \tag{27}$$

The $\mathbf{c}$-estimator is given by solving

$$\sum (y_i - \boldsymbol{\theta} \cdot \boldsymbol{x}_i)\,(x_i + y_i\boldsymbol{\theta} + \mathbf{c}) = 0. \tag{28}$$

The optimal vector $\mathbf{c}^*$ is given by

$$\mathbf{c}^* = \sigma^2 V_\xi^{-1} \bar{\boldsymbol{\xi}}, \tag{29}$$

where V_ξ is the covariance matrix of $\boldsymbol{\xi}$. The optimal $\mathbf{c}^*$ can be estimated from the data.

5. Information geometry of estimating functions

We have introduced a simple family $f_c(x, y; \theta)$ of unbiased estimating functions and adaptive choice of c, in the present simple case of estimating linear relations. We have shown that there is an estimator better than TLS in the case of errors in variables.

However, many problems remain: How can one find unbiased estimating functions, or does an unbiased estimating function really exist? When they exist, what is the best estimating function? How can we apply the adaptive method of choosing a good estimating function in the general case?

These questions are answered by information geometry of estimating functions in semiparametric statistical models in [3]. Information geometry [1, 4] is a differential geometrical theory of the manifold of probability distributions, and is applicable to a wide area of stochastic phenomena. We explain hereafter information geometry of semiparametric statistical models in a very intuitive way.

Let us consider a semiparametric statistical model $M = \{p(x; \theta, s)\}$, where x is a random variable (it is a pair (x, y) in the case of a linear relation), θ is a parameter of interest, and s is the nuisance parameter of function degrees of freedom. The model M is a submanifold included in the set $S = \{p(x)\}$ of all the probability density functions which are mutually absolutely continuous. It is not easy to consider the exact mathematical framework of differential geometry of the function space of probability distributions. See Pistone and Rogantin [9] for exact treatment. We only describe intuitive explanations.

We first consider the manifold of all the probability density functions $p(x) > 0$. Let ε be a small positive number, and let $r(x)$ be a small deviation of $p(x)$ in the direction of $r(x)$, giving $p(x)\{1 + \varepsilon r(x)\}$. Then, r satisfies

$$E_p[r(x)] = 0 \tag{30}$$

Let T_p be the set of all such r . This is the set of all the directions of deviations of p and is hence regarded as the tangent space of S at p. We introduce the inner product of $r_1(x)$ and $r_2(x)$ by

$$\langle r_1, r_2 \rangle = E_p[r_1(x) r_2(x)] . \tag{31}$$

Then, S becomes a Hilbert space with respect to this inner product.

The semiparametric model M is a submanifold in S. At a point $p = p(x; \theta, s)$, the tangent space T_p includes

$$u(x; \theta, s) = \frac{d}{d\theta} \log p(x; \theta, s) \tag{32}$$

which is the direction due to a slight change in θ, because we have

$$p(x; \theta + \varepsilon, s) = p(x; \theta, s)\{1 + \varepsilon u(x; \theta, s)\} . \tag{33}$$

This is called the score function. We can also consider a change in the nuisance function. Let $k(\xi, t)$ be a curve parameterized by t of the nuisance distribution $s(\xi)$, where $k(\xi, 0) = s(\xi)$. Then,

$$v(x; \theta, k) = \frac{d}{dt} \log p\{x; \theta,\ k(\xi, t)\} \tag{34}$$

represents the direction of change in s along the function $k(\xi, t)$. The set of $v[k]$ is an infinite-dimensional space, because there are infinitely many functions $k(\xi, t)$. The set of all such functions forms a subspace of T_p called the nuisance subspace T_p^N. The tangent space is a direct sum of the subspace T_p^U spanned by the score function u, the nuisance tangent space T_p^N, and the subspace T_p^A orthogonal to these, called the ancillary subspace,

$$T_p = T_p^U \oplus T_p^N \oplus T_p^A. \tag{35}$$

Let us first consider a simple special case, where the nuisance parameter $s(\xi)$ is specified by a finite number of parameters $\boldsymbol{\eta} = (\eta_1, \ldots, \eta_m)$. The statistical model $M' = \{p(x; \theta), s(\boldsymbol{\eta})\}$ includes only a finite number of parameters $(\theta, \boldsymbol{\eta})$. Hence, the joint maximum likelihood estimator $(\hat{\theta}, \hat{\boldsymbol{\eta}})$ is asymptotically consistent and Fisher efficient. In such a case, let $u^E(x, \theta, \boldsymbol{\eta})$ be the efficient score which is the projection of the score $u(x)$ to the subspace orthogonal to T_p^N. Then,

$$g^E = \langle u^E, u^E \rangle \tag{36}$$

gives the efficient Fisher information, and the best asymptotic estimator satisfies

$$\mathbf{E}\left[\left(\hat{\theta} - \theta_0\right)^2\right] \approx \frac{1}{ng^E}. \tag{37}$$

The estimator is obtained from the estimating equation

$$\sum u^E (x_i, \theta) = 0 \tag{38}$$

provided $s(\xi, \boldsymbol{\eta})$ is estimated and the projection is given here.

In the general case, the true nuisance function $s(\xi)$ does not belong to the class $\{s(\xi, \boldsymbol{\eta})\}$. In this case, u^E is not in general an unbiased estimating function. Hence, when s is misspecified, the estimator is not necessarily unbiased nor efficient. It is important to find unbiased estimating functions. We can prove that an unbiased estimating function $f(x, \theta)$, satisfying

$$\mathbf{E}_{\theta, s}\left[f(x, \theta)\right] = 0 \tag{39}$$

for all s should be orthogonal to the nuisance subspace at any point (θ, s). In order to elucidate the orthogonality conditions common to all s, we define the two shifts operators of a tangent vector:

$$\overset{(e)}{\prod}{}_s^{s'} r(x) = r(x) - \mathbf{E}_{\theta, s'}\left[r(x)\right] \tag{40}$$

called the e-parallel shift of r from s to s', and

$$\overset{(m)}{\prod}{}_{s}^{s'} = \frac{p(x;\theta,s)}{p(x;\theta,s')}r(x) \tag{41}$$

called the m-parallel shift of r from s to s'. The important fact is that the two parallel shifts keep the inner product invariant,

$$\left\langle \overset{(e)}{\prod}{}_{s}^{s'} r_1, \overset{(m)}{\prod}{}_{s}^{s'} r_2 \right\rangle_{\theta,s'} = \langle r_1(x), r_2(x)\rangle_{\theta,s'} \,. \tag{42}$$

This is given rise to by the duality of affine connections in information geometry.

The efficient score calculated at (θ, s) is in general not orthogonal to the nuisance tangent space at different (θ, s'). We want to make it orthogonal to any s. We then have an unbiased estimating function. By using the parallel shift, if we shift m-parallelly all the tangent vectors of the nuisance tangent spaces $T^N_{\theta,s'}$ at all s' to s, and then project the score function at (θ, s) to the subspace orthogonal to all such shifted nuisance subspaces, we obtain an unbiased estimating function. Its e-parallel shift is also an unbiased estimating function.

By such consideration, we obtain all the estimating functions when they exist. The optimal one is the projected score $u^I(x;\theta,s)$ of $u(x;\theta,s)$, when the true nuisance parameter is s. It is important that, even when we misspecify $s(\xi)$, it is still an unbiased estimating function, giving a good estimator. The efficient score $u^E(x;\theta,s)$ does not have this property.

In some special cases, the m-parallel shift of $T^N_{\theta,s'}$ from s' to s is the same as $T^N_{\theta,s}$. In such a case, the efficient score u^E itself is an unbiased estimating function. The misspecification of s is not serious in such a case, and we can use any rough estimator of s. The mixture model in general has such a nice property, including the present case [6].

In general, given a parametric family of unbiased estimating function $f(x,\theta,\boldsymbol{\eta})$, by choosing an adequate $\boldsymbol{\eta}$, we have a good estimator. To this end, we use a simple parametric model of the nuisance function $s(\xi,\boldsymbol{\eta})$. The family usually does not include the true nuisance function $s(\xi)$. It does not in general give a good result, if we use the simple statistical model $M = \{p(x;\theta), s(\boldsymbol{\eta})\}$ for estimation. However, if we use a family of unbiased estimators, and plug any rough estimate of $\hat{\boldsymbol{\eta}}$ in it, it gives a good result. See [3] for a detailed analysis of the estimating functions in terms of information geometry.

6. Conclusion

We have used a simple model of estimating a linear relation in an errors-in-variables model. It is shown that neither the maximum likelihood nor the total least squares method gives the optimal estimator. We have presented a simple method of improving the estimator.

The theory is based on information geometry of estimation functions. We have touched upon its theoretical framework. This idea can be used for developing a more general model of estimation. For example, it is used in Independent Component Analysis by Amari and Cardoso [2].

References

[1] S. Amari, *Differential-Geometrical Methods in Statistics*, Springer Lecture Notes in Statistics, vol. 28, Springer-Verlag, New York, 1985.

[2] S. Amari and J.-F. Cardoso, Blind source separation — semiparametric statistical approach, *IEEE Trans. on Signal Processing*, 45:2692–2700, 1997.

[3] S. Amari, and M. Kawanabe, Information geometry of estimating functions in semiparametric statistical models, *Bernoulli*, 3:29–54, 1997.

[4] S. Amari, and H. Nagaoka, *Methods of Information Geometry*, AMS and Oxford University Press, 2000.

[5] J. M. Begun, W. J. Hall, W. M. Huang, and J. A. Wellner, Information and asymptotic efficiency in parametric-nonparametric models, *Annals of Statistics*, 11:432–452, 1983.

[6] P. J. Bickel, C. A. J. Klaassen, Y. Ritov, and J. A. Wellner, *Efficient and Adaptive Estimation for Semiparametric Models*, Johns Hopkins University Press, Baltimore, 1994.

[7] V. P. Godambe, Conditional likelihood and unconditional optimum estimating equations, *Biometrica*, 63:277–284, 1976.

[8] J. Neyman, and E. L. Scott, Consistent estimates based on partially consistent observations, *Econometrica*, 32:1–32, 1948.

[9] Pistone, G. and Rogantin, M.-P., The exponential statistical manifold: Mean parameters, orthogonality, and space transformation, *Bernoulli*, 5:721–760, 1999.

UNIFYING LEAST SQUARES, TOTAL LEAST SQUARES AND DATA LEAST SQUARES

Christopher C. Paige
School of Computer Science, McGill University,
Montreal, Quebec, Canada, H3A 2A7
paige@cs.mcgill.ca

Zdeněk Strakoš
Institute of Computer Science, Academy of Sciences of the Czech Republic,
Pod Vodárenskou věží 2, 182 07 Praha 8, Czech Republic
strakos@cs.cas.cz

Abstract The standard approaches to solving overdetermined linear systems $Ax \approx b$ construct minimal corrections to the vector b and/or the matrix A such that the corrected system is compatible. In ordinary least squares (LS) the correction is restricted to b, while in data least squares (DLS) it is restricted to A. In scaled total least squares (Scaled TLS) [15], corrections to both b and A are allowed, and their relative sizes depend on a parameter γ. Scaled TLS becomes total least squares (TLS) when $\gamma = 1$, and in the limit corresponds to LS when $\gamma \to 0$, and DLS when $\gamma \to \infty$.

In [13] we presented a particularly useful formulation of the Scaled TLS problem, as well as a new assumption that guarantees the existence and uniqueness of meaningful Scaled TLS solutions for all parameters $\gamma > 0$, making the whole Scaled TLS theory consistent. This paper refers to results in [13] and is mainly historical, but it also gives some simpler derivations and some new theory. Here it is shown how any linear system $Ax \approx b$ can be reduced to a minimally dimensioned core system satisfying our assumption. The basics of practical algorithms for both the Scaled TLS and DLS problems are indicated for either dense or large sparse systems.

Keywords: scaled total least squares, ordinary least squares, data least squares, core problem, orthogonal reduction, singular value decomposition.

S. Van Huffel and P. Lemmerling (eds.), Total Least Squares and Errors-in-Variables Modeling, 25–34.

1. Introduction

Two useful approaches to solving the overdetermined linear system

$$Ax \approx b, \quad A \text{ an } n \text{ by } k \text{ matrix}, b \text{ an } n\text{-vector}, \quad b \notin \mathcal{R}(A), \tag{1}$$

are ordinary least squares (LS, or OLS, see for example [1], [8, §5.3]) and total least squares (TLS, see [6, 7], also [1, §4.6], [8, §12.3], [11]). In LS we seek (we use $\|\cdot\|$ to denote the vector 2-norm)

$$\text{LS distance} \quad \equiv \quad \min_{r,x} \|r\| \quad \text{subject to} \quad Ax = b - r. \tag{2}$$

In TLS, G and r are sought to minimize the Frobenius (F) norm in

$$\text{TLS distance} \quad \equiv \quad \min_{r,G,x} \|[r, G]\|_F \quad \text{s. t.} \quad (A + G)x = b - r. \tag{3}$$

The opposite case to LS is the data least squares problem (DLS), see [9]. In DLS the correction is allowed only in A

$$\text{DLS distance} \quad \equiv \quad \min_{G,x} \|G\|_F \quad \text{subject to} \quad (A + G)x = b. \tag{4}$$

All these approaches can be unified by considering the following very general scaled TLS problem (Scaled TLS), see the paper [15] by B. D. Rao, who called it "weighted TLS": for a given $\gamma > 0$,

$$\text{Scaled TLS distance} \quad \equiv \quad \min_{\tilde{r},\tilde{G},\tilde{x}} \|[\tilde{r}\gamma, \tilde{G}]\|_F \quad \text{s. t.} \quad (A+\tilde{G})\tilde{x} = b-\tilde{r}. \tag{5}$$

Here the relative sizes of the corrections in A and b are determined by the real parameter $\gamma > 0$. As $\gamma \to 0$ the Scaled TLS solution approaches the LS solution, when $\gamma = 1$ (5) coincides with the TLS formulation, and as $\gamma \to \infty$ it approaches DLS. The case $\gamma \to 0$ is not completely obvious, since setting $\gamma = 0$ in (5) leads to $\tilde{G} = 0$ but allows *arbitrary* $\tilde{r}$. However consideration of very small γ should at least partially convince the reader that the LS solution is obtained. The case $\gamma = 1$ is obvious, and we see that $\gamma \to \infty$ requires $\tilde{r} \to 0$, leading to DLS. For more on Scaled TLS and DLS see also [2]. Scaling by a diagonal matrix was considered in [7], and this motivated later researchers, leading eventually to the Scaled TLS formulation in [15]. The paper [4] considered the case where only some of the columns of the data matrix are contaminated, and this also suggested a way of treating LS as well as TLS in the one formulation.

The formulation of the Scaled TLS problem that we use is slightly different from that in (5). For any positive bounded γ, substitute in (5) $r \equiv \tilde{r}\gamma$, $x \equiv \tilde{x}$ and $G \equiv \tilde{G}$ to obtain the new formulation:

$$\text{Scaled TLS distance} \quad \equiv \quad \min_{r,G,x} \|[r, G]\|_F \quad \text{s. t.} \quad (A+G)x\gamma = b\gamma-r. \tag{6}$$

We call the $x = x(\gamma)$ that minimizes this distance the *Scaled TLS solution* of (6). In analogy with (3), we call $x(\gamma)\gamma$ the *TLS solution* of (6). In (6) we could have written x instead of $x\gamma$. We chose the present form so that for positive bounded γ, the Scaled TLS solution $x = x(\gamma)$ of (6) is identical to the solution $\tilde{x}$ of (5). Thus (5) and (6) have identical distances and solutions for positive bounded γ. Therefore our results and discussions based on (6) apply fully to the Scaled TLS problem (5).

In [13, §6] we showed for (6) in the general case of complex data that as $\gamma \to 0$, $x(\gamma)$ becomes the LS solution x of (2), (Scaled TLS distance)/γ becomes the LS distance. As $\gamma \to \infty$, $x(\gamma)$ becomes the DLS solution x of (4), and the Scaled TLS distance becomes the DLS distance. The convergence of the Scaled TLS problem to the LS problem has been described in [15], and essentially in [7], for the real case.

We found that the development of our results was more simple and intuitive using the formulation (6) rather than (5). In particular, all the known TLS theory and algorithms can be applied directly to (6). The equivalence of (6) and (5) is extremely useful. This equivalence was pointed out to us by Sabine Van Huffel [10] after she read an earlier version of our work based on (6). We have not seen it stated in the literature, but it is implicit in the paper by Rao [15].

In (6), γ simply scales the right-hand side vector b (and the Scaled TLS solution $x = x(\gamma)$). Thus it is appropriate to call the formulation (6) the *Scaled* TLS problem, rather than the "weighted" TLS problem as was done in [15]. This also avoids the possibility of confusing the meaning of "weighted" here with its different meaning in "weighted least squares".

Using γ can have a statistical significance. Suppose that the elements of A are known to have independent zero-mean random errors of equal standard deviation δ_A. Suppose also that the elements of b have been observed with independent zero-mean random errors of equal standard deviation δ_b, and that the errors in b and A are independent. Then taking $\gamma = \delta_A/\delta_b$ in (6) will ensure that all the errors in that model have equal standard deviation (and so variance), and (6) is the ideal formulation for providing estimates. This agrees with the limiting behaviour described above, for clearly if $\delta_A = 0$ and $\delta_b \neq 0$, then LS is the correct choice, while if $\delta_A \neq 0$ and $\delta_b = 0$, then DLS is the correct choice. However (6) can also be useful outside any statistical context, and then γ does not have the above interpretation, see for example [14] which is summarized in our other contribution in this book.

In all these formulations, if $b \in \mathcal{R}(A)$, then zero distance can be obtained via a direct solution. Otherwise TLS, and so Scaled TLS solutions can be found via the singular value decomposition (SVD). Let $\sigma_{\min}(\cdot)$ denote the smallest singular value of a given matrix. To be precise,

$\sigma_{\min}(M)$ will denote the j-th largest singular value of an n by j matrix M, and will be zero if $n < j$. The interlacing property for the eigenvalues of $[A, b]^T[A, b]$ and of A^TA [16, Ch2, §47, pp. 103–4] tells us that $\sigma_{\min}([A, b]) \le \sigma_{\min}(A)$. When

$$\sigma_{\min}([A, b]) < \sigma_{\min}(A) \tag{7}$$

the n by k matrix A must have rank k, the unique solution of the TLS problem (3) is obtained from scaling the right singular vector of $[A, b]$ corresponding to $\sigma_{\min}([A, b])$, and the norm of the TLS correction satisfies $\min_{r,G,x} \|[r, G]\|_F = \sigma_{\min}([A, b])$, (see for example [8, §12.3]).

However when $\sigma_{\min}([A, b]) = \sigma_{\min}(A)$, the theory and solution methods became complicated, see for example the discussions on nongeneric problems in [11]. For this and other reasons we argued in [13] that (7) should not be used as a basis for the TLS theory.

A similar argument to that following (7) shows that when

$$\sigma_{\min}([A, b\gamma]) < \sigma_{\min}(A) \quad \text{for a given } \gamma > 0, \tag{8}$$

the Scaled TLS distance in (6) is $\sigma_{\min}([A, b\gamma])$, but we also showed in [13] that this should not be used as a basis for the Scaled TLS theory.

In the general case, let $\mathcal{U}_{\min}$ be the left singular vector subspace of A corresponding to $\sigma_{\min}(A)$. We showed in [13] that a satisfactory condition for building the theory of the TLS, DLS and Scaled TLS formulations for solving (1) is the γ-independent criterion:

$$\text{the } n \times k \text{ matrix } A \text{ has rank } k, \text{ and } \; b \not\perp \mathcal{U}_{\min}. \tag{9}$$

We showed in [13, Thm.3.1] that this implies

$$\sigma(\gamma) \equiv \sigma_{\min}([A, b\gamma]) < \sigma_{\min}(A) \quad \text{for all } \gamma \ge 0, \tag{10}$$

which of course implies (7) and (8). The stronger condition (9) is the simplest one. It can be checked using direct computations, see Sectinon 3, while the others each apparently require two SVDs.

A crucial property of the criterion (9) is that *any* linear system $Ax \approx b$ can in theory be reduced to a "core" problem satisfying (9). We show how this can be done by direct computations that can be usefully applied to all Scaled TLS and DLS problems.

This paper is necessarily short, and can be considered as an introduction to [13] which contains the full theory that has been omitted here. That paper presented a new and thorough analysis of the theoretical foundations of the Scaled TLS problem, and of its relationships to the LS and DLS problems. Here we mention some of those results, but concentrate mainly on the concept of the core problem.

The rest of the paper is organized as follows. Section 2 indicates why the formulations (3)–(6) are incomplete without the criterion (9). Section 3 shows how to handle the completely general Scaled TLS problem by reducing it to a core problem that satisfies an even stronger criterion than (9) — one which ensures the core problem is irreducible in the sense of containing no information that is redundant or irrelevant to the solution. Section 4 discusses how Scaled TLS problems can be solved computationally, and describes a simple solution to the DLS problem. Section 5 summarizes the advances and outlines the philosophy.

2. Conditions for meaningful solutions

The problem formulations (3)–(6) are not good for solving $Ax \approx b$ in certain cases. It was shown in [13, §7] that (3)–(6) are not good when n by k A does not have rank k. The formulations should at least demand the solution vectors be orthogonal to the null space. It is preferable to eliminate the null space, so now let us assume A has rank k.

We argue that (3)–(6) are best restricted to problems of the form (1) satisfying (9). Suppose the data can be transformed so that

$$[\,\tilde{b} \,\|\, \tilde{A}\,] = P^T [\, b \,\|\, AQ\,] = \left[\begin{array}{c||c|c} b_1 & A_{11} & 0 \\ \hline 0 & 0 & A_{22} \end{array}\right], \tag{11}$$

where P and Q are orthogonal. The approximation problem $Ax \approx b$ then represents two *independent* approximation problems:

$$A_{11}x_1 \approx b_1, \qquad A_{22}x_2 \approx 0, \qquad x \equiv Q\begin{bmatrix} x_1 \\ x_2 \end{bmatrix}, \tag{12}$$

in that the solution to each of these has no effect upon, and can be found independently of, the other. In this case the non-core problem $A_{22}x_2 \approx 0$ has the solution $x_2 = 0$, and only $A_{11}x_1 \approx b_1$ need be solved.

If $b \perp \mathcal{U}_{\min}$, see (9), then orthogonal P and Q clearly exist giving (11) where A_{22} contains all the singular values of A equal to $\sigma_{\min}(A)$. Then it was shown in [13, §7] that (3)–(6) applied directly to the combined problem $Ax \approx b$ can give meaningless solutions. But even when they give meaningful solutions these minimum singular values are irrelevant, and should be removed from the problem, since rounding errors effectively introduce a nonzero vector below b_1 in (11), and so cause these irrelevant singular values to contaminate the solution. Although (2) in theory gives $x_2 = 0$, this last comment suggests we might gain by insisting on (9) for LS too.

The criterion (9) leads to a clear and consistent theory, and ensures that the minimum singular value of A is relevant to the solution. Fortunately there is an elegant transformation which produces the minimally dimensioned core problem obeying (9) from any given $[b, A]$.

3. The core problem within $Ax \approx b$

Here we answer the following question. Given any n by k matrix A and n-vector b, how can the problem $Ax \approx b$ be split into two independent problems as in (11) and (12), giving a trivial problem $A_{22}x_2 \approx 0$ of maximal dimensions, and the minimally dimensioned core problem $A_{11}x_1 \approx b_1$ satisfying (9)? We can set $x_2 = 0$, and then each of the formulations (2)–(6) gives a unique meaningful solution to $A_{11}x_1 \approx b_1$.

We answer this by choosing orthogonal matrices P and Q to produce the following real bidiagonal matrix, see for example [8, §5.4.3–5, pp. 251–254]. In the usual case of $n \times k$ A with $n > k$ we obtain, (where the bottom 0 either represents a zero vector or is nonexistent):

$$[\tilde{b}, \tilde{A}] \equiv P^T [\, b \mid A \,] \begin{bmatrix} 1 & \\ & Q \end{bmatrix} = \left[\begin{array}{c|ccccc} \beta_1 & \alpha_1 & & & & \\ & \beta_2 & \alpha_2 & & & \\ & & \cdot & \cdot & & \\ & & & \beta_k & \alpha_k & \\ & & & & \beta_{k+1} & \\ & & & & 0 & \end{array}\right]. \qquad (13)$$

Remember the Scaled TLS solution requires some knowledge of the SVD of $[b\gamma, A]$. This SVD can quickly be computed from this bidiagonal form for any choice of γ, see for example [8, §8.6.2, pp. 452–456].

There are two ways this algorithm can terminate prematurely: with a $\beta_j = 0$, giving (11) and (12) with $A_{11}x_1 = b_1$ being a compatible system; or with $\alpha_j = 0$, giving an incompatible system $A_{11}x_1 \approx b_1$ in (12).

The computations described in [8, §5.4.3–5, pp. 251–254] are designed for dense matrices. If we have large sparse $[b, A]$, then we could consider the iterative bidiagonalization of Golub and Kahan in [5]. This iterative bidiagonalization is the basis for the LSQR algorithm in [12] which solves large sparse LS (as well as consistent) problems. In theory after j and a half steps this produces the first $j + 1$ columns of P, the first j columns of Q, and the leading $j + 1$ by $j + 1$ block of the right-hand side in (13). Björck [1, §7.6.5, pp.310-311] suggested applying this iterative bidiagonalization to the TLS problem, see also [3, §4.1]. Now we see this approach is also applicable to solving the Scaled TLS problem, as well as (at least in theory) delivering the core problem, for any large sparse linear system $Ax \approx b$. The adaptation of LSQR for solving large sparse Scaled TLS or DLS problems using finite precision computations will be further investigated. See Section 4 for the DLS solution using (13).

The main theoretical importance of the reduction (13) is that if $\beta_j\alpha_j \neq 0$, $j = 1, \ldots, k$, then our criterion (9) holds for the reduced bidiagonal matrix. We now prove this in an extended version of [13, Thm.8.1].

Theorem 1. *Suppose n by k A has SVD $A = \sum_{i=1}^{k} u_i\sigma_i v_i^T$, and there exist orthogonal matrices P and Q giving (13) with*

$$\beta_j\alpha_j \neq 0, \quad j = 1, \ldots, k. \tag{14}$$

Then we have a stronger condition than (9) for this b and A:

$$rank(A) = k; \qquad b^T u_i \neq 0, \quad i = 1, \ldots, k, \tag{15}$$

and no nontrivial split of the form (11) can be obtained with orthogonal P and Q, showing $Ax \approx b$ is the minimally dimensioned core problem. Also the k singular values of A are distinct and nonzero; the $k+1$ singular values of $[b, A]$ are distinct, and all nonzero if and only if $\beta_{k+1} \neq 0$.

Proof. $\tilde{A}$ and A have the same singular values, as do $[\tilde{b}, \tilde{A}]$ and $[b, A]$, and $\tilde{A} = P^T AQ$ has the SVD $\tilde{A} = \sum_{i=1}^{k} \tilde{u}_i\sigma_i\tilde{v}_i^T \equiv \sum_{i=1}^{k} P^T u_i\sigma_i v_i^T Q$, so

$$b^T u_i = b^T P P^T u_i = \tilde{b}^T \tilde{u}_i, \quad i = 1, \ldots, k.$$

Write $\tilde{A} \equiv [b_1, A_1]$, then $\tilde{A}^T\tilde{A}$ is $k \times k$ tridiagonal with nonzero next to diagonal elements, and $A_1^T A_1$ remains when the first row and column are deleted. Thus the eigenvalues of $A_1^T A_1$ strictly separate those of $\tilde{A}^T\tilde{A}$, see [16, Ch.5, §37, p.300], and the singular values of A_1 strictly separate those of $\tilde{A}$. Thus $\tilde{A}$, and so A, has distinct singular values. A similar argument holds for $[b, A]$. A clearly has rank k, and $[b, A]$ has rank $k+1$ if and only if $\beta_{k+1} \neq 0$. Suppose σ is a singular value of $\tilde{A}$ with singular vectors u and v such that

$$\tilde{b}^T u = \beta_1 e_1^T u = 0, \quad u\sigma = \tilde{A}v, \quad \sigma v^T = u^T\tilde{A}, \quad \|u\| = \|v\| = 1,$$

then $0 = e_1^T u\sigma = e_1^T\tilde{A}v = \alpha_1 e_1^T v$, and $e_1^T v = 0$. Writing $v = \left[\begin{smallmatrix} 0 \\ q \end{smallmatrix}\right]$ shows

$$\tilde{A}v = A_1 q = u\sigma, \quad u^T A_1 = \sigma q^T, \quad \|u\| = \|q\| = 1,$$

so σ is also a singular value of A_1. This is a contradiction since the singular values of A_1 strictly separate those of $\tilde{A}$, so (15) holds.

Finally if (11) could exist with nontrivial A_{22}, then b would be orthogonal to a left singular vector subspace of A, which (15) has proven is impossible. □

Thus we need not derive results for the most general possible $[b\gamma, A]$. We can instead assume (9). Any more general $Ax \approx b$ problem can be reduced to a core problem that satisfies (15) (and so (9)) by applying the reduction (13) and stopping at the first zero β_j or α_j. Suppose the resulting core data is $[b_1, A_{11}]$, see (11). Then the theorem also showed that A_{11} has no multiple singular values, so any singular value repeats must appear in A_{22}.

In theory we need not insist on (15), because a problem only satisfying (9) will in theory give the same solution and distance as it would if it were reduced to one satisfying (15). But in practice it is preferable to carry out the reduction (13) leading to (15), see Section 4.

4. Computing Scaled TLS and DLS solutions

In order to compute either Scaled TLS solutions or the DLS solution for given data $[b, A]$, we recommend first carrying out a reduction of the form (13) to the core problem in Section 3 — unless there are clear reasons for not doing so. The reasons for doing so are hard to reject. For general data we will not know if the formulations (2)–(6) have unique meaningful solutions, but the reduction will give us a subproblem for which this is so. Even if we know the original data satisfies (9), it is (from the computational point of view) highly preferable to remove all the irrelevant information from our data as early in the solution process as possible, and this is exactly what the transformation (13) does. In any case we still need some sort of SVD of the data, and this will usually first perform a reduction as costly as that in (13). But (13) allows us to find the SVD of $[b\gamma, A]$ easily for different choices of γ and so is the obvious choice. There are excellent fast and accurate algorithms for finding all or part of the SVD of (13) with β_1 replaced by $\beta_1\gamma$. We can find just the smallest singular value and its singular vectors, from which the solution vector $x(\gamma)$ can be simply attained. If we have some idea of the accuracy of our data, then when we use numerically reliable orthogonal transformations in (13), we will have a good idea of what element of (13) (if any) we can set to zero to stop the computation as soon as possible. Thus the crucial decisions can be made *before* any SVD computations are carried out. This is more efficient, but it is almost certainly more reliable to make such decisions from (independent) orthogonal transformations of the original data than from the elements of singular vectors, (see for example [11, p.23]). The remaining computations for Scaled TLS are fairly obvious. Finally (13) leads to a solution to the DLS problem (4), which we now describe. The theory here is simpler than that in [13, §9].

Suppose that the core part $[\tilde{b}, \tilde{A}]$ of the transformed $[b, A]$ has the form in (13) with (14). We will solve the DLS problem for this reduced, or core data. Now Theorem 1 proved (9) holds. If $\beta_{k+1} = 0$ the DLS distance is zero, and the solution is obvious. Otherwise, writing

$$[\tilde{b}|\tilde{A}] \equiv \left[\begin{array}{c|c} \beta_1 & \alpha_1 e_1^T \\ \hline 0 & A_2 \end{array}\right] \equiv P^T[b|AQ], \quad \tilde{G} \equiv \left[\begin{array}{c} g^T \\ \hline G_2 \end{array}\right] \equiv P^T G Q, \quad \tilde{x} \equiv Q^T x,$$

we see for this reduced data that the DLS problem (4) becomes

$$\min_{g,G_2,\tilde{x}} \{\|g\|^2 + \|G_2\|_F^2\} \quad \text{s. t.} \quad \left[\begin{array}{c|c} \beta_1 & \alpha_1 e_1^T + g^T \\ 0 & A_2 + G_2 \end{array}\right] \left[\begin{array}{c} -1 \\ \tilde{x} \end{array}\right] = 0.$$

Since β_1 is nonzero, $\tilde{x} \neq 0$, and the minimum $\|G_2\|_F$ in $(A_2 + G_2)\tilde{x} = 0$ is $\sigma_{\min}(A_2)$, with $\tilde{x}$ proportional to the right singular vector v of A_2 corresponding to $\sigma_{\min}(A_2)$. But then $e_1^T v \neq 0$ (otherwise $\sigma_{\min}(A_2)$ would also be a singular value of $\tilde{A}$) and we can take $g = 0$ so that

$$\tilde{x}_D = v\beta_1/(\alpha_1 e_1^T v), \qquad \sigma_D = \sigma_{\min}(A_2), \tag{16}$$

are the DLS solution and distance in (4) for the reduced data $[\tilde{b}, \tilde{A}]$. The smallest singular value and its right singular vector of the nonsingular bidiagonal matrix A_2 are relatively easy to find, see for example [8, §8.6.2, pp. 452–456].

5. Summary and conclusions

The philosophy behind our approach is radically different from that of previous TLS, Scaled TLS or DLS work known to us. The Scaled TLS formulation (6) makes it easy to analyze and solve the Scaled TLS problem (it shows the Scaled TLS problem is just the TLS problem with its right-hand side b scaled by γ, so all the TLS artillery is available). But more importantly than that, the approach of reducing a problem $Ax \approx b$ to its "core" problem (Section 3) and solving that core problem simplifies our understanding of the area. It also simplifies the development of algorithms, while unifying the theoretical problems in the area. Crucial to all this is the (γ-independent) criterion (9) for Scaled TLS (also TLS, DLS and even possibly LS) problems, that was introduced in [13]. The key is that *any* Scaled TLS (or LS or TLS or DLS) problem can in theory be transformed by *direct* orthogonal transformations into two independent problems: a (possibly nonexistent) trivial problem, and a core problem, where the core problem automatically satisfies (9). Solving the core problem then solves the original problem. Thus no complicated conditions such as (7) or (9) need be tested, and no special cases need be treated. All the decisions can be made by examining the sizes of elements in the orthogonally transformed data. Both theory and computations can thus be unified, simplified and clarified.

Acknowledgments

This work was supported by NSERC of Canada Grant OGP0009236 and by the GA AS CR under grant A2030801. Part of this work was performed while Z. Strakos was visiting Emory University, Atlanta, GA.

References

[1] Å. Björck, *Numerical Methods for Least Squares Problems*, SIAM Publications, Philadelphia PA, 1996.

[2] G. Cirrincione, *A Neural Approach to the Structure of Motion Problem*, PhD thesis, LIS INPG Grenoble, 1998.

[3] R. D. Fierro, G. H. Golub, P. C. Hansen and D. P. O'Leary, Regularization by truncated total least squares, *SIAM J. Sci. Comput.*, 18:1223–1241, 1997.

[4] G. H. Golub, A. Hoffman and G. W. Stewart, A generalization of the Eckart-Young-Mirsky matrix approximation theorem, *Lin. Alg. Appl.*, 88/89:317–327, 1987.

[5] G. H. Golub and W. Kahan, *Calculating the singular values and pseudo-inverse of a matrix*, J. SIAM, Series B, Numer. Anal., 2:205–224, 1965.

[6] G. H. Golub and C. Reinsch, Singular value decomposition and least squares solutions, *Numerische Mathematik*, 14:403-420, 1970. Also in *Handbook for Automatic Computation Vol. 2: Linear Algebra*, by J. H. Wilkinson and C. Reinsch, (eds.), pp. 134-151, Springer, New York, 1971.

[7] G. H. Golub and C. F. Van Loan, An analysis of the total least squares problem, *SIAM J. Numer. Anal.*, 17:883–893, 1980.

[8] ——, *Matrix Computations*, The Johns Hopkins University Press, Baltimore MD, third ed. 1996.

[9] R. D. D. Groat and E. M. Dowling, The data least squares problem and channel equalization, *IEEE Trans. Signal Processing*, 42(1):407–411, 1993.

[10] S. Van Huffel. Personal communication, June 1999.

[11] S. Van Huffel and J. Vandewalle, *The Total Least Squares Problem: Computational Aspects and Analysis*, SIAM, Philadelphia, 1991.

[12] C. C. Paige and M. A. Saunders, LSQR: An algorithm for sparse linear equations and sparse least squares, *ACM Trans. Math. Software*, 8:43–71, 1982.

[13] C. C. Paige and Z. Strakoš, Scaled total least squares fundamentals, to appear in *Numerische Mathematik.*

[14] C. C. Paige and Z. Strakoš, Bounds for the least squares distance using scaled total least squares, to appear in *Numerische Mathematik.*

[15] B. D. Rao, Unified treatment of LS, TLS and truncated SVD methods using a weighted TLS framework, In: S. Van Huffel (editor), *Recent Advances in Total Least Squares Techniques and Errors-in-Variables Modelling*, pp. 11–20, SIAM Publications, Philadelphia PA, 1997.

[16] J. Wilkinson, *The Algebraic Eigenvalue Problem*, Clarendon Press, Oxford, 1965.

BOUNDS FOR THE LEAST SQUARES RESIDUAL USING SCALED TOTAL LEAST SQUARES

Christopher C. Paige
School of Computer Science, McGill University
Montreal, Quebec, Canada, H3A 2A7
paige@cs.mcgill.ca

Zdeněk Strakoš
Institute of Computer Science, Academy of Sciences of the Czech Republic,
Pod Vodárenskou věží 2, 182 07 Praha 8, Czech Republic
strakos@cs.cas.cz

Abstract The standard approaches to solving overdetermined linear systems $Ax \approx b$ construct minimal corrections to the data to make the corrected system compatible. In ordinary least squares (LS) the correction is restricted to the right hand side b, while in scaled total least squares (Scaled TLS) [10, 9] corrections to both b and A are allowed, and their relative sizes are determined by a real positive parameter γ. As $\gamma \to 0$, the Scaled TLS solution approaches the LS solution. Fundamentals of the Scaled TLS problem are analyzed in our paper [9] and in the contribution in this book entitled *Unifying least squares, total least squares and data least squares.*

This contribution is based on the paper [7]. It presents a theoretical analysis of the relationship between the sizes of the LS and Scaled TLS *corrections* (called the LS and Scaled TLS distances) in terms of γ. We give new upper and lower bounds on the LS distance in terms of the Scaled TLS distance, compare these to existing bounds, and examine the tightness of the new bounds.

This work can be applied to the analysis of iterative methods which minimize the residual norm [8, 5].

Keywords: ordinary least squares, scaled total least squares, singular value decomposition, linear equations, least squares residual.

S. Van Huffel and P. Lemmerling (eds.), Total Least Squares and Errors-in-Variables Modeling, 35–44.

1. Introduction

Consider an overdetermined approximate linear system

$$Ax \approx b, \quad A \text{ an } n \text{ by } k \text{ matrix}, \quad b \text{ an } n\text{-vector}, \quad b \notin \mathcal{R}(A), \tag{1}$$

where $\mathcal{R}(M)$ denote the range (column space) of a matrix M. In LS we seek (we use $\|\cdot\|$ to denote the vector 2-norm)

$$\text{LS distance} \quad \equiv \quad \min_{r,x} \|r\| \quad \text{subject to} \quad Ax = b - r. \tag{2}$$

In Scaled TLS, for a given parameter $\gamma > 0$, x, G and r are sought to minimize the Frobenius (F) norm in

$$\text{Scaled TLS distance} \;\equiv\; \min_{r,G,x} \|[r, G]\|_F \quad \text{s. t.} \quad (A+G)x\gamma = b\gamma - r. \tag{3}$$

We call the $x = x(\gamma)$ which minimizes this distance the Scaled TLS solution of (3). Here the relative sizes of the corrections G and r in A and $b\gamma$ are determined by the real scaling parameter $\gamma > 0$. As $\gamma \to 0$ the Scaled TLS solution approaches the LS solution. The formulation (3) is studied in detail in [9]. We present an introduction to and refine some results of [9] in our contribution *Unifying least squares, total least squares and data least squares* presented in this book. Here we follow the notation introduced there. In applications γ can have a statistical interpretation, see for example [9, §1], but here we regard γ simply as a variable.

Scaled TLS solutions can be found via the singular value decomposition (SVD). Let $\sigma_{\min}(\cdot)$ denote the smallest singular value of a matrix, and let P_k be the orthogonal projector onto the left singular vector subspace of A corresponding to $\sigma_{\min}(A)$. The bounds presented here will assume

$$\text{the } n \times (k+1) \text{ matrix } [A, b] \text{ has rank } k+1, \text{ and } \quad P_k b \neq 0. \tag{4}$$

We showed in [9, (3.7)] that this implied

$$0 < \sigma(\gamma) \equiv \sigma_{\min}([A, b\gamma]) < \sigma_{\min}(A) \quad \text{for all } \gamma > 0. \tag{5}$$

In this case the unique solution of the Scaled TLS problem (3) is (in theory) obtained from scaling the right singular vector of $[A, b\gamma]$ corresponding to $\sigma_{\min}([A, b\gamma])$, and the norm of the Scaled TLS correction satisfies, for a given $\gamma > 0$ (see for example [9, (1.9)], or [3, §12.3] when $\gamma = 1$),

$$\text{Scaled TLS distance in (3)} \quad = \quad \sigma_{\min}([A, b\gamma]). \tag{6}$$

The paper [7] and the presentation of the bounds in this contribution are greatly simplified by only dealing with problems where (4) holds. The assumption (4) is equivalent to that in [9, (1.10)] plus the restriction $b \notin \mathcal{R}(A)$, which eliminates the theoretically trivial case $b \in \mathcal{R}(A)$. It is sufficient to note here that nearly all practical overdetermined problems will already satisfy (4), but any overdetermined (and incompatible) problem that does not can be reduced to one that does, see [9, §8], and the bounds presented here with this assumption will be applicable to the original problem.

It is known that (see for example [9, (6.3)])

$$\lim_{\gamma\to 0} \frac{\text{Scaled TLS distance in (3)}}{\gamma} = \lim_{\gamma\to 0} \frac{\sigma_{\min}([A, b\gamma])}{\gamma} = \|r\|, \text{ the LS distance in (2),} \tag{7}$$

but here we examine the relationship between these distances for *any* $\gamma > 0$. This will bound the rate at which these quantities approach each other for small γ, as well as provide bounds on the LS distance in terms of $\sigma_{\min}([A, b\gamma])$, and *vice versa*, for all $\gamma > 0$. It will in general simplify the presentation to assume $\gamma > 0$, since when $\gamma = 0$ is meaningful, the values will be obvious.

Van Huffel and Vandewalle [6] derived several useful bounds for TLS versus LS (the $\gamma = 1$ case). Our results extend some of these to the case of general $\gamma > 0$, as well as provide new bounds.

The contribution is organized as follows. In Section 2 we present our main result, in particular, bounds on the least squares residual norm $\|r\|$ (LS distance) in terms of the scaled total least squares distance $\sigma_{\min}([A, b\gamma])$. We show how good these bounds are, and how varying γ gives important insights into the asymptotic relationship between the LS and Scaled TLS distances. In Section 3 we compare our bounds to previous results. In Section 4 we analyze the ratio of the minimal singular values of $[A, b\gamma]$ and A which determines the tightness of the presented bounds.

2. Main result

Our main result relating the LS distance $\|r\|$ to the Scaled TLS distance $\sigma_{\min}([A, b\gamma])$ is formulated in the following theorem, see [7, Theorem 4.1 and Corollary 6.1].

Theorem 1. *Given a scalar $\gamma > 0$, and an n by $k+1$ matrix $[A, b]$, use $\sigma(\cdot)$ to denote singular values and $\|\cdot\|$ to denote 2-norms. If r and x*

solve $\min_{r,x} \|r\|$ *subject to* $Ax = b - r$, *and (4) holds, then*

$$0 < \theta(\gamma) \equiv \frac{\sigma_{\min}([A, b\gamma])}{\sigma_{max}(A)} \leq \delta(\gamma) \equiv \frac{\sigma_{\min}([A, b\gamma])}{\sigma_{\min}(A)} < 1, \tag{8}$$

and we have bounds on the LS residual norm $\|r\|$ *in terms of the Scaled TLS distance* $\sigma_{\min}([A, b\gamma])$*:*

$$\begin{aligned} \lambda_r \equiv \sigma_{\min}([A, b\gamma])\{\gamma^{-2} + \|x\|^2\}^{\frac{1}{2}} &< \sigma_{\min}([A, b\gamma])\,\{\gamma^{-2} + \frac{\|x\|^2}{1 - \theta(\gamma)^2}\}^{\frac{1}{2}} \\ \leq \|r\| \leq \mu_r &\equiv \sigma_{\min}([A, b\gamma])\{\gamma^{-2} + \frac{\|x\|^2}{1 - \delta(\gamma)^2}\}^{\frac{1}{2}}. \end{aligned} \tag{9}$$

Equivalently,

$$\begin{aligned} \lambda_\sigma \equiv \|r\|/\{\gamma^{-2} + \frac{\|x\|^2}{1 - \delta(\gamma)^2}\}^{\frac{1}{2}} &\leq \sigma_{\min}([A, b\gamma]) \\ \leq \|r\|/\{\gamma^{-2} + \frac{\|x\|^2}{1 - \theta(\gamma)^2}\}^{\frac{1}{2}} &\leq \mu_\sigma \equiv \|r\|/\{\gamma^{-2} + \|x\|^2\}^{\frac{1}{2}}. \end{aligned} \tag{10}$$

In addition to that, $\delta(\gamma)$ *is bounded as*

$$\frac{\gamma\|r\|}{\|[A, b\gamma]\|} \leq \delta(\gamma) \leq \frac{\gamma\|r\|}{\sigma_k([A, b\gamma])} \leq \frac{\gamma\|r\|}{\sigma_{\min}(A)}. \tag{11}$$

We see that the difference between the upper and the lower bounds in (9) depends on the size of $(1 - \delta(\gamma)^2)^{-1}$. If $\delta(\gamma) \ll 1$, then this difference will be very small. Bounds in (11) give us some indication of the size of $\delta(\gamma)$. We see from (11) that if $\gamma\|r\|$ is small compared with $\sigma_k([A, b\gamma])$ then $\delta(\gamma) \ll 1$, but if $\gamma\|r\|$ is not small compared with $\|[A, b\gamma]\|$ then $\delta(\gamma)$ cannot be small. If $[A, b\gamma]$ is well-conditioned in the sense that $\sigma_{\min}([A, b\gamma])$ is not too much smaller than $\|[A, b\gamma]\|$, then (11) gives us a very good idea of $\delta(\gamma)$. We will study $\delta(\gamma)$ in more detail in Section 4.

A crucial aspect of Theorem 1 is that it gives both an upper and a lower bound on the minimum residual norm $\|r\|$, or on $\sigma_{\min}([A, b\gamma])$, which is the Scaled TLS distance in (3). The weaker lower bound in (9), or upper bound in (10), is sufficient for many uses, and is relatively easy to derive, but the upper bound in (9), or lower bound in (10), is what makes the theorem strong.

The following corollary [7, Corollary 4.2] examines the *tightness* of the bounds (9)–(10), to indicate just how good they can be. In fact it shows that *all* the relative gaps go to zero (as functions of the scaling parameter γ) at least as fast as $O(\gamma^4)$.

Corollary 1. *Under the same conditions as in Theorem 1, with* $\sigma \equiv \sigma(\gamma) \equiv \sigma_{\min}([A, b\gamma])$*, the notation in (9)–(10), and*

$$\begin{aligned} \eta_r &\equiv (\|r\| - \lambda_r)/\|r\|, \quad \eta_\sigma \equiv (\sigma - \lambda_\sigma)/\sigma, \\ \zeta_r &\equiv (\mu_r - \lambda_r)/\|r\|, \quad \zeta_\sigma \equiv (\mu_\sigma - \lambda_\sigma)/\sigma, \end{aligned} \tag{12}$$

we have the following bounds

$$\begin{aligned} &0 < \eta_r \leq \zeta_r, \qquad 0 < \eta_\sigma \leq \zeta_\sigma, \\ &0 < \zeta_r, \zeta_\sigma < \frac{\gamma^2\|x\|^2}{2 + \gamma^2\|x\|^2} \cdot \frac{\delta(\gamma)^2}{1 - \delta(\gamma)^2} \to 0 \quad \text{as} \quad \gamma \to 0, \end{aligned} \tag{13}$$

where the upper bound goes to zero at least as fast as $O(\gamma^4)$.

Thus when $\delta(\gamma) \ll 1$, or γ is small, the upper and lower bounds in (9)–(10) are not only very good, but very good in a *relative* sense, which is important for small $\|r\|$ or $\sigma_{\min}([A, b\gamma])$. We see Corollary 1 makes precise a nice theoretical observation with practical consequences — small γ ensures very tight bounds (9) on $\|r\|$. In particular, for small γ we see

$$\|r\| \;\approx\; \lambda_r \equiv \sigma_{\min}([A, b\gamma])\,\{\gamma^{-2} + \|x\|^2\}^{\frac{1}{2}}, \tag{14}$$

and the relative error is bounded above by $O(\gamma^4)$.

When $\delta(\gamma) < 1$, [6, Thm. 2.7] showed (for $\gamma = 1$) the closed form TLS solution $x\gamma = x(\gamma)\gamma$ of (3) is

$$x(\gamma)\gamma = \{A^TA - \sigma^2_{\min}([A, b\gamma])I\}^{-1}A^Tb\gamma,$$

and with $r_{ScaledTLS} \equiv b\gamma - Ax(\gamma)\gamma$, [6, (6.19)] showed (for $\gamma = 1$)

$$\|r_{ScaledTLS}\| = \sigma_{\min}([A, b\gamma])(1 + \|x(\gamma)\gamma\|^2)^{\frac{1}{2}}. \tag{15}$$

Relation (14) can be seen to give an analogue of this for the LS solution: since $r\gamma = b\gamma - Ax\gamma$ in (2), the bounds (9), (11) and (13) show a strong relationship between $\gamma\|r\|$ and $\sigma_{\min}([A, b\gamma])$ for small $\delta(\gamma)$, γ, $\|r\|$ or $\|x\|/(1 - \delta(\gamma)^2)$:

$$\gamma\|r\| \;\approx\; \sigma_{\min}([A, b\gamma])\,\{1 + \gamma^2\|x\|^2\}^{\frac{1}{2}}. \tag{16}$$

The assumption $P_k b \neq 0$ in (4) is not necessary for proving the bounds (9)–(10). From the proof of Theorem 1 in [7] it is clear that these bounds only require $\delta(\gamma) < 1$. However $\delta(\gamma) < 1$ does not guarantee $P_k b \neq 0$. When $P_k b = 0$, $\|r\|$ contains no information whatsoever about $\sigma_{\min}(A)$, while the bounds do. By assuming $P_k b \neq 0$ we avoid this inconsistency. Moreover, we will consider various values of the parameter γ, and so we prefer the theorem's assumption to be independent of γ.

We end this section by a comment on possible consequences of Theorem 1 for understanding methods for large Scaled TLS problems. For small $\delta(\gamma)$, γ, $\|r\|$ or $\|x\|^2/(1-\delta(\gamma)^2)$, (10) with (11) and (13) show

$$\sigma_{k+1}^2([A, b\gamma]) \approx \frac{\gamma^2\|r\|^2}{1+\gamma^2\|x\|^2} = \|[A, b\gamma]\begin{bmatrix}-x\gamma\\1\end{bmatrix}\|^2 / \|\begin{bmatrix}-x\gamma\\1\end{bmatrix}\|^2;$$

so the Scaled TLS distance is well approximated using the Rayleigh quotient corresponding to the unique LS solution of $Ax\gamma = b\gamma - r\gamma$. As pointed out by Åke Björck in a personal communication, this may help to explain the behaviour of algorithms proposed in [1].

3. Comparison with previous bounds

The best previously published bounds relating LS and TLS distances appear to be those of Van Huffel and Vandewalle [6]. The relevant bounds of that reference, and a new bound, can be derived from (9), and we present them as a corollary (cf. [7, Corollary 5.1]).

Corollary 2. *Under the same conditions and assumptions as in Theorem 1, with* $\sigma(\gamma) \equiv \sigma_{\min}([A, b\gamma])$, $\delta(\gamma) \equiv \sigma_{\min}([A, b\gamma])/\sigma_{\min}(A)$,

$$\begin{aligned}\frac{\sigma_{\min}([A, b\gamma])}{\gamma} &\leq \frac{\sigma_{\min}([A, b\gamma])}{\gamma}\{1 - \frac{\sigma_{\min}^2([A, b\gamma])}{\|A\|^2} + \frac{\|b\|^2\gamma^2}{\|A\|^2}\}^{\frac{1}{2}} \\ &\leq \|r\| \leq \frac{\sigma_{\min}([A, b\gamma])}{\gamma}\{1 - \delta(\gamma)^2 + \frac{\|b\|^2\gamma^2}{\sigma_{\min}(A)^2}\}^{\frac{1}{2}}. \qquad (17)\end{aligned}$$

When $\gamma = 1$ the weaker lower bound and the upper bound in (17) are the equivalents for our situation of (6.34) and (6.35) in [6]. The stronger lower bound seems new. A slightly weaker upper bound was derived in [4, (2.3)]. Experimental results presented, e.g., in [7] demonstrate that our bounds in (9) can be significantly better than those in (17). The relationship of these bounds is, however, intricate. While (17) was in [7, Corollary 5.1] derived from (9), it is not *always* true that the latter is tighter. When $\delta(\gamma) \approx 1$ and $\|r\| \approx \|b\|$, it is possible for the upper bound in (17) to be smaller than that in (9). But in this case $\sigma_{\min}([A, b\gamma]) \approx \sigma_{\min}(A)$, and then the upper bound in (17) becomes the trivial $\|r\| \lesssim \|b\|$. Summarizing, when the upper bound in (17) is tighter than the upper bound in (9), the former becomes trivial and the latter is irrelevant.

The bounds (17) and (9) differ because the easily available $\|x\|$ in (9) was replaced by its upper and lower bounds to obtain (17). But there is another reason (9) is preferable to (17). The latter bounds require knowledge of $\sigma_{\min}(A)$, as well as $\sigma_{\min}([A, b\gamma])$. Admittedly (8) shows we also need these to know $\delta(\gamma)$ exactly, but, assuming that (4) holds, we know $\delta(\gamma) < 1$, and is bounded away from 1 always (see Theorem 2 in the following section). In fact there are situations where we know $\delta(\gamma) \ll 1$. Thus (9) is not only simpler and often significantly stronger than (17), it is more easily applicable.

4. Tightness parameter

The results presented above show the crucial role of the parameter $\delta(\gamma) = \sigma_{\min}([A, b\gamma])/\sigma_{\min}(A)$. It represents a ratio of the smallest singular value of the matrix appended by a column (here $[A, b\gamma]$) to the smallest singular value of the original matrix (here A). Though the definition is simple, the nature of $\delta(\gamma)$ is very subtle and its behaviour is very complicated.

Let the $n \times k$ matrix A have rank k and singular values σ_i with singular value decomposition (SVD)

$$A = U_A \Sigma V^T, \quad \Sigma \equiv \text{diag}(\sigma_1, \ldots, \sigma_k), \quad \sigma_1 \geq \ldots \geq \sigma_k > 0. \qquad (18)$$

Here U_A is $n \times k$ matrix, $U_A^T U_A = I_k$, Σ is $k \times k$, and $k \times k$ V is orthogonal. Let

$$a \equiv (\alpha_1, \ldots, \alpha_k)^T \equiv [u_1, \ldots, u_k]^T b = U_A^T b. \qquad (19)$$

The elements of a are the components of the vector of observations b in the directions of the left singular vectors of the data matrix A.

Assume (4) holds. Then using the notation in (18)–(19), $0 < \sigma(\gamma) < \sigma_k \equiv \sigma_{\min}(A)$ holds for all $\gamma > 0$, and the Scaled TLS distance in (3) is $\sigma(\gamma) \equiv \sigma_{\min}([A, b\gamma])$, which is the smallest positive solution of

$$0 = \psi_k(\sigma(\gamma), \gamma) = \gamma^2 \|r\|^2 - \sigma(\gamma)^2 - \gamma^2 \sigma(\gamma)^2 \sum_{i=1}^{k} \frac{|\alpha_i|^2}{\sigma_i^2 - \sigma(\gamma)^2}. \qquad (20)$$

Moreover, if (4) holds and $\gamma > 0$, then $0 < \delta(\gamma) < 1$, and $\delta(\gamma)$ increases as γ increases, and decreases as γ decreases, strictly monotonically. This was derived in [9, §4]. With $\gamma = 1$, (20) was derived in [2], see also [6, Thm. 2.7, & (6.36)]. These latter derivations assumed the weaker condition $\sigma_{\min}([A, b]) < \sigma_{\min}(A)$, and so do not generalize to Scaled TLS for all $\gamma > 0$, see [9].

Our bounds containing the factor $(1 - \delta(\gamma)^2)^{-1}$ would be useless if $\delta(\gamma) = 1$ and of limited value when $\delta(\gamma) \approx 1$. The following theorem [7, Theorem 3.1] shows that when (4) holds, $\delta(\gamma)$ is *bounded away from unity* for all γ, giving an upper bound on $(1 - \delta(\gamma)^2)^{-1}$. It is important that these bounds exist, but remember they are worst case bounds, and give no indication of the sizes of $\delta(\gamma)$ or $(1-\delta(\gamma)^2)^{-1}$ for the values of γ we will usually be interested in.

Theorem 2. *With the notation and assumptions of (18)–(20), let $n \times k$ A have singular values $\sigma_1 \geq \cdots \geq \sigma_j > \sigma_{j+1} = \cdots = \sigma_k > 0$. Then since (4) holds,*

$$\|P_k b\|^2 = \sum_{i=j+1}^{k} |\alpha_i|^2 > 0, \tag{21}$$

$$\delta(\gamma)^2 \equiv \frac{\sigma_{\min}^2([A, b\gamma])}{\sigma_k^2} \leq \frac{\|r\|^2}{\|P_k b\|^2 + \|r\|^2} < 1 \quad \textit{for all } \gamma \geq 0, \tag{22}$$

$$(1-\delta(\gamma)^2)^{-1} \leq 1 + \|r\|^2/\|P_k b\|^2 \quad \textit{for all } \gamma \geq 0, \tag{23}$$

where P_k is described just before (4).

This shows that when (4) holds, $\delta(\gamma)$ is bounded away from unity, so $\sigma_{\min}([A, b\gamma])$ is bounded away from $\sigma_{\min}(A)$, for all γ.

The inequality (22) has a useful explanatory purpose. We cannot have $\delta(\gamma) \approx 1$ unless $P_k b$, the projection of b onto the left singular vector subspace of A corresponding to $\sigma_{\min}(A)$, is very small compared to r. It is straightforward to show that replacing A by

$$\tilde{A} = A - \sum_{i=j+1}^{k} u_i \sigma_{\min}(A) v_i^T$$

in (2) increases the square of the LS residual by $\|P_k b\|^2$, thus giving a small relative change when $P_k b$ is small compared to r. This confirms that the criterion (4) (see also [9, (1.10)]) is exactly what is needed. When $P_k b = 0$ the smallest singular value $\sigma_{\min}(A)$ has no influence to the solution of the LS problem and should be eliminated from our considerations. When $P_k b$ is small, elimination of $\sigma_{\min}(A)$ (replacing of A by $\tilde{A}$) has little effect on the LS solution.

We will finish this contribution by a short note illustrating the conceptual and technical complications which arise when the assumption (4) is not used. First we must analyze when $\delta(\gamma) = 1$. The necessary and sufficient conditions for $\delta(\gamma) = 1$ were given in [9, Theorem 3.1]. Here we will explain the main idea in relation to the secular equation (20). Let $n \times k$ A have singular values $\sigma_1 \geq \cdots \geq \sigma_j > \sigma_{j+1} = \cdots = \sigma_k > 0$. When $\delta(\gamma) = 1$, b has no components in the left singular vector subspace of A corresponding to $\sigma_{\min}(A)$, $P_k b = 0$, $\alpha_{j+1} = \cdots = \alpha_k = 0$ and the matrix with the appended column $[A, b\gamma]$ has $k - j$ singular values equal to $\sigma_{\min}(A)$. The singular values of $[A, b\gamma]$ different from those of A are solutions $\sigma(\gamma)$ of the deflated secular equation, see [11, Ch2, §47, pp. 103–104],

$$0 = \psi_j(\sigma(\gamma), \gamma) = \gamma^2 \|r\|^2 - \sigma(\gamma)^2 - \gamma^2 \sigma(\gamma)^2 \sum_{i=1}^{j} \frac{|\alpha_i|^2}{\sigma_i^2 - \sigma(\gamma)^2}, \quad (24)$$

where the summation term is ignored if all singular values of A are equal. Note that $\psi_j(0, \gamma) > 0$, so that $\delta(\gamma) = 1$ requires that $\psi_j(\sigma_k, \gamma) \geq 0$ (if $\psi_j(\sigma_k, \gamma) < 0$, then the deflated secular equation (24) must have a positive solution σ less than $\sigma_{\min}(A)$ which contradicts the condition $\delta(\gamma) = 1$).

It is interesting to note that for the particular choice of $\gamma = \sigma_k / \|r\|$, the condition $\psi_j(\sigma_k, \gamma) \geq 0$ is equivalent to $\alpha_1 = \cdots = \alpha_j = 0$, i.e. $U_A^T b = 0$ and $r = b$. In the other words, $\delta(\gamma) < 1$ for $\gamma < \sigma_k / \|b\|$ (the last column of the matrix $[A, b\gamma]$ has for $\gamma < \sigma_k / \|b\|$ norm less than $\sigma_{\min}(A)$), and for the choice $\gamma_b \equiv \sigma_k / \|b\|$ the condition $\delta(\gamma_b) = 1$ is equivalent to the fact that in the LS problem (2) the LS solution $x = 0$ is trivial and $r = b$. When this particular γ_b is used with (17), we obtain (see also [5, Section 2])

$$\delta(\gamma_b) \|b\| \leq \|r\| \leq \delta(\gamma_b) \|b\| \{2 - \delta(\gamma_b)^2\}^{\frac{1}{2}}. \quad (25)$$

The results presented here have been successfully applied outside the Errors-in-Variables Modeling field for analysis of convergence and numerical stability of Krylov subspace methods, see [8], [5].

5. Conclusion

Summarizing, our contribution (which is based on [7]) shows new bounds for the LS residual norm $\|r\| = \min_x \|b - Ax\|$ in terms of the Scaled TLS distance $\sigma_{\min}([A, b\gamma])$, and presents several important corollaries describing the tightness of the bounds and their dependence on the parameter γ. The bounds were seen to be very good when $\sigma_{\min}([A, b\gamma])$ was sufficiently smaller than $\sigma_{\min}(A)$. When $\sigma_{\min}([A, b\gamma]) \approx \sigma_{\min}(A)$, it is shown that the smallest singular value $\sigma_{\min}(A)$ and its singular vectors did not play a significant role in the solution of the LS problem.

Acknowledgments

This work was supported by NSERC of Canada Grant OGP0009236 and by the GA AS CR under grant A2030801. Part of this work was performed while Zdenek Strakos was visiting Emory University, Atlanta, GA, U.S.A.

References

[1] Å. Björck, P. Heggerness, and P. Matstoms, Methods for large scale total least squares problems, *SIAM J. Matrix Anal. Appl.*, 22:413–442, 2000.

[2] G. H. Golub and C. F. Van Loan, An analysis of the total least squares problem, *SIAM J. Numer. Anal.*, 17:883–893, 1980.

[3] ——, *Matrix Computations*, The Johns Hopkins University Press, Baltimore MD, third ed. 1996.

[4] A. Greenbaum, M. Rozložník and Z. Strakoš. Numerical behavior of the modified Gram-Schmidt GMRES implementation. *BIT*, 37(3):706–719, 1997.

[5] J. Liesen, M. Rozložník and Z. Strakoš. On Convergence and Implementation of Residual Minimizing Krylov Subspace Methods, to appear in *SIAM J. Sci. Comput.*

[6] S. Van Huffel and J. Vandewalle, *The Total Least Squares Problem: Computational Aspects and Analysis*, SIAM, Philadelphia, 1991.

[7] C. C. Paige and Z. Strakoš. Bounds for the least squares distance using scaled total least squares, to appear in *Numerische Mathematik.*

[8] C. C. Paige and Z. Strakoš. Residual and backward error bounds in minimum residual Krylov subspace methods to appear in *SIAM J. Sci. Comput.*

[9] C. C. Paige and Z. Strakoš, Scaled total least squares fundamentals, to appear in *Numerische Mathematik.*

[10] B. D. Rao, Unified treatment of LS, TLS and truncated SVD methods using a weighted TLS framework, In: S. Van Huffel (editor), *Recent Advances in Total Least Squares Techniques and Errors-in-Variables Modelling*, pp. 11–20, SIAM Publications, Philadelphia PA, 1997.

[11] J. Wilkinson, *The Algebraic Eigenvalue Problem*, Clarendon Press, Oxford, 1965.

Part II

TOTAL LEAST SQUARES ALGORITHMS

RECENT DEVELOPMENTS IN RANK REVEALING AND LANCZOS METHODS FOR TLS-RELATED PROBLEMS

Ricardo D. Fierro

Department of Mathematics, California State University, San Marcos, CA 92096 USA

fierro@csusm.edu

Per Christian Hansen

Informatics and Mathematical Modelling Building 321,
Technical University of Denmark, DK-2800, Lyngby, Denmark

pch@imm.dtu.dk

Abstract The singular value decomposition (SVD) of a data matrix is useful for providing rank and subspace information as well as solving and analyzing total least squares (TLS) problems. Rank-revealing decompositions, herein referred to as UTV decompositions, and Lanczos methods have been proposed as substitutes for the SVD in various applications through the years. These type of methods are appropriate for solving TLS-related problems because they can efficiently and reliably provide an orthogonal basis that accurately approximates the needed numerical row space or null space.

The development and availability of reliable and robust codes for TLS algorithms always lags the pace of the literature. We discuss some recent developments in the use of UTV decompositions and Lanczos methods for solving TLS-related problems along with related Matlab codes.

Keywords: singular value decomposition, total least squares, subspace estimation, rank revealing decomposition, updating, downdating, Hankel TLS, Hankel/Toeplitz matrix.

S. Van Huffel and P. Lemmerling (eds.), Total Least Squares and Errors-in-Variables Modeling, 47–56.

1. Introduction

The singular value decomposition (SVD) of a data matrix is useful for providing rank and subspace information as well as solving and analyzing total least squares (TLS) problems. This orthogonal decomposition is an important analytical tool and the algorithm is highly regarded, but for many applications the SVD requires too much storage, computation time, or cannot be implemented efficiently for situations that involve adding new data or deleting old data. Rank revealing decompositions, herein referred to as UTV decompositions, and Lanczos methods have been proposed and analyzed as substitutes for the SVD in various applications through the years. These types of methods are appropriate for solving TLS-related problems because they can efficiently and reliably provide an orthogonal basis that accurately approximates the needed numerical row space or null space. Here, TLS-related problems are ones that involve working with some orthogonal matrix V which is block-partitioned as

$$V = \begin{bmatrix} V_{11} & V_{12} \\ V_{21} & V_{22} \end{bmatrix} \qquad (1)$$

and subsequently make use of $-V_{12}V_{22}^{\dagger}$ or $(V_{11}^T)^{\dagger}V_{21}^T$.

The development and availability of reliable codes for TLS-related algorithms always lags the pace of the literature. One reason is that independent researchers tend to focus on different TLS-related applications, thus the demand for specialized TLS algorithms grows slowly rather than rapidly. Another reason is that quality codes suitable for public use involve additional and more careful programming for reliability and robustness.

We discuss some recent developments in the use of UTV decompositions and Lanczos methods for solving TLS-related problems along with related codes [9, 10]. Hopefully this work will stimulate a collection of reliable MATLAB codes for solving TLS-related problems and thereby encourage others to consider using or adapting these algorithms for their applications, as well as contributing new algorithms.

In Section 1 we consider algorithms for recursive TLS problems, particularly exponential windowing and rectangular windowing. In Section 2 we consider subspace estimation for TLS-related problems involving Toeplitz matrices so that fast matrix-vector multiplications can be used efficiently. As expected, the core of the iterative algorithm we consider relies on the Lanczos method.

2. Modifying the ULV decomposition

The URV and ULV decompositions [21, 22], each considered a rank revealing decomposition, were introduced as fast alternatives to the eigendecomposition and singular value decomposition in subspace-based signal processing methods. Here we consider aspects of the ULV decomposition because the details for the URV decomposition are similar. The codes we discuss in this section are available in a recent Matlab collection called "UTV Tools" [9, 10] for computing and modifying rank revealing UTV decompositions. They can be used, for example, in direction-of-arrival estimation of time-varying signals from an array of sensors using TLS-ESPRIT [18, 20]. This particular algorithm computes an orthogonal matrix V in (1), and then the eigenvalues of $-V_{12}V_{22}^{-1}$ or $(V_{11}^T)^{-1}V_{21}^T$ are related to the direction-of-arrival estimation of the signals.

2.1 Exponential windowing

Some signal processing applications use a procedure called "exponential windowing" to process data. Here, a new sample of signals is appended as a row w^T in a matrix βA where β is the weighting factor which is multiplied to existing rows to dampen the influence of old data. The goal is to use the ULV decomposition of $A = ULV^T$, where U is an $m \times n$ orthonormal matrix, L is an $n \times n$ lower triangular matrix, and V is an $n \times n$ orthogonal matrix, to obtain the ULV decomposition of the modified matrix

$$\bar{A} = \begin{bmatrix} \beta A \\ w^T \end{bmatrix} = \bar{U}\bar{L}\bar{V}^T.$$

The ULV updating procedure for computing the new ULV decomposition $\bar{A} = \bar{U}\bar{L}\bar{V}^T$ is described in [22] and implemented in **ulv_up.m**; the calling sequence may look as follows:

```
[p,L,V,U,vec] = ulv_up(p,L,V,U,w,beta,tol_rank).
```

The inputs are: `p`, numerical rank of A revealed in the $n \times n$ matrix L; `L`, `V`, `U`, the ULV factors such that $A = ULV^T$; `w`, the new row added to A; `beta`, the weighting factor in $(0, 1]$; and `tol_rank`, the rank decision tolerance. The outputs are: `p`, the numerical rank of $\bar{A}$; `L`, `V`, `U`, the ULV factors such that $\bar{A} = \bar{U}\bar{L}\bar{V}^T$; `vec`, a 5×1 vector with elements that describe the quality of the decomposition, such as `v(1)` $\approx \|\bar{L}(p+1\colon n, 1\colon p)\|$.

Matrices U and $\bar{U}$ are not needed in some applications, and the updating procedure can be accomplished safely without it. For $\beta = 1$, the numerical rank can either remain the same or increase by one. For $0 < \beta < 1$, the rank can either remain the same, increase by one, or decrease by one, depending on the rank tolerance `tol_rank`. Condition estimation is performed with **ccvl.m**, which is based on the generalized LINPACK condition estimator [4].

The ULV-based null space approximation $\bar{V}(:\,, p+1\colon n)$, where p is the numerical rank of $\bar{A}$, can be refined by reducing the norm of the off-diagonal block of $\bar{L}$ below a prescribed tolerance `tol_ref`, which can be included as an additional input parameter. In the ith step this refinement is done by shiftless QR iterations on $\bar{L}(i, 1\colon i)$ using Givens rotations [20, 21, 22].

2.2 Rectangular windowing

As an alternative to exponential windowing, some signal processing applications use a two-step procedure called "rectangular windowing" to process data. The first step involves updating, using **ulv_up.m**, and here this means adding a new row of data as described in Section 2.1 with $\beta = 1$. Hence we work with the current matrix $\bar{A} = [A^T w]^T = \bar{U}\bar{L}\bar{V}^T$. The second step involves the nontrivial task of downdating, using **ulv_dw.m**, and here this means finding a ULV decomposition $\bar{A}(2\colon m, 1\colon n) = \hat{U}\hat{L}\hat{V}^T$ from $\bar{A} = \bar{U}\bar{L}\bar{V}^T$. The combined ULV updating/downdating procedure for computing the new ULV decomposition using the rectangular windowing method is implemented in **ulv_win.m**, and the calling sequence may look as follows:

```
[p,L,V,U,vec] = ulv_win(p,L,V,U,w,alg_type,tol_rank).
```

The inputs and outputs of **ulv_win** are similar to **ulv_up.m**, except for the input `alg_type`. This is due to the various ways the downdating step can be implemented, depending on the availability of $\bar{U}$. For `alg_type` $= 1$ or 2, $\bar{U}$ is unavailable and A is required, and $\bar{U}(1, :\,)$ can be obtained using the LINPACK/CSNE (corrected semi-normal equations) method as described in [19] and implemented in **ulv_csne.m**. The downdating algorithm ignores the rank revealing structure in $\bar{L}$ for `alg_type` $= 1$ (see [2, 19]), but makes use of the structure as much as possible for `alg_type` $= 2$, which results in a slightly more accurate but also more complicated variant (see [3]). For `alg_type` $= 3$, the matrix $\bar{U}$ is maintained, hence we follow the LINPACK approach, where the Modified Gram-Schmidt algorithm with reorthogonalization as described in [5] and implemented in **mgsr.m** is used to generate the expanded orthonormal matrix $[\bar{U},\ q]$.

3. Lanczos-based subspace estimation

Applications which involve measured signals considered to be a mixture of pure signals plus noise include noise suppression in speech processing (e.g., Jensen et al. [13] and metabolite quantification in medical diagnosis (e.g., Barkhuijsen et al. [1]). Subspace estimation methods in signal processing require the computation of an orthogonal basis for a subspace associated with a structured matrix that is derived from the sampled data [16]. Approximate or inexact signal subspaces typically perform just as well as SVD-based signal subspaces due to the noise in the data [7, 8, 11]. Krylov subspace techniques are very effective for this purpose when matrix-vector multiplications can be performed quickly.

3.1 New algorithm: TKSVD

To motivate our new algorithm referred to as TKSVD [9], we consider a TLS application in one-dimensional time-domain signal processing. In a variant [6, 24, 25] of the HSVD method [15], the N uniformly sampled data points y_ℓ ($\ell = 0, 1, \ldots, N-1$) are arranged to form the $m \times n$ Toeplitz matrix

$$A = \begin{bmatrix} y_{m-1} & y_m & \cdots & y_{N-1} \\ y_{m-2} & y_{m-1} & \cdots & y_{N-2} \\ \vdots & \vdots & \ddots & \vdots \\ y_0 & y_1 & \cdots & y_{n-1} \end{bmatrix} \qquad (N = m + n - 1). \qquad (2)$$

The measurement y_ℓ ($\ell = 0, 1, \ldots, N-1$) is assumed to be a sum of a pure signal y_ℓ^{pure}, characterized by

$$y_\ell^{pure} = \sum_{p=1}^{k} a_p e^{j\phi_p} e^{(-d_p + j2\pi f_p)\, t_\ell},$$

and white Gaussian noise e_ℓ, that is, $y_\ell = y_\ell^{pure} + e_\ell$. Here, $t_\ell = \ell\Delta t + t_0$, where Δt is the sampling interval, t_0 is the time between the effective time origin and the first data point to be included in the analysis. The parameter estimates (frequencies f_p, damping factors d_p, amplitudes a_p, and phase shifts ϕ_p) are related to the TLS solution to $\bar{V}(1\colon n-1, 1\colon k)E \approx \bar{V}(2\colon n, 1\colon k)$, for $k \ll min(m, n)$ [24, 25], where the columns of $\bar{V}(:, 1\colon k)$ represent the right singular vectors corresponding to the k dominant singular values of A. Therefore, a Lanczos-based method that computes an approximation $V(:, 1\colon k)$ to $\bar{V}(:, 1\colon k)$ is a natural choice.

The algorithm implemented in **tksvd.m** blends the Lanczos method, reorthogonalization, and explicit restarts for computing $V(:, 1\colon k)$. The core of the algorithm is the symmetric Lanczos method [12, p. 480] applied to $A^T A$ in product form, and basically consists of two phases.

Phase 1. In this phase, the Lanczos vector q_i is orthogonalized against previously generated Lanczos vectors $q_1, \ldots, q_{i-1}$, hence it requires complete reorthogonalization. This is not a serious drawback for the application we have in mind because $k \ll \min(m, n)$ and $m + n - 1 = 2^p$ for some small whole number p, usually $8 \leq p \leq 10$. The Lanczos algorithm runs a fixed number `max_iter` of steps, say `max_iter` $= k$ or $2k$, and then the residuals are computed to identify the converged Ritz values $\lambda_1, \ldots, \lambda_{k_1}$ and Ritz vectors $v_1, \ldots, v_{k_1}$. Generally, $k_1 < k$ and a restart is required. To restart the process, TKSVD orthogonalizes the unconverged Ritz vector associated with the smallest residual against all the converged Ritz vectors, with normalization. Every new Lanczos vector q_i is orthogonalized against previously generated Lanczos vectors $q_1, \ldots, q_{i-1}$ as well as all converged Ritz vectors $v_1, \ldots, v_{k_2}$. After a few restarts, we finally get $k_2 = k$.

Phase 2. There is no guarantee that the k converged Ritz pairs $(\lambda_1, v_1), \ldots (\lambda_k, v_k)$ are the correct ones, so additional iterations are required for possible exchanges. TKSVD restarts in the usual way, except now it monitors the dominant Ritz values $\lambda_1^{(i)}$ of the tridiagonal Rayleigh quotient matrices T_i and terminates the Lanczos process once they appear to be converging. Then the residuals are computed to identify converged Ritz values, say $\theta_1, \ldots, \theta_{k_3}$. TKSVD terminates if $\lambda_k > \max\{\theta_1, \ldots, \theta_{k_3}\}$. Otherwise, the Ritz pairs associated with the k largest Ritz values $\lambda_1, \ldots, \lambda_k, \ldots, \theta_1, \ldots, \theta_{k_3}$ are selected, and the algorithm restarts and monitors the dominant Ritz values in the usual way, and eventually captures the correct k dominant Ritz pairs. This approach takes advantage of the ability of the Lanczos method to converge quickly to the dominant Ritz value [14]. Hence, Phase 2 makes the algorithm reliable because it recovers close or multiple singular values that would ordinarily be skipped by the basic Lanczos method.

Given a user's tolerance δ, the test used to identify converged Ritz pairs produces a relative residual

$$\rho \equiv \|A^T U_k - V(:,1\colon k) R_k^T\| / \|A\| < \delta \,, \tag{3}$$

where $U_k R_k$ is the skinny QR factorization of $AV(:,1\colon k)$. The MATLAB routine is called by `[Uk,Rk,Vk,vec] = tksvd(colA,rowA,k,options)`.

3.2 Numerical results

Now we give the results of some experiments based on Data Set 002 from the BioSource Database [23], which consists of voxels that are part of an image of a human brain. The analysis of Nuclear Magnetic Resonance (NMR) data is very difficult and requires expertise in NMR spectroscopy to determine the chemical properties of the physical matter. Therefore, we provide numerical results without regard to its relation to chemical properties. All computations were performed with MATLAB on a PC with a Pentium III 450 MHz processor and 128 MB of memory. For typical data in this database, the first few singular values of A decay rapidly, but the remaining singular values decay slowly, making them more "difficult" to capture.

In the first experiment, we consider the voxel represented by the complex signal vector in the file `csi-4-7.mat`. Table 1 summarizes the results of computing $V(:,1:k)$ for $N = 1024$, $m = 513$, $n = 512$, $\delta = 10^{-5}$, and various model orders k. Here, $\hat{\rho}$ is an *a posteriori* estimate of the relative residual ρ in (3).

Table 1. Summary of numerical results for the original signal in the MRS data file `csi-4-7.mat` for $N = 1024$, $m = 513$, $n = 512$, various model orders k, and accuracy tolerance $\delta = 10^{-5}$. $\hat{\rho}$ is an a posteriori estimate of the relative residual ρ.

k	`max_iter`	restarts	ρ	$\hat{\rho}$	flops	time
5	5	5	1.70×10^{-6}	1.70×10^{-6}	1.09×10^{7}	0.39s
5	10	2	3.41×10^{-7}	3.41×10^{-7}	9.93×10^{6}	0.32s
10	10	6	2.08×10^{-7}	3.91×10^{-7}	3.04×10^{7}	0.99s
10	20	1	2.82×10^{-7}	2.82×10^{-7}	1.66×10^{7}	0.49s
15	15	9	1.30×10^{-6}	2.12×10^{-6}	7.83×10^{7}	2.23s
15	30	3	1.10×10^{-6}	1.19×10^{-6}	6.56×10^{7}	1.70s
20	20	10	9.03×10^{-7}	1.18×10^{-6}	1.23×10^{8}	3.35s
20	40	3	4.77×10^{-7}	6.46×10^{-7}	1.72×10^{8}	3.68s
25	25	13	1.23×10^{-6}	1.50×10^{-6}	2.27×10^{8}	5.88s
25	50	3	5.56×10^{-7}	6.34×10^{-7}	2.89×10^{8}	5.71s

The numerical results in Table 1 depend on the starting vector in TKSVD, among other parameters, and the data set. In the second experiment, we considered a few other data sets as indicated in Table 2. For each signal, we report large-sample 95% confidence intervals for the expected flop count and CPU time using 100 samples, where TKSVD used a random starting vector each time. In all the samples we chose $N = 1024$, $m = 513$, $n = 512$, orders $k = 15$ and 30, accuracy tolerance $\delta = 10^{-5}$, and `max_iter` $= 2k$.

Table 2. 95% confidence intervals for expected flop count and CPU time (seconds) of TKSVD for 100 samples. For each signal, $N = 1024$, $m = 513$, $n = 512$, and $\delta = 10^{-5}$.

csi-4-7.mat	**flops**	**CPU time**
$k = 15$	$(7.43 \times 10^7, 7.99 \times 10^7)$	(1.93, 2.05)
$k = 30$	$(3.52 \times 10^8, 4.11 \times 10^8)$	(8.41, 9.37)
csi-6-8.mat		
$k = 15$	$(4.32 \times 10^7, 4.55 \times 10^7)$	(1.19, 1.24)
$k = 30$	$(3.37 \times 10^8, 3.69 \times 10^8)$	(6.79, 7.29)
csi-9-6.mat		
$k = 15$	$(6.60 \times 10^7, 7.02 \times 10^7)$	(1.66, 1.75)
$k = 30$	$(4.36 \times 10^8, 4.63 \times 10^8)$	(9.08, 9.52)
csi-8-7.mat		
$k = 15$	$(7.85 \times 10^7, 8.47 \times 10^7)$	(1.91, 2.04)
$k = 30$	$(3.87 \times 10^8, 4.28 \times 10^8)$	(8.21, 8.87)
csi-10-10.mat		
$k = 15$	$(4.39 \times 10^7, 4.64 \times 10^7)$	(1.18, 1.23)
$k = 30$	$(2.21 \times 10^8, 2.66 \times 10^8)$	(5.00, 5.75)
csi-5-5.mat		
$k = 15$	$(7.26 \times 10^7, 7.88 \times 10^7)$	(1.87, 2.00)
$k = 30$	$(3.15 \times 10^8, 3.68 \times 10^8)$	(6.98, 7.86)

To compare the numerical results in Tables 1 and 2, we carefully implemented a MATLAB routine for bidiagonalizing a 513×512 complex Toeplitz matrix using Householder transformations and accumulating only the right transformations. This approach, which represents the first phase of the SVD algorithm, required 2.83×10^9 flops and 384.97 seconds, while MATLAB's "economy size" built-in SVD routine required 5.78×10^9 flops and 139.68 seconds.

Acknowledgments

The authors are grateful to Leentje Vanhamme and Sabine Van Huffel at the Dept. of Electrical Engineering (ESAT) at K.U. (Leuven, Belgium) for helpful discussions concerning the BioSource Database [23].

References

[1] H. Barkhuijsen, R. D. Beer, and D. V. Ormondt. Improved Algorithm for Non-iterative Time-Domain Model Fitting to Exponentially Damped Magnetic Resonance Signals. *J. Magn. Reson.*, 30:553–557, 1987.

[2] J. L. Barlow, P. A. Yoon, and H. Zha. An Algorithm and a Stability Theory for Downdating the ULV Decomposition. *BIT*, 36:14–40, 1996.

[3] A.W. Bojanczyk and J.M. Lebak. Downdating a ULLV Decomposition of Two Matrices, in J.G. Lewis (Ed.), *Applied Linear Algebra*, SIAM, Phil., 1994.

[4] A.K. Cline, A.R. Conn, and C.F. Van Loan. Generalizing the LINPACK Condition Estimator. in J.P. Hennart (Ed.), *Numerical Analysis*, Lecture Notes in Mathematics, 909: 73–83, Springer, 1982.

[5] J.W. Daniel, W.B. Gragg, L. Kaufman and G.W. Stewart. Reorthogonalization and Stable Algorithms for Updating the Gram-Schmidt QR Factorization, *Math. Comp.*, 30:772–795, 1976.

[6] R. de Beer and D. van Ormondt. *Analysis of NMR Data Using Time Domain Fitting Procedures.* in NMR Basic Principles and Progress, Vol. 26, Springer-Verlag, Berlin, 1992.

[7] L. Eldén and E. Sjöström. Fast Computation of the Principal Singular Vectors of Toeplitz Matrices Arising in Exponential Data Modelling of a Toeplitz Matrix. *Signal Processing*, 50:151–164, 1996.

[8] R.D. Fierro. Lanczos, Householder Transformations, and Implicit Deflation for Fast and Reliable Dominant Singular Subspace Computation. *Num. Lin. Alg. with Applic.*, 8:245–264, 2001.

[9] The software is available from the authors' home pages:
R.D. Fierro: http://www.csusm.edu/Math/fierro/RDF.html.
P.C. Hansen: http://www.imm.dtu.dk/~pch.

[10] R.D. Fierro, P.C. Hansen, and P.S.K. Hansen. UTV Tools: Matlab Templates for Rank Revealing UTV Decompositions. *Numer. Algo.*, 20:165–194, 1999.

[11] R.D. Fierro, L. Vanhamme, and S. Van Huffel. *Total Least Squares Algorithms Based on Rank-Revealing Complete Orthogonal Decompositions.* in Recent Advances in Total Least Squares Techniques and Errors-in-Variables Modeling, S. Van Huffel, ed., SIAM Proceedings series, 1997, pp. 99–116.

[12] G. Golub and C. V. Loan. *Matrix Computations.* John Hopkins University Press, 1996.

[13] S.H. Jensen, P.C. Hansen, S.D. Hansen, and J.Aa. Sørensen. Reduction of Broad-band Noise in Speech by Truncated QSVD. *IEEE Trans. Audio Speech Proc.*, 3:439–448, 1995.

[14] J. Kuczynski and H. Wozniakowski. Estimating the Largest Eigenvalue by the Power and Lanczos Algorithm with a Random Start. *SIAM J. Matrix Anal. Appl.*, 13:1094–1122, 1992.

[15] S. Kung, K. Arun, and B. Rao. State-space and Singular Value Decomposition-based Approximation Methods for the Harmonic Retrieval Problem. *J. Opt. Soc. Am.*, 73:1799–1811, 1983.

[16] F. T. Luk and D. Vandevoorde. Decomposing a Signal into a Sum of Exponentials in R.H. Chan, T.F. Chan, and G.H.Golub (Eds.) *Iterative Methods in Scientific Computing*, Springer, 1997.

[17] M. Moonen, P. Van Dooren, and J. Vandewalle A Note on Efficient Numerically Stabilized Rank-One Eigenstructure Updating. *IEEE Trans. on Signal Proc.*, 39:1911–1913, 1991.

[18] B. Ottersten and T. Kailath. Direction-of-Arrival Estimation for Wide-Band Signals Using the ESPRIT Algorithm. *IEEE Trans. on Acoustics, Speech, and Signal Proc.*, 38:317–327, 1990.

[19] H. Park and L. Eldèn. Downdating the Rank Revealing URV Decomposition. *SIAM J. Matrix Anal. Appl.*, 16:138–155, 1995.

[20] K.J. Ray Liu, D. O'Leary, G.W. Stewart, and Y.J. Wu. URV ESPRIT for Tracking Time-Varying Signals. *IEEE Trans. on Signal Proc.*, 42:3441–3448, 1994.

[21] G.W. Stewart. An Updating Algorithm for Subspace Tracking. *IEEE Trans. on Signal Proc.*, 40:1535–1541, 1992.

[22] G.W. Stewart. Updating a Rank-Revealing ULV Decomposition. *SIAM J. Matrix Anal. Appl*, 14:494–499, 1993.

[23] Data Set 002 from the BioSource Database can be accessd at http://www.esat.kuleuven.ac.be/sista/members/biomed/data002.htm.

[24] S. Van Huffel, L. Aerts, J. Bervoets, J. Vandewalle, C. Decanniere, and P. Van Hecke. Improved Quantitative Time-Domain Analysis of NMR Data by Total Least Squares. in *Proc. 6th Eur. Sign. Process. Conf. (EUSIPCO)*, vol. III, 1992, pp. 1721–1724.

[25] S. Van Huffel, H. Chen, C. Decanniere, and P. V. Hecke. Algorithm for Time-domain NMR Data Fitting Based on Total Least Squares. *J. Magn. Reson.*, 110:228–237, 1994.

[26] L. Vanhamme, R. Fierro, S. Van Huffel, and R. D. Beer. Fast Removal of Residual Water in Proton Spectra. *J. Magn. Reson.*, 132:197–203, 1998.

A REGULARIZED TOTAL LEAST SQUARES ALGORITHM

Hongbin Guo and Rosemary A. Renaut

Department of Mathematics, Arizona State University, Tempe, AZ 85287 1804

{hb_guo, renaut}@asu.edu

Abstract Error-contaminated systems $Ax \approx b$, for which A is ill-conditioned, are considered. Such systems may be solved using Tikhonov-like regularized total least squares (R-TLS) methods. Golub et al, 1999, presented a direct algorithm for the solution of the Lagrange multiplier formulation for the R-TLS problem. Here we present a parameter independent algorithm for the approximate R-TLS solution. The algorithm, which utilizes the shifted inverse power method, relies only on a prescribed estimate for the regularization constraint condition and does not require the specification of other regularization parameters. An extension of the algorithm for nonsmooth solutions is also presented.

Keywords: total least squares, regularization.

1. Introduction

We consider the development of algorithms for the solution of the Tikhonov type regularization of the total least squares (R-TLS) problem

$$\min\|(E,f)\|_F \quad \text{subject to} \quad (A+E)x = b+f \quad \text{and} \quad \|Lx\| \leq \delta. \qquad (1)$$

Here δ is a regularization parameter, and $L \in \mathbb{R}^{l\times n}$ defines a (semi)norm on the solution which is frequently chosen to approximate the first or second derivative operator. The utility of regularization for TLS, some of its properties, and its verification numerically via a parameter dependent direct algorithm were presented in [5]. Their algorithm, however, requires the development of a systematic approach for parameter selection. An alternative approach which solves a quadratic eigenvalue problem was presented by Golub at the workshop. Here we develop an iterative algorithm utilizing the shifted inverse power method for which the only prescribed parameter, δ, should be determined *a priori* from knowledge of the underlying physical model. The algorithm then de-

S. Van Huffel and P. Lemmerling (eds.), Total Least Squares and Errors-in-Variables Modeling, 57–66.

termines both the Lagrange multipliers and estimates a R-TLS solution concurrently.

In Section 2 we derive the algorithm which is then extended in Section 3 when regularization information on subdomains is available. Numerical results are presented in Section 4.

2. Derivation of the algorithm

The Lagrange multiplier formulation for the R-TLS problem (1), given by

$$\mathcal{L}(E,f,x,\xi,\mu) = \|(E,f)\|_F^2 + \xi^T(Ax+Ex-b-f) + \mu(\|Lx\|^2 - \delta^2),$$

where ξ and μ are to be determined, [9], can be simplified by the Kuhn-Tucker conditions to yield the equations

$$\mu L^T Lx + A^T f - \|f\|^2 x = 0, \tag{2}$$

$$(1+\|x\|^2)f = Ax - b, \tag{3}$$

$$\mu \geq 0, \;\; \mu(x^T L^T Lx - \delta^2) = 0. \tag{4}$$

Let x^* and μ^* denote the values of the solution x and parameter μ at the optimal solution of (1). If $\mu^* = 0$, the constraint $\|Lx\| \leq \delta$ is inactive, (e.g. see [9] Chapter 11.3), δ has been chosen too large, and $x^* = x_{TLS}$, where x_{TLS} is the solution of the TLS problem without regularization. Thus in the development of the algorithm we explicitly assume that the regularization parameter δ is provided such that $\|Lx\| \leq \delta$ is active. Equivalently, by (4), this requires

$$x^T L^T Lx - \delta^2 = 0, \tag{5}$$

and Theorem 2.1 of [5] can be applied to characterize the R-TLS solution.

Theorem 1. *For regularization problem* (1), *if the constraint is active, the solution* x^* *satisfies*

$$(A^T A + \lambda_I I + \lambda_L L^T L)x^* = A^T b, \tag{6}$$

$$\mu > 0, \;\; (x^*)^T L^T Lx^* - \delta^2 = 0, \tag{7}$$

where

$$\lambda_I = -\frac{\|Ax^* - b\|^2}{1+\|x^*\|^2}, \tag{8}$$

$$\lambda_L = \mu(1+\|x^*\|^2), \tag{9}$$

$$\mu = -\frac{1}{\delta^2}\left(\frac{b^T(Ax^*-b)}{1+\|x^*\|^2} + \frac{\|Ax^*-b\|^2}{(1+\|x^*\|^2)^2}\right). \tag{10}$$

This result can be further extended to characterize the R-TLS solution in terms of an eigenpair for an augmented system describing equations (2)–(4).

Theorem 2. *If the constraint is active, the solution x^* of regularization problem (1) satisfies:*

$$B(x^*)\begin{bmatrix} x^* \\ -1 \end{bmatrix} = -\lambda_I \begin{bmatrix} x^* \\ -1 \end{bmatrix}, \tag{11}$$

where

$$B(x^*) = \begin{bmatrix} A^TA + \lambda_L(x^*)L^TL, & A^Tb \\ b^TA, & -\lambda_L(x^*)\delta^2 + b^Tb \end{bmatrix},$$

and λ_L is determined by (9) *and (10).*

Conversely, if the pair $(\hat{x}^T, -1), -\hat{\lambda}$ is an eigenvector-eigenvalue pair for matrix $B(\hat{x})$,

$$B(\hat{x})\begin{bmatrix} \hat{x} \\ -1 \end{bmatrix} = -\hat{\lambda} \begin{bmatrix} \hat{x} \\ -1 \end{bmatrix} \tag{12}$$

and $\mu(\hat{x}) \neq 0$, then

$$\hat{x}^TL^TL\hat{x} = \delta^2, \ \hat{\lambda} = -\frac{\|A\hat{x} - b\|^2}{1 + \|\hat{x}\|^2},$$

and, furthermore, $\hat{x}$ satisfies (6)

Now, as in [5], we observe, from the relation $(E, f) = (-fx^T, f) = f(-x^T, 1)$ and (3), that

$$\|(E, f)\|_F^2 = \|f\|^2(1 + \|x\|^2) = \frac{\|Ax - b\|^2}{1 + \|x\|^2} = -\lambda_I.$$

Thus determining the minimum $\hat{\lambda}(\hat{x})$ which satisfies the eigenvalue equation (11) and $\mu(\hat{x}) > 0$ is equivalent to solving (1). Moreover, the Lagrange parameters are explicitly prescribed provided that an estimate for δ has been made available. In particular, because of the continuous dependence of the matrix spectrum on its entries, we have shown, by including the constraint within (11), that the R-TLS solution may be obtained utilizing a approach similar to Rayleigh quotient iteration (RQI) for the TLS problem [2], modified for matrix $B = B(x)$.

Our initial results use the shifted inverse power method to obtain an approximate value for $\lambda_I(x^*)$ and x^*. We denote by $\lambda_I^{(k)}$, $\lambda_L^{(k)}$, the values for $\lambda_I(x)$, $\lambda_L(x)$, given by (8) and (9), respectively, with $x = x^{(k)}$, where throughout the superscript k denotes the values at the kth iteration. The estimate of the normalized eigenvector of B is given by

$$z^{(k)} = ((x^{(k)})^T, -1)/\sqrt{1 + \|x^{(k)}\|^2}, \tag{13}$$

and we define the residual at step k by $\rho^{(k)} = \|B(x^{(k)})z^{(k)} + \lambda_I(x^{(k)})z^{(k)}\|$. We set a tolerance τ, on the residual ρ such that the iteration stops when $\rho^{(k)}/|\lambda_I^{(k)}| < \tau$ is obtained. In fact, by the Cauchy-Schwartz inequality, and $\|z^{(k)}\| = 1$,

$$\begin{aligned} \rho^{(k)} &\geq |(z^{(k)})^T(B(x^{(k)}) + \lambda_I(x^{(k)}))z^{(k)}| \\ &= |\mu(x^{(k)})(\|Lx^{(k)}\|^2 - \delta^2)|. \end{aligned} \tag{14}$$

Thus, the residual $\rho^{(k)}$ also provides an upper estimate for the violation of the constraint condition (4) and, if $z^{(k)}$ is sufficiently close to an eigenvector of $B(x^{(k)})$, then the inequality in (14) is replaced by equality.

In order to start the iteration we shall assume the calculation of an initial estimate $x^{(0)}$ from the regularized LS problem with a small regularization parameter, [7]. While it remains to determine a strategy for finding the optimal initial estimate of x, the choice $x^{(0)} = x_{RLS}$ is sufficient for the purposes of the verification of the algorithm when $L \neq I$. In the latter case, we note by [5], that the regularized LS and TLS solutions are identical for $\delta < \|x_{LS}\|$ and thus this initial choice is inappropriate.

Algorithm 1.

(i) Given $x^{(0)}$, calculate $\lambda_I^{(0)}$,$\lambda_L^{(0)}$ and $z^{(0)}$ from (8), (9) and (13).

(ii) Iterate to convergence, for $k = 0, 1, 2, \ldots$

(a) solve $(B(x^{(k)}) + \lambda_I^{(k)} I)y = z^{(k)}$.

(b) $x^{(k+1)} = -y(1:n)/y(n+1)$.

(c) Update $\lambda_I^{(k+1)}$ and $\lambda_L^{(k+1)}$.

(d) $z^{(k+1)} = y/\|y\|$.

(e) Update $\rho^{(k+1)}$.

(f) If $\frac{\rho^{(k+1)}}{|\lambda_I^{(k+1)}|} < \tau$ and $\mu(x^{k+1}) > 0$, break.

end for.

Remark 1. If $y(n+1) = 0$, we can not update x. In this case, we generate a new $z^{(k)}$ by combining $z^{(k)}$ and y, e.g. $y = y/\|y\|_2$, $z^{(k)} = (z^{(k)} + y)/\|z^{(k)} + y\|_2$.

Remark 2. The Kuhn-Tucker conditions for the original convex objective and constraint functions, (1), are both necessary and sufficient. The reduced conditions, (2)–(4), are, however, only necessary, and uniqueness of the solution is not guaranteed. Clearly, should a different starting value generate an alternative solution, then the one with minimal eigenvalue is chosen for the R-TLS problem. We note that this difficulty with convergence is also encountered with the RQI for the TLS problem, [2].

Remark 3. The efficiency of the algorithm clearly depends on the cost of solving the system $(B(x^{(k)}) + \lambda_I^{(k)} I)y = z^{(k)}$.

3. Non-smooth solutions

We now suppose that the global problem domain can be decomposed such that known local constraints can be provided for each subdomain. Specifically, we suppose that the domain is partitioned as

$$x = \begin{bmatrix} x^{(1)} \\ \vdots \\ x^{(p)} \end{bmatrix}, \quad \text{and that} \quad \|L_i x^{(i)}\| \leq \delta_i, \quad i = 1, 2, \ldots, p,$$

where $\|L_i x^{(i)}\| \leq \delta_i$, and L is a block-diagonal matrix,

$$L = \text{diag}(L_1, L_2, \ldots, L_p).$$

This may well provide for a more accurate estimate of the global TLS solution, particularly in cases for which the solution is smooth on some region, but less smooth on another and may be especially significant when there are jumps between subdomains.

We define the set of active constraints via

$$\mathcal{S} = \{i : \|L_i x^{(i)}\| \leq \delta_i, \text{ is active constraint}\},$$

and denote the local solution by $\overline{x^{(i)}}^T = [0, \ldots, {x^{(i)}}^T, \ldots, 0]$. Then we have

$$(A^T A + \lambda_I I + \lambda_L L^T L)x = A^T b,$$

$$\mu_i \geq 0, \ \sum_{i=1}^{p} \mu_i (x^{(i)T} L_i^T L_i x^{(i)} - \delta_i^2) = 0,$$

where

$$\lambda_I = -\frac{\|Ax - b\|^2}{1 + \|x\|^2},$$

$$\lambda_L = \text{diag}(\lambda_L^{(i)}) \in \mathbb{R}^{n \times n}, \ \lambda_L^{(i)} = \mu_i(1 + \|x\|^2), \ i = 1, 2, \cdots, p,$$

$$\mu_i(1 + \|x\|^2)\delta_i^2 = \overline{x^{(i)}}^T A^T b - \lambda_I \|x^{(i)}\|^2 - \overline{x^{(i)}}^T A^T A x, \ \text{if } i \in \mathcal{S},$$

$$\mu_i = 0, \ \text{if } i \notin \mathcal{S}.$$

From

$$\lambda_I\|x\|^2 = b^T Ax - \|Ax\|^2 - \sum_{i\in\mathcal{S}} \lambda_L^{(i)}\delta_i^2,$$

and

$$\lambda_I(1+\|x\|^2) = -b^T b - \|Ax\|^2 + 2b^T Ax,$$

we have

$$\lambda_I = b^T Ax - b^T b + \sum_{i\in\mathcal{S}} \lambda_L^{(i)}\delta_i^2.$$

Thus again (11) is satisfied for matrix

$$B = \begin{bmatrix} A^T A + \lambda_L(x)L^T L, & A^T b \\ b^T A, & b^T b - \sum_{i\in\mathcal{S}} \lambda_L^{(i)}\delta_i^2 \end{bmatrix},$$

and a similar algorithm can be constructed.

4. Numerical results

It has already been demonstrated in [5] that R-TLS is able to compute better results than Tikhonov's method. Moreover, it is generally accepted that the ordinary TLS solution does not differ significantly from the ordinary LS solution for *small* levels of noise, [3], [10]. Thus, the purpose of our numerical experiments is not to compare the effectiveness of the R-TLS concept with the Tikhonov or LS solutions. Rather, we seek to validate the presented parameter-independent R-TLS algorithm. We report results for three examples; one of these, `ilaplace` is taken from the regularization tools package, [7], and is tested statistically. The second example is designed to evaluate the behavious of the algorithm subject to different choices for the parameter δ, and the last is designed to evaluate the modified R-TLS algorithm which was presented in Section 3 for the case of a non-smooth solution. Moreover, since the purpose of the presentation is validation rather than efficiency, for all calculations we have used the MATLAB operator "\" to solve $(B+\lambda_I(x)I)y = z$.

In the design of our experiments we wish to test the robustness of the algorithm to perturbations in both the system matrix and the right hand side. In particular, we note that in most cases, the condition of the TLS problem is mainly determined by the perturbation of the system matrix, rather than the perturbation of the right hand side. Thus we suppose that the exact system $Ax^* = b$ is perturbed such that we solve the system

$$\overline{A}x_{RTLS} = \overline{b},\ \overline{A} = A + \sigma E,\ \overline{b} = b + \sigma e. \tag{15}$$

Here the entries in the error matrices E and e, obtained using the Matlab function `randn`, are normally distributed with zero mean and unit standard variation, and are normalized such that $\|E\|_F = \|e\|_2 = 1$. Different error levels are considered by chosing the parameter σ to be 1, 0.1, 0.01. 0.001, *cf.* [5]. It is obvious that the solution of the perturbed system, (15), x_{RTLS} depends on the choice for E and e, and that agreement with the exact solution x^* for a single case does not predict the overall behaviour of the algorithm. Thus, our results are presented as histograms for the relative error, $\|x_{RTLS} - x^*\|/\|x^*\|$, in the solution over 500 independent simulations of the same example. This approach compares with the approach presented in [3] apart from one important difference. In [3] the relative error is measured relative to an *optimal* solution, x_{opt}. Their algorithms for regularization are all dependent on certain parameter choices, either number of iterations, number of terms used in an SVD expansion, or a penalty parameter. Thus for each given algorithm there is a set of possible solutions dependent on the choices from these parameters. The optimal solution is defined to be the solution from this set of possible solutions which is closest to the exact solution. For our presented algorithm this is not an issue and we simply compare the obtained solution with the exact solution.

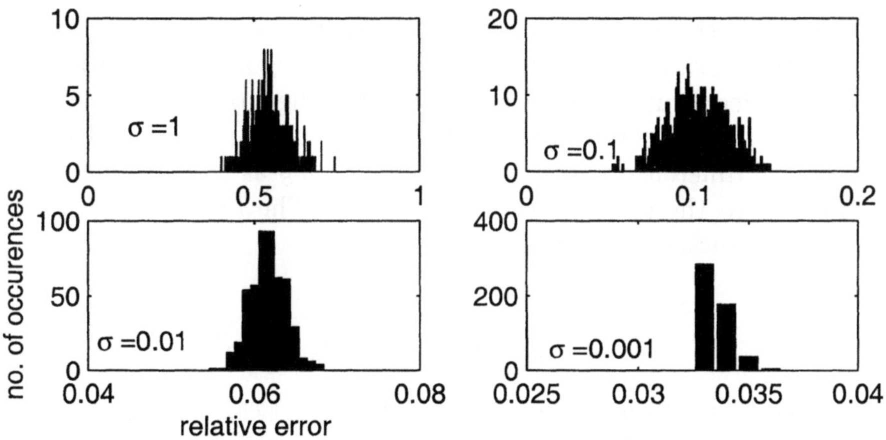

Figure 1. Example 1: The four histograms illustrate the statistical distribution of relative error. $\|x_{RTLS} - x^*\|/\|x^*\|$.

Example 1. Function `ilaplace` [5], [7], the inverse Laplace transform problem, is used to generate matrix A with order 65×64, and vectors x^*, $b = Ax^*$, but modified such that the scaling $\|A\|_F = \|Ax^*\|_2 = 1$ is adopted. The operator $L \in \mathbb{R}^{(n-1)\times n}$ approximates the first-derivative operator, $\delta = 0.9\|Lx^*\|$ bounds $\|Lx\|$, and we chose tolerance $\tau = 10^{-4}$ in the algorithm. Numerical results shown in Figure 1 demonstrate that the solutions are reasonable.

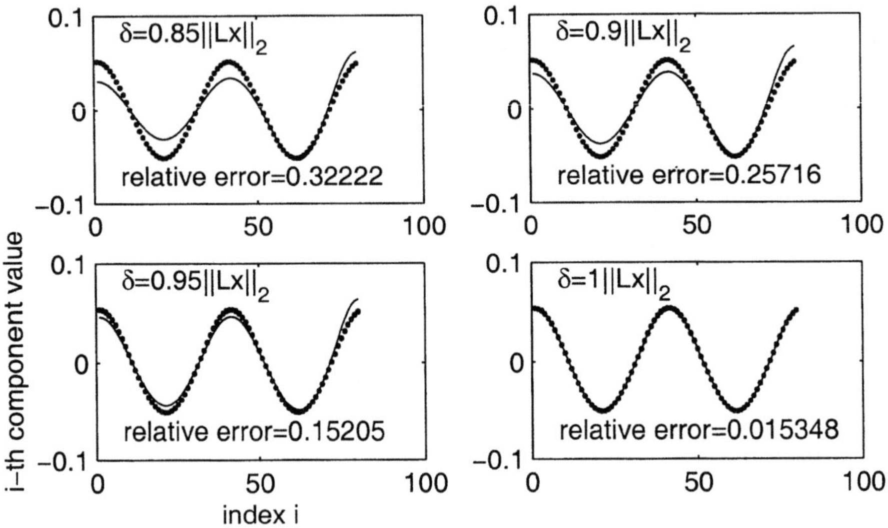

Figure 2. Example 2: The exact TLS solutions and RTLS solutions are given by the dotted lines and solid lines, respectively.

Example 2. In the former example the exact TLS solution is not known and it is therefore not possible to compare the regularized solution and non-regularized solution. Thus to evaluate the effect of the regularization we construct an ill-conditioned system with known TLS solution, and measure the difference between x_{TLS} and x_{RTLS} as the parameter δ is taken closer to the exact value for $\|Lx_{TLS}\|$. The $m \times (n+1)$ matrix $[A, b]$ is constructed as follows:

$$\begin{aligned} [A,b] &= U\Sigma V^T,\ \Sigma^T = [\ \Sigma_1,\ \ 0\ ] \\ U &= I_m - 2xx^T,\ V = I_n - 2yy^T, \end{aligned}$$

where vectors x and y are normalized to size 1, $\|x\|_2 = \|y\|_2 = 1$, with entries $x(i) = \sin(4\pi i/m), i = 0, 1, \dots, m-1$ and $y(j) = \cos(4\pi j/n)$, $j = 0, 1, \dots, n-1$. We use $m = 100$, $n = 80$ and take the singular values in the diagonal matrix of ordered singular values, Σ_1 to be geometrically spaced between 1 and 10^{-40}, [3]. The operator L again approximates the first order derivative and we again use tolerance $\tau = 10^{-4}$. Figure 2 illustrates the results x_{RTLS} as compared to $x_{TLS} = -V(1:n, n+1)/V(n+1, n+1)$ for the regularization choices $\delta = \theta\|Lx_{TLS}\|$, $\theta = 0.8, 0.85, 0.9$ and 1.0. We see that as θ becomes closer to 1.0, so that δ is closer to $\|Lx_{TLS}\|$, the solution x_{RTLS} tends to x_{TLS}.

Example 3. This example is designed to test the modified algorithm for the case in which the solution is not expected to be smooth and we have

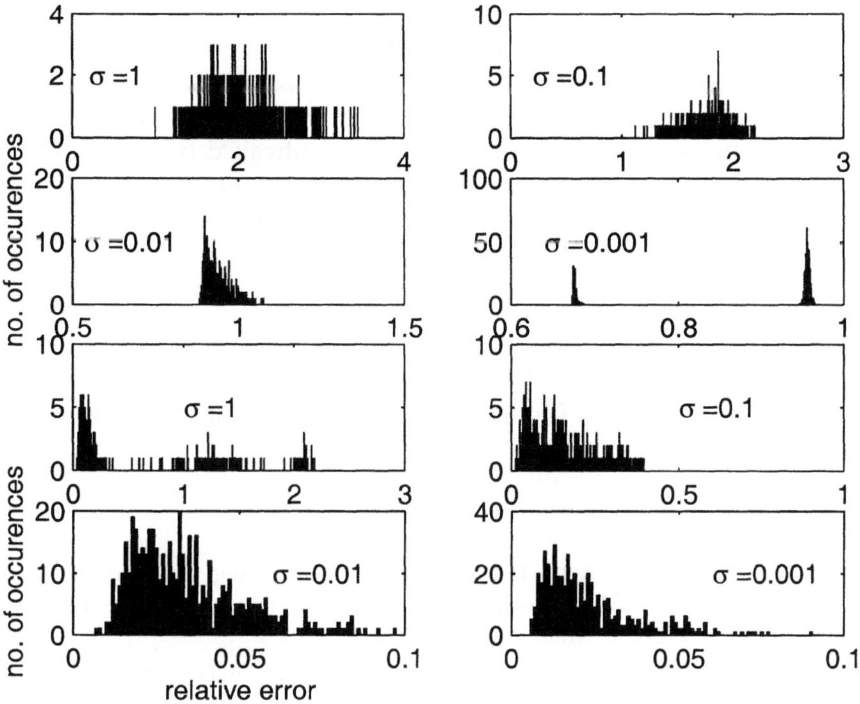

Figure 3. Example 3: upper and lower histograms show results for global constraint $||Lx|| \leq \delta$, and subdomain constraints $||L_i x|| \leq \delta_i$, resp.

information about the positions at which a lack of smoothness will occur. This example uses the system matrix defined from *shaw*, [7], but with exact solution x^* given by

$$x^*(i) = \begin{cases} \cos((i-1) * \pi/32) & 1 \leq i \leq 32, \\ \cos((i-33) * \pi/32) & 33 \leq i \leq 64, \end{cases}$$

and the scaling such that $||b|| = ||Ax^*||_2 = 1$. The global algorithm which uses $\delta = ||Lx^*||_2$, and tolerance $\tau = 10^{-8}$, and is illustrated in the upper four histograms of Figure 3, cannot provide an accurate estimate for x^*. On the other hand, with the subdomain parameters defined according to the values on each subdomain, $\delta_i = ||L_i x_i^*||_2,\ i = 1,2$, where L_1 and L_2 are first-derivative operators for each subdomain, and the tolerance is just $\tau = 10^{-4}$, we see from the lower four histograms of Figure 3 that better results are obtained.

5. Conclusions

The presented parameter independent algorithm for solving the regularized TLS problem works well for ill-conditioned problems with moderate error, provided only that an estimate for the degree of smoothness of the solution can be prescribed based on expected physical behaviour of the solution. Future work is planned in order to make the algorithm practical, to study the convergence behaviour and to take advantage of system structure.

Acknowledgments

This work has been supported in part by the Arizona Alzheimer's Disease Research Center which is funded by the Arizona Department of Health Services.

References

[1] R. C. Allen, Jr., W. R. Boland, V. Faber, and G. M. Wing. Singular values and condition numbers of galerkin matrices arising from linear integral equations of the first kind. *J. Math. Anal. Appl.*, 109:564–590, 1985.

[2] A. Björck, P. Heggernes, and P. Matstoms. Methods for large scale total least squares problems. *SIAM J. Matrix Anal. Appl.*, 22(2):413–429, 2000.

[3] R. D. Fierro, G. H. Golub, P. C. Hansen, and D. P. O'Leary. Regularization by truncated total least squares. *SIAM J. Sci. Comput.*, 18:1223–1241, 1997.

[4] G. H. Golub and C. F. Van Loan. *Matrix Computations, 3rd ed.* Johns Hopkins University Press, Baltimore, 1996.

[5] G. H. Golub, P. C. Hansen, and D. P. O'Leary. Tikhonov regularization and total least squares. *SIAM J. Matrix Anal. Appl.*, 21:185–194, 1999.

[6] G. H. Golub and C. F. Van Loan. An analysis of the total least squares problem. *SIAM J. Numer. Anal.*, 17:883–893, 1980.

[7] P. C. Hansen. Regularization tools: A matlab package for analysis and solution of discrete ill-posed problems. *Numer. Alg.*, 6:1–35, 1994.

[8] S. Van Huffel and J. Vandewalle. *The total least squares problem: computational aspects and analysis.* SIAM, Philadelphia, 1991.

[9] D. G. Luenberger. *Linear and nonlinear programming.* Addison-Wesley publishing company, second edition, 1984.

[10] G. W. Stewart. On the invariance of perturbed null vectors under column scaling. *Numer. Math.*, 44:61–65, 1984.

THE PARAMETRIC QUADRATIC FORM METHOD FOR SOLVING TLS PROBLEMS WITH ELEMENTWISE WEIGHTING

Amedeo Premoli
Dipartimento di Elettronica e Informazione, Politecnico di Milano
Piazza Leonardo da Vinci 32, I-20133 Milano, Italy

Maria Luisa Rastello
Istituto Elettrotecnico Nazionale G. Ferraris
Strada delle Cacce 91, I-10135 Torino, Italy
rastello@ien.it

Abstract This paper considers the Total Least Squares (TLS) problem with Elementwise Weighting (EW), removing the assumption of identical variances of data and observation errors while preserving their statistical independence. To solve them we approximate the gradient of the weighted Frobenius norm of errors by the gradient of an *ad hoc* Parametric Quadratic Form and present a fast iterative algorithm, exhibiting quadratic convergence. Some numerical examples show the capabilities of this algorithm in solving EW-TLS problems.

Keywords: elementwise-weighted total least squares, parametric quadratic form.

1. Introduction

The problem of *linear parameter estimation* consists in the solution of an overdetermined system of linear equations

$$A\,x \approx b \tag{1}$$

where $A \in \mathbb{R}^{m\times n}$, with $n < m$, is the *data matrix*, $b \in \mathbb{R}^m$ is the *observation vector* (univariate problem) and $x \in \mathbb{R}^n$ is the unknown vector; hereafter, $\mathcal{I} \equiv \{1, 2, \ldots, m\}$ and $\mathcal{J} \equiv \{1, 2, \ldots, n\}$.

In the *Ordinary Least Squares* (OLS) problems there is the underlying assumption that independently distributed (i.d.) errors affect only the vector b. Methods for estimating the effect of this assumption on the

S. Van Huffel and P. Lemmerling (eds.), Total Least Squares and Errors-in-Variables Modeling, 67–76.

OLS solution are given in [1]. The *Total Least Squares* (TLS) problems, introduced in [2], take into account i.d. errors in both data and observations, under the assumption of identical variances (i.i.d. errors). Their solutions have been found by using the *Singular Value Decomposition* (SVD), as pointed out in [3] and [4]. A full review on general TLS problems can be found in [5]. The case opposite, but less popular, to the OLS problem is the *Data Least Squares* (DLS) problem [6], where i.i.d. errors affect only the $m \times n$ elements of A.

In most applications, the above assumptions on error variances are unrealistic. The elements of A and b can be affected by i.d. errors with zero averages but different variances, because they represent, in general, various physical quantities and therefore they can be measured with different accuracy. The formulation and solution of these problems are of wide interest in many measurement applications, as evidenced in [7] [8]. The *Elementwise-Weighted Total-Least-Squares* (EW-TLS) problem, considered here, is more general than the above *Least Squares* (LS) problems [9], because it assumes that both data and observation errors, Δa_{ij} and Δb_i, are independently but not identically distributed with zero-average and variances σ_{ij}^2 $\forall\ ij \in \mathcal{I} \times \mathcal{J}$ and σ_{ib}^2 $\forall\ i \in \mathcal{I}$, respectively. In [10] and [11] the consistence of EW-TLS estimators is discussed.

This paper proposes an original — to authors' knowledge — algorithm solving general EW-TLS problems. It is based on the minimization of the weighted Frobenius norm (WFN) of the errors, expressed in terms of the sum of ratios of quadratic forms. The gradient of the WFN is approximated by the gradient of a parametric quadratic form (PQF). The minima of a sequence of these PQFs converge to the minimizer of the WFN of errors within few iterations. Some numerical examples show the performance of the algorithm.

2. Statement of EW-TLS problem

We introduce the WFN of data and observation errors

$$F(x, \Delta A, \Delta b) = \sum_{i \in \mathcal{I}} \left(\Delta a_i^{\mathrm{T}} S_i^{-1} \Delta a_i + \Delta b_i^2 \sigma_{ib}^{-2} \right) , \tag{2}$$

where $\Delta a_i = [\Delta a_{i1}\ \Delta a_{i2}\ \cdots\ \Delta a_{in}]^{\mathrm{T}}$ and $S_i \equiv \mathrm{diag}\left(\sigma_{i1}^2, \sigma_{i2}^2, \ldots, \sigma_{in}^2\right)$ ($\forall\ i \in \mathcal{I}$). Solving EW-TLS problems consists in finding the optimal values $\breve{x}$, $\Delta \breve{A}$ and $\Delta \breve{b}$ of of x, ΔA and Δb minimizing $F(x, \Delta A, \Delta b)$

$$\min_{\langle x, \Delta A, \Delta b \rangle} F(x, \Delta A, \Delta b) \text{ subject to } (A - \Delta A)\ x = b - \Delta b\ . \tag{3}$$

Note that the values $\check{x}$, $\Delta\check{A}$ and $\Delta\check{b}$ are invariant with respect to an arbitrary scaling factor in the variances. When $\sigma_{ij} = 0$, the corresponding error Δa_{ij} is excluded from the WFN (2), because it is zero. If $S_i = 0$ and $\sigma_{ib}^2 = 0$, the i-th scalar equation in Eq. (1) becomes exact and "$\approx$" must be replaced by "=". The possible equations with $S_i = 0$ and $\sigma_{ib}^2 = 0$ play the role of "exact" constraints on x. In principle, these cases can be treated as limit cases of the EW-TLS problems, but their solutions require rather different numerical approaches, and their investigation is out of the aim of the present paper.

By choosing specific structures of standard deviations σ_{ij} and σ_{ib} , the above formulation of EW-TLS problems includes :

(i) Ordinary Least Squares (OLS) problems : $\sigma_{ij} = 0 \ \forall \ ij \in \mathcal{I} \times \mathcal{J}$ and σ_{ib} are nonzero and invariant $\forall \ i \in \mathcal{I}$.

(ii) Ordinary Weighted Least Squares (OWLS) problems : $\sigma_{ij} = 0$ $\forall \ ij \in \mathcal{I} \times \mathcal{J}$ and σ_{ib} are nonzero and varying $\forall \ i \in \mathcal{I}$.

(iii) Total Least Squares (TLS) problems : $\sigma_{ij} \ \forall \ ij \in \mathcal{I} \times \mathcal{J}$ and σ_{ib} $\forall \ i \in \mathcal{I}$ are nonzero and coincident.

(iv) Data Least Squares (DLS) problems : $\sigma_{ij} \ \forall \ ij \in \mathcal{I} \times \mathcal{J}$ are nonzero and coincident and $\sigma_{ib} = 0 \ \forall \ i \in \mathcal{I}$.

(v) Relative Error Total Least Squares (RE-TLS) problems [9] : σ_{ij} $\propto |a_{ij}| \ \forall \ ij \in \mathcal{I} \times \mathcal{J}$ and $\sigma_{ib} \propto |b_i| \ \forall \ i \in \mathcal{I}$.

(vi) Weighted Total Least Squares (WTLS) problems [2] [12] : $\sigma_{ij}^2 =$ $\sigma_{ib} \, \sigma'_j \ \forall \ ij \in \mathcal{I} \times \mathcal{J}$, where σ_{ib} $(i \in \mathcal{I})$ and σ'_j $(j \in \mathcal{J})$ can assume any value.

On the other hand, EW-TLS problems can be regarded as particular Structured-TLS problems [13] [14]. However, in the latter the number of uncertain free parameters is upper constrained by the number of the $m \times (n+1)$ elements of system (1) and is, in practice, much fewer than it.

With reference to Eq. (3), we substitute $\Delta b = \Delta A \, x - A \, x + b$ into $F(x, \Delta A, \Delta b)$ and obtain a new expression of the WFN

$$F(x, \Delta A) = \sum_{i \in \mathcal{I}} \left[\Delta a_i^{\mathrm{T}} S_i^{-1} \Delta a_i + \left(x^{\mathrm{T}} \Delta a_i - x^{\mathrm{T}} a_i + b_i \right)^2 \sigma_{ib}^{-2} \right] . \qquad (4)$$

To minimize the above WFN, we determine, for a given value of x, the m gradients of $F(x, \Delta A)$ with respect to the vectors $\Delta a_1, \Delta a_2, \ldots, \Delta a_m$

$$\frac{\partial F(x, \Delta A)}{\partial \Delta a_i} = 2 \left[S_i^{-1} \Delta a_i + \left(x^{\mathrm{T}} \Delta a_i - x^{\mathrm{T}} a_i + b_i \right) \sigma_{ib}^{-2} x \right] \quad \forall \ i \in \mathcal{I} \, .$$

Setting them to zero, we obtain m systems of n algebraic equations in the unknown n-vectors Δa_1, Δa_2, ..., Δa_m

$$\left[S_\imath^{-1} + x\, x^{\mathrm{T}} \sigma_{\imath b}^{-2}\right] \Delta a_\imath \;=\; \left(x^{\mathrm{T}}\, a_\imath - b_\imath\right)\, \sigma_{\imath b}^{-2}\, x \quad \forall\, \imath \in \mathcal{I}\,. \tag{5}$$

Each of the $n \times n$ matrices in Eqs. (5) coincides with the sum of a diagonal matrix and a unit-rank matrix : their inverses are given by

$$\left[S_\imath^{-1} + x\, x^{\mathrm{T}} \sigma_{\imath b}^{-2}\right]^{-1} \;=\; S_\imath - S_\imath\, x\, [S_\imath\, x]^{\mathrm{T}} / \left(x^{\mathrm{T}}\, S_\imath\, x + \sigma_{\imath b}^2\right) \qquad \imath \in \mathcal{I}$$

and the solutions of systems (5) are

$$\Delta a_\imath \;=\; \left[\left(x^{\mathrm{T}}\, a_\imath - b_\imath\right) / \left(x^{\mathrm{T}}\, S_\imath\, x + \sigma_{\imath b}^2\right)\right] S_\imath\, x \qquad \imath \in \mathcal{I}\,. \tag{6}$$

Hereafter, we assume that $x^{\mathrm{T}}\ S_\imath\ x + \sigma_{\imath b}^2 > 0$, for any $\imath$ and $\forall x$ in a neighborhood of the minimizer of $F(x, \Delta A)$. In practice, this limitation is not heavy. Note that, for any pair "$\imath\jmath$" with $\sigma_{\imath\jmath} = 0$, $\Delta a_{\imath\jmath} = 0$, as expected.

By substituting the values $\Delta a_\imath$ in Eqs. (6) into $F(x, \Delta A)$ (see Eq. 4), we obtain a new expression of the WFN, depending only on x

$$F(x) \doteq \sum_{\imath\in\mathcal{I}} \left(\frac{x^{\mathrm{T}} a_\imath - b_\imath}{x^{\mathrm{T}}\ S_\imath\ x + \sigma_{\imath b}^2}\right)^2 \left(x^{\mathrm{T}} S_\imath x + \sigma_{\imath b}^2\right) = \sum_{\imath\in\mathcal{I}} \frac{\left(x^{\mathrm{T}}\ a_\imath - b_\imath\right)^2}{x^{\mathrm{T}}\ S_\imath\ x + \sigma_{\imath b}^2}\,. \tag{7}$$

Finally, the constrained minimization of $F(x, \Delta A, \Delta b)$ in Eq. (3) can be substituted by the less troublesome unconstrained minimization of $F(x)$. Note that $F(x)$ can be assumed to be non-negative $\forall x$, since all $(a_\imath^{\mathrm{T}}\ x - b_\imath)^2$ are non-negative and all $x^{\mathrm{T}}\ S_\imath\ x + \sigma_{\imath b}^2$ are assumed to be positive.

3. Parametric quadratic form (PQF)

The gradient of $F(x)$ in Eq. (7) with respect to x results to be

$$\frac{\partial F(x)}{\partial x} \;=\; \sum_{\imath\in\mathcal{I}} \frac{1}{x^{\mathrm{T}}\ S_\imath\ x + \sigma_{\imath b}^2} \frac{\partial \left(a_\imath^{\mathrm{T}}\ x - b_\imath\right)^2}{\partial x} -$$

$$- \sum_{\imath\in\mathcal{I}} \left(\frac{a_\imath^{\mathrm{T}}\ x - b_\imath}{x^{\mathrm{T}}\ S_\imath\ x + \sigma_{\imath b}^2}\right)^2 \frac{\partial \left(x^{\mathrm{T}}\ S_\imath\ x + \sigma_{\imath b}^2\right)}{\partial x}\,.$$

In a sufficiently small neighborhood $\mathcal{D}^*$ of a generic value x^* the gradient can be approximated by

$$\left.\frac{\partial F(x)}{\partial x}\right|_{x\in\mathcal{D}^*} \simeq \sum_{\imath\in\mathcal{I}} \frac{1}{x^{*\mathrm{T}}\ S_\imath\ x^* + \sigma_{\imath b}^2} \frac{\partial \left(a_\imath^{\mathrm{T}}\ x - b_\imath\right)^2}{\partial x} -$$

$$-\sum_{i\in\mathcal{I}} \left(\frac{a_i^{\mathrm{T}}\, x^* - b_i}{x^{*\mathrm{T}}\, S_i\, x^* + \sigma_{ib}^2} \right)^2 \frac{\partial \left(x^{\mathrm{T}}\, S_i\, x + \sigma_{ib}^2 \right)}{\partial x} \, .$$

This approximated gradient can be regarded as the exact gradient of the form

$$Q(x, x^*) = \sum_{i\in\mathcal{I}} \left[\frac{\left(a_i^{\mathrm{T}}\, x - b_i\right)^2}{x^{*\mathrm{T}}\, S_i\, x^* + \sigma_{ib}^2} - \frac{\left(x^{\mathrm{T}}\, S_i\, x + \sigma_{ib}^2 \right) \left(a_i^{\mathrm{T}}\, x^* - b_i \right)^2}{\left(x^{*\mathrm{T}}\, S_i\, x^* + \sigma_{ib}^2 \right)^2} \right] .$$

It can be rewritten as a parametric quadratic form (PQF) :

$$Q(x, x^*) = x^{\mathrm{T}} A^{\mathrm{T}} Z^* A x - 2\, x^{\mathrm{T}} A^{\mathrm{T}} Z^* b + b^{\mathrm{T}} Z^* b - x^{\mathrm{T}} G^* x - \gamma_b^* \, , \qquad (8)$$

where the matrices Z^* and G^* are

$$Z^* \equiv \operatorname{diag}\left(\zeta_1^*, \zeta_2^*, \ldots, \zeta_m^*\right) \text{ with } \zeta_i^* = \frac{1}{x^{*\mathrm{T}}\, S_i\, x^* + \sigma_{ib}^2} \qquad i \in \mathcal{I} \, ,$$

$$G^* \equiv \operatorname{diag}\left(\gamma_1^*, \gamma_2^*, \ldots, \gamma_n^*\right) \text{ with } \gamma_j^* = \sum_{i\in\mathcal{I}} \frac{\sigma_{ij}^2 \left(a_i^{\mathrm{T}}\, x^* - b_i \right)^2}{\left(x^{*\mathrm{T}}\, S_i\, x^* + \sigma_{ib}^2 \right)^2} \qquad j \in \mathcal{J} \, ,$$

and the scalar γ_b^* is

$$\gamma_b^* = \sum_{i\in\mathcal{I}} \frac{\sigma_{ib}^2 \left(a_i^{\mathrm{T}}\, x^* - b_i \right)^2}{\left(x^{*\mathrm{T}}\, S_i\, x^* + \sigma_{ib}^2 \right)^2} \, .$$

If the Hessian matrix of $Q(x, x^*)$ is positive definite, $Q(x, x^*)$ is strictly convex and its equilibrium point is its (unique) minimum. If x^* coincides with the minimizer $\check{x}$ of $F(x)$ in Eq. (7), $Q(x, x^*)$ is necessarily positive definite and preserves the positive definiteness also in a suitable neighborhood of $\check{x}$. Then the zeroing of the gradient of $Q(x, x^*)$ in Eq. (8), supplies the value x^{**} approximating the minimizer $\check{x}$:

$$x^{**} = \left[A^{\mathrm{T}}\, Z^*\, A - G^* \right]^{-1} A^{\mathrm{T}}\, Z^*\, b \, . \qquad (9)$$

4. Iterative PQF algorithm

We present an iterative algorithm solving EW-TLS problems: it is based on the minimization of a sequence of PQFs, each of them approximating $F(x)$ in a different point. This sequence starts from a given value x^{**} of x. Under suitable conditions, the successive solutions of these PQFs converge to the global minimum of $F(x)$ in Eq. (7).

In detail, the algorithm steps are as follows

(i) Read the $m \times n$ matrix A , with variances σ_{ij}^2 $(ij \in \mathcal{I} \times \mathcal{J})$, the m-vector b , with variances σ_{ib}^2 $(i \in \mathcal{I})$, and the maximum allowed number of iterations K.

(ii) Set the iteration counter $k = 0$ and choose the starting value x^{**}.

(iii) Increase $k = k + 1$.

(iv) $x^* = x^{**}$

(v) If $k = K$, the algorithm does not converge. End.

(vi) Calculate $Z^* \equiv \text{diag}\,(\zeta_1^*, \zeta_2^*, \ldots, \zeta_m^*)$

$$\text{with} \quad \zeta_i^* = \frac{1}{x^{*\mathrm{T}}\, S_i\, x^* + \sigma_{ib}^2} \qquad i \in \mathcal{I} .$$

(vii) Calculate $G^* \equiv \text{diag}\,(\gamma_1^*, \gamma_2^*, \ldots, \gamma_n^*)$

$$\text{with} \quad \gamma_j^* = \sum_{i \in \mathcal{I}} \left(\frac{\sigma_{ij}(a_i^{\mathrm{T}}\, x^* - b_i)}{x^{*\mathrm{T}}\, S_i\, x^* + \sigma_{ib}^2} \right)^2 \qquad j \in \mathcal{J} .$$

(viii) Calculate $x^{**} \;=\; \left[A^{\mathrm{T}}\, Z^*\, A - G^*\right]^{-1}\, A^{\mathrm{T}}\, Z^*\, b$.

(ix) Calculate $\|\Delta x\| = [x^{**} - x^*]^{\mathrm{T}}\, [x^{**} - x^*]\,/n$.

(x) If $\|\Delta x\| \geq \epsilon_0^2 \|x^*\|$, go to step iii.

(xi) The value x^{**} in step viii approximates the minimizer $\check{x}$ with a relative accuracy less than ϵ_0 . End.

The convergence is controlled by the distance $\|\Delta x\|$ between two successive values x^* and x^{**} of x : when $\|\Delta x\| < \epsilon_0^2 \|x^*\|$ in step x, the algorithm converges to an equilibrium point of $F(x)$. Moreover, if the PQF defined in this equilibrium point is positive definite, preliminary simulations suggest that this point is likely to be the global minimum. Parameter K, assigned by the user, truncates the iterations in step v, when the convergence is not reached.

5. Performance of PQF algorithm

Some simulations were performed to test the performance of the iterative PQF algorithm in solving EW-TLS problems. In all examples the starting point x^{**} was chosen coincident with the minimizer of the corresponding OWLS problem, obtained from the EW-TLS problem by zeroing the $m \times n$ variances σ_{ij}^2 of data and preserving the m variances

$\sigma_{\imath b}^2$ of observations. For $m \times n = 32 \times 10$ the algorithm reached the solution x^* with an accuracy $\|\Delta x\|$ better than $\epsilon_0^2 \|x^*\| = 10^{-12}\|x^*\|$ within no more than $K = 8$ iterations, showing quadratic convergence. However, an exhaustive investigation on the influence of the starting point x^{**} on the convergence of the algorithm is still in progress. Some results on this aspect are reported in [10] and [11].

Successively, the PQF algorithm was employed to evaluate the capabilities of the EW-TLS formulation in solving general LS problems. To this aim we generated N noisy matrices A and vectors b as follows:

(i) We generated a noiseless matrix $\breve{A}$, and determined the corresponding vector $\breve{b}$ so that the overdetermined system (1) admits an exact solution $\breve{x}$: i.e. $\breve{b} = \breve{A}\,\breve{x}$.

(ii) We added to each element $\breve{a}_{\imath\jmath}$ and $\breve{b}_\imath$ an independently distributed error with a standard deviation proportional to $\breve{\sigma}_{\imath\jmath}$ and $\breve{\sigma}_{\imath b}$, respectively:

$$a_{\imath\jmath} = \breve{a}_{\imath\jmath} + \breve{\sigma}_{\imath\jmath}\,\theta_{\imath\jmath} \quad \imath\jmath \in \mathcal{I} \times \mathcal{J} \quad ; \quad b_\imath = \breve{b}_\imath + \breve{\sigma}_{\imath b}\,\theta_{\imath b} \quad \imath \in \mathcal{I} \qquad (10)$$

where $\theta_{\imath\jmath}$ ($\imath\jmath \in \mathcal{I} \times \mathcal{J}$) and $\theta_{\imath b}$ ($\imath \in \mathcal{I}$) are identically and independently distributed random values with zero-average.

By choosing m and n even, the standard deviations $\breve{\sigma}_{\imath\jmath}$ and $\breve{\sigma}_{\imath b}$ were set to :

$$\begin{array}{ccccccccccc} & & 1 & & n/2 & n/2{+}1 & & n & & & \\ & & \downarrow & & \downarrow & \downarrow & & \downarrow & & & \\ 1 & \longrightarrow & \breve{\sigma}_{\rm a} & \cdots & \breve{\sigma}_{\rm a} & \breve{\sigma}_{\rm c} & \cdots & \breve{\sigma}_{\rm c} & & \breve{\sigma}_{\rm b} & \\ & & \cdots & \cdots & \cdots & \cdots & \cdots & \cdots & & \cdots & \\ m/2 & \longrightarrow & \breve{\sigma}_{\rm a} & \cdots & \breve{\sigma}_{\rm a} & \breve{\sigma}_{\rm c} & \cdots & \breve{\sigma}_{\rm c} & ; & \breve{\sigma}_{\rm b} & (11) \\ m/2+1 & \longrightarrow & \breve{\sigma}_{\rm c} & \cdots & \breve{\sigma}_{\rm c} & \breve{\sigma}_{\rm a} & \cdots & \breve{\sigma}_{\rm a} & & \breve{\sigma}_{\rm b} & \\ & & \cdots & \cdots & \cdots & \cdots & \cdots & \cdots & & \cdots & \\ m & \longrightarrow & \breve{\sigma}_{\rm c} & \cdots & \breve{\sigma}_{\rm c} & \breve{\sigma}_{\rm a} & \cdots & \breve{\sigma}_{\rm a} & & \breve{\sigma}_{\rm b} & \end{array}$$

In all examples $\breve{\sigma}_{\rm a} = 0.1$, $\breve{\sigma}_{\rm c} = 0.001$ and $\breve{\sigma}_{\rm b} = 0.1$.

The standard deviations $\sigma_{\imath\jmath}$ ($\imath\jmath \in \mathcal{I} \times \mathcal{J}$) and $\sigma_{\imath b}$ ($\imath \in \mathcal{I}$), constituting the weights of the WFN in Eq. (2), were chosen following the same criterion adopted for $\breve{\sigma}_{\imath\jmath}$: they are set equal to three different values $\sigma_{\rm a}$, $\sigma_{\rm c}$ and $\sigma_{\rm b}$, according to scheme (11). Five simulated examples were characterized by specific values of $\sigma_{\rm a}$, $\sigma_{\rm c}$ and $\sigma_{\rm b}$, as follows :

(a) $\sigma_{\rm a} = \sigma_{\rm c} = 0.001$ and $\sigma_{\rm b} = 0.1$: these values are close to those of an OLS problem, because the standard deviations of data are much lower that those of observations, according to item i in Section 2.

(b) $\sigma_{\mathrm{a}} = \sigma_{\mathrm{c}} = \sigma_{\mathrm{b}} = 0.1$: these values identify an ordinary TLS problem according to item iii in Section 2.

(c) $\sigma_{\mathrm{a}} = \sigma_{\mathrm{b}} = 0.1$ and $\sigma_{\mathrm{c}} = 0.01$: these values are near but not coincident with the corresponding $\breve{\sigma}_{\mathrm{a}}$, $\breve{\sigma}_{\mathrm{b}}$ and $\breve{\sigma}_{\mathrm{c}}$.

(d) $\sigma_{\mathrm{a}} = \sigma_{\mathrm{b}} = 0.1$ and $\sigma_{\mathrm{c}} = 0.001$: these values are coincident with the corresponding standard deviations $\breve{\sigma}_{\mathrm{a}}$, $\breve{\sigma}_{\mathrm{b}}$ and $\breve{\sigma}_{\mathrm{c}}$.

(e) $\sigma_{\mathrm{a}} = 0.001$ and $\sigma_{\mathrm{c}} = \sigma_{\mathrm{b}} = 0.1$: the structure is as in (11), but $\sigma_{\mathrm{a}} = \breve{\sigma}_{\mathrm{c}}$ and $\sigma_{\mathrm{c}} = \breve{\sigma}_{\mathrm{a}}$.

Each example was simulated $N = 200$ times by choosing N different realizations of the stochastic noise θ_{ij} and θ_{ib} in Eqs. (10). We calculated the standard deviations $\hat{\sigma}_{ij}$ and $\hat{\sigma}_{ib}$ of the $m \times (n+1)$ noisy elements a_{ij} and b_i and the corresponding standard deviations $\bar{\sigma}_{ij}$ and $\bar{\sigma}_{ib}$ of the optimal errors $\Delta\breve{a}_{ij}$ and $\Delta\breve{a}_{ib}$ supplied by the PQF algorithm in the N simulations. Finally we introduce the figure of merit :

$$\alpha = \frac{1}{m(n+1)} \left[\sum_{ij \in \mathcal{I} \times \mathcal{J}} (\bar{\sigma}_{ij} - \hat{\sigma}_{ij})^2 + \sum_{i \in \mathcal{I}} (\bar{\sigma}_{ib} - \hat{\sigma}_{ib})^2 \right] ,$$

where the standard deviations were properly normalized.

Moreover, we introduced the global figure of merit β, defined as the average value of the n standard deviations of the elements of x supplied by the algorithm in the N stochastic simulations. Tables 1 and 2 report values of α and β of the five examples for two different dimensions $m \times n$ of the system of equations.

Table 1. Figures of merit α and β for $m \times n \equiv 16 \times 4$.

	σ_{a}	σ_{c}	σ_{b}	α	β
(a)	0.001	0.001	0.1	0.0238	0.0104
(b)	0.1	0.1	0.1	0.0128	0.0105
(c)	0.1	0.01	0.1	0.0071	0.0084
(d)	0.1	0.001	0.1	0.0071	0.0084
(e)	0.001	0.1	0.1	0.0221	0.0157

Results in Tables 1 and 2 show that the solutions supplied by EW-TLS and ordinary TLS formulations are quite different both in terms of α and β : for instance compare example (b) with example (d). Moreover, high values of both α and β denote the disagreement between variances

Table 2. Figures of merit α and β for $m \times n \equiv 32 \times 10$.

	σ_a	σ_c	σ_b	α	β
(a)	0.001	0.001	0.1	0.00898	0.0273
(b)	0.1	0.1	0.1	0.00307	0.0275
(c)	0.1	0.01	0.1	0.00196	0.0212
(d)	0.1	0.001	0.1	0.00197	0.0212
(e)	0.001	0.1	0.1	0.00582	0.0459

σ_{ij}^2 and $\breve{\sigma}_{ij}^2$: compare examples (a), (b) and (e) with examples (c) and (d). When this disagreement is limited to those variances with small values, α and β do not vary remarkably : compare example (d) with example (e).

6. Conclusions

The proposed iterative algorithm solves Total-Least-Squares problems, in which errors are assumed to be elementwise independently distributed but with different variances (Elementwise-Weighting). It is based on the introduction of an *ad hoc* Parametric Quadratic Form, whose gradient approximates that of the weighted Frobenius norm of the errors. Preliminary results show that the algorithm exhibits quadratic convergence. Further numerical simulations are in progress for confirming these results.

Acknowledgments

The authors are indebted to Sabine Van Huffel, Alexander Kukush, Philippe Lemmerling and Ivan Markovsky for some stimulating talks and suggestions.

References

[1] S.D. Hodges and P.G. Moore. Data uncertainties and least square regression. *Applied Statistics*, 21:185–195, 1972.

[2] G.H. Golub and C.F. Van Loan. An analysis of the total least squares problem. *SIAM Journal of Numerical Analysis*, 17:883–893, 1980.

[3] G.H. Golub and C. Reinsch. Singular value decomposition and least squares solutions. *Numer. Math.*, 14:403–420, 1970.

[4] G.H. Golub. Some modified eigenvalue problems. *SIAM Review*, 15:318–344, 1973.

[5] S. Van Huffel and J. Vandewalle. *The Total Least Squares Problem: Computational Aspects and Analysis.* Frontiers in Applied Mathematics. SIAM Philadelphia, 1991.

[6] R.D. Degroat and E. Dowling. The data least squares problem and channel equalization. *IEEE Trans. on Signal Processing*, 41(1): 407–411, 1993.

[7] A. Premoli, M.L. Rastello and G. Cirrincione. A new approach to total least-squares techniques for metrological applications. In P. Ciarlini, M.G. Cox, F. Pavese and D. Richter, editors, *Advanced Mathematical Tools in Metrology II, Series on Advances in Mathematics for Applied Sciences*40:206–215, World Scientific, 1996.

[8] G. Cirrincione, S. Van Huffel, A. Premoli and M.L. Rastello. Iteratively reweighted total least squares algorithms for different variances in observations and data. In P. Ciarlini, M.G. Cox, E. Felipe, F. Pavese and D. Richter, editors, *Advanced Mathematical Tools in Metrology V, Series on Advances in Mathematics for Applied Sciences*57:77–84, World Scientific, 2001.

[9] B. De Moor. Structured total least squares and L_2 approximation problems. *Linear Algebra and its Applications*, 188/189:163–205, 1993.

[10] A. Kukush and S. Van Huffel. Consistency of elementwise-weighted total least squares estimator in a multivariate errors-in-variables model $AX = B$. Submitted for publication, June 2001.

[11] A. Kukush, I. Markovsky and S. Van Huffel. On consistent estimators in linear and bilinear multivariate errors-in-variables models. In *this volume.*

[12] B.D. Rao. Unified treatment of LS, TLS and truncated SVD methods using a weighted TLS framework. In S. Van Huffel, editor, *Recent Advances in Total Least Squares Techniques and Errors-in-Variables Modeling*, 11–20, SIAM, 1997.

[13] P. Lemmerling, S. Van Huffel and B. De Moor. Structured total least squares problems: formulation, algorithm and applications. In S. Van Huffel, editor, *Recent Advances in Total Least Squares Techniques and Errors-in-Variables Modeling*, 215–223, SIAM, 1997.

[14] B. De Moor. Linear system identification, structured total least squares and the Riemannian SVD. In S. Van Huffel, editor, *Recent Advances in Total Least Squares Techniques and Errors-in-Variables Modeling*, 225–238, SIAM, 1997.

Part III

STRUCTURED TOTAL LEAST SQUARES PROBLEMS

STRUCTURED TOTAL LEAST SQUARES

Analysis, Algorithms and Applications

Philippe Lemmerling and Sabine Van Huffel
ESAT-SISTA/COSIC, K.U.Leuven, Arenberg 10, B-3001 Leuven-Heverlee, Belgium
{Philippe.Lemmerling, Sabine.VanHuffel}@esat.kuleuven.ac.be

Abstract In this paper an overview is given of the Structured Total Least Squares (STLS) approach and its recent extensions. The Structured Total Least Squares (STLS) problem is a natural extension of the Total Least Squares (TLS) problem when constraints on the matrix structure need to be imposed. Similar to the ordinary TLS approach, the STLS approach can be used to determine the parameter vector of a linear model, given some noisy measurements. In many signal processing applications, the imposition of this matrix structure constraint is necessary for obtaining Maximum Likelihood (ML) estimates of the parameter vector.

A basic algorithmic scheme leading to fast implementations of the STLS approach through the exploitation of the low displacement rank structure is presented. An example illustrates the possible benefit of using the STLS approach.

Keywords: structured total least squares, Hankel/Toeplitz matrix, displacement rank.

1. Introduction

Many data fitting techniques give rise to an overdetermined set of linear equations

$$Ax \approx b,$$

where $A \in \mathbb{R}^{m\times n}$, $b \in \mathbb{R}^{m\times 1}$ contain the measured data and $x \in \mathbb{R}^{n\times 1}$ is the so-called parameter vector that characterizes a linear model.

One way of determining the latter parameter vector is by solving the Total Least Squares (TLS) [12, 25] problem formulation:

$$\min_{\Delta A,\Delta b,x} \|[\Delta A\ \Delta b]\|_F^2 \tag{1}$$
$$\text{such that } (A+\Delta A)x = b+\Delta b.$$

S. Van Huffel and P. Lemmerling (eds.), Total Least Squares and Errors-in-Variables Modeling, 79–91.

As explained in [25], the resulting parameter vector x will be a Maximum Likelihood (ML) estimate only if the measurement errors for each observation[1] are independently and identically normally distributed with common covariance matrix proportional to the identity matrix and zero mean. This implies that the measurement errors on samples belonging to two different observations (i.e. samples belonging to two different rows of $[A\ b]$) should be independent.

In many signal processing applications this last condition is violated. A typical example occurs when the "observed" data matrix $[A\ b]$ is a Toeplitz matrix constructed from an observed data vector:

$$[A\ b] = \begin{bmatrix} s(n+1) & s(n) & \cdots & s(2) & s(1) \\ s(n+2) & s(n+1) & \ddots & \ddots & s(2) \\ \vdots & \ddots & \ddots & \ddots & \vdots \\ s(m+n) & s(m+n-1) & \cdots & s(m+1) & s(m) \end{bmatrix}, \tag{2}$$

where $s \in \mathbb{R}^{(m+n)\times 1}$ is the observed data vector. Clearly the measurement errors on samples belonging to different rows are no longer independent, since the same observations occur in several rows.

Intuitively (for a proof see [2]) it is clear that in order to find a ML estimate of the parameter vector x, $[\Delta A\ \Delta b]$ should have the same structure as $[A\ b]$. This yields the formulation of the so-called Structured Total Least Squares (STLS) problem:

$$\begin{aligned} &\min_{\Delta A, \Delta b, x} \|[\Delta A\ \Delta b]\|_X^2 \\ &\text{such that } (A+\Delta A)x = b + \Delta b, \\ &\text{and } [\Delta A\ \Delta b] \text{ has the same structure as } [A\ b], \end{aligned} \tag{3}$$

where $\|.\|_X$ is an appropriate 2-norm defined on the entries of $[\Delta A\ \Delta b]$. Summarizing we can say that in the STLS formulation matrix structure constraints are added to the TLS problem formulation in order to increase the statistical accuracy of the estimated parameter vector. Notice that this is a first level at which we exploit structure in order to obtain statistically more accurate results[2]. Further on we will take advantage of the structure present on a second level, namely the level of algorithms for solving STLS problems: the matrices involved in these algorithms inherit partly the structure that appears in the previously mentioned matrix structure constraints.

[1] One observation corresponds to one row of the data matrix $[A\ b]$.

[2] Note however that the matrix structure constraints can be added for reasons other than statistical ones. The latter is demonstrated in [13].

The paper is organized as follows. In the next section an overview is given of the existing approaches for solving the STLS problem and some recent extensions are presented as well. Section 3 presents the basic algorithm that allows the exploitation of the structure at the second level i.e. the algorithmic level. An example illustrating the benefit of the use of the STLS approach is described in Section 4.

2. Overview of STLS approaches

In this section an overview is given of several STLS approaches. In the first subsection only STLS problems in which a linear matrix structure needs to preserved, are considered. In the second subsection two extensions are discussed: an extension towards STLS problems in which nonlinear structures need to be preserved and an extension in which a general cost function instead of the L_2 cost function in (3) is minimized. In the remainder of the paper we adopt a Matlab like notation for vectors and matrices.

2.1 Linearly structured matrices

In recent years many different formulations have been proposed for the STLS problem involving linearly structured matrices: the Constrained Total Least Squares (CTLS) approach [1, 2], the Structured Total Least Norm (STLN) approach [21, 24] and the Riemannian Singular Value Decomposition (RiSVD) approach [6, 7]. All these approaches start more or less from a formulation similar to (3), but the final formulation for which an algorithm is developed might be quite different. The STLN approach is the most straightforward one. It basically solves the STLS problem as a nonlinearly constrained optimization problem, i.e. (3) is solved. The other two approaches first derive an equivalent formulation for which an algorithm is then developed. This will be discussed in the following paragraphs while showing the analogies to solving the TLS problem.

The CTLS approach transforms the original STLS problem (3) into the following optimization problem[3]:

$$\min_{y \in \mathbb{R}^{(n+1)\times 1}} y^T S^T D_y^{-1} S y, \; y(n+1) = -1, \tag{4}$$

with $D_y \equiv H_y W^{-1} H_y^T$, with $W \in \mathbb{R}^{k\times k}$ a weighting matrix, $x = y(1:n)$, $S \equiv [A \; b]$, $\Delta S \equiv [\Delta A \; \Delta b]$, $H_y \equiv [T_1 y \; T_2 y \cdots T_k y] \in \mathbb{R}^{m\times k}$ where k is

[3]Note that in the remainder of the paper only structures having nonsingular D_y matrices are considered. This is not an overly stringent condition, since many popular structures such as Hankel and Toeplitz matrices belong to this class.

the number of different entries in S (and thus also in ΔS, since S and ΔS have the same structure) and $T_i, i = 1, \ldots, q$ are the so-called "basis matrices", i.e. they are used to construct the linearly structured matrix S starting from the vector $s \in \mathbb{R}^{k\times 1}$ that contains the *different* elements of S: $S = \sum_{i=1}^{k} s(i)T_i$.

Note that the objective function in (4) looks like a Rayleigh quotient. The difference with a classical Rayleigh quotient is the introduction of the matrix D_y, which depends on y and on the specific structure that needs to be preserved in the STLS problem. If no structure were imposed on ΔS in (3), we would obtain (1) but above all, D_y would[4] become $\|y\|_2^2$ and (4) would yield the eigenvector corresponding to the smallest eigenvalue of $S^T S$ (since then (4) would indeed correspond to the minimization of the well-known Rayleigh quotient). Normally, we would use the SVD of S in order to determine the solution of the TLS problem. More specifically we would determine the right singular vector corresponding to the smallest singular value of S, but in theory we could also look for the eigenvector corresponding to the smallest eigenvalue of $S^T S$. This is exactly what is done through the Rayleigh quotient minimization. Intuitively we can thus summarize the previous as follows. The solution of the STLS problem is found by minimizing a Rayleigh quotient-like cost function, in which the specific structure to be preserved in the STLS problem is reflected by the specific form of D_y.

The RiSVD approach is derived by using the technique of Lagrange multipliers. The result is the following equivalent problem formulation

Find the triplet (u, τ, v) corresponding to the smallest τ such that

$$Sv = D_v u\tau \quad u^T D_v u = 1 \tag{5}$$

$$S^T u = D_u v\tau \quad v^T D_u v = 1, \quad v^T v = 1, \tag{6}$$

with $D_v \equiv H_v W^{-1} H_v^T$, $D_u \equiv H_u W^{-1} H_u^T$, $W \in \mathbb{R}^{k\times k}$ a weighting matrix, $H_v \equiv [T_1 v\; T_2 v \cdots T_k v] \in \mathbb{R}^{m\times k}$, $H_u \equiv [T_1^T u\; T_2^T u \cdots T_k^T u] \in \mathbb{R}^{n\times k}$, where k is again the number of different entries in S and $T_i, i = 1, \ldots, k$ are the so-called "basis matrices", i.e. they are used to construct the linearly structured matrix S starting from the vector $s \in \mathbb{R}^{k\times 1}$ that contains the *different* elements of S: $S = \sum_{i=1}^{k} s(i)T_i$.

Note that the equations (5) and (6) are very similar to the classical SVD equations. The difference with the classical SVD equations is the introduction of the matrices D_u and D_v, which depend on respectively u and v and on the specific structure that needs to be preserved in the STLS problem. If no structure were imposed on ΔS in (3), we would

[4]Simply write out the formula for D_y in case S is unstructured.

obtain (1) but above all, we would[5] obtain the classical SVD equations in (5) and (6). This should come as no surprise since the SVD is the standard way for solving the TLS problem. Intuitively we can summarize the previous as follows. The STLS problem is solved by determining the right singular vector corresponding to the smallest singular value of a nonlinear SVD (5)-(6), in which the specific structure to be preserved in the STLS problem is reflected by the specific form of D_u and D_v.

2.2 Recent extensions

The previous subsection only considered STLS problems in which a linear matrix structure is preserved. However, important applications in signal processing contain STLS problems in which a nonlinear matrix structure (e.g. a Vandermonde-like structure) needs to be preserved. It appears e.g. in subspace-based parameter estimation algorithms (such as e.g. ESPRIT [19, 22, 23, 5]). Extensions towards nonlinearly structured matrices exist both in the STLN [20] and the RiSVD framework [14]. Further details can be found in these references.

In [8], the STLN framework for linearly structured matrices is extended towards general cost functions instead of the L_2 cost function in (3). This is particularly useful in case some prior knowledge on the probability density function of the measurement noise is present, since the cost function in the STLS problem formulation is related to this probability density function. Furthermore, the so-called misfit versus latency extension does not only include measurement errors (the so-called misfit), i.e. ΔA and Δb in (3), but also a latency term (which then shows up in the cost function and the constraints of (3)) that refers to unmeasured inputs (for details see [15]).

3. STLS algorithms

Before describing the basic algorithm, we reformulate (3) as follows:

$$\min_{\Delta s, x} 0.5 \Delta s^T W \Delta s \quad \text{such that } (A + \Delta A)x = b + \Delta b, \tag{7}$$

where $\Delta s \in \mathbb{R}^{k \times 1}$ contains the k *different* elements of $[\Delta A\ \Delta b]$ and $W \in \mathbb{R}^{k \times k}$ is a weighting matrix. Note that by taking $W = 2diag([1\ 2 \cdots n + 1 \cdots n+1 \cdots 2\ 1]) \in \mathbb{R}^{(m+n)\times(m+n)}$, it becomes clear that (7) is an extension of (1) since the former problem imposes an additional constraint on the structure of $[\Delta A\ \Delta b]$ by expressing its entries as linear functions

[5]Simply write out the formulas for D_u and D_v in case S is unstructured.

of Δs. Also observe that the correction $[\Delta A\ \Delta b]$ obtained in (1) is typically unstructured as it is the result of a Singular Value Decomposition of the data matrix $[A\ b]$.

In most signal processing applications $\|\Delta s\|_2^2$ has to be minimized and therefore we choose $W = I_{m+n}$ in (7), where $I_k \in \mathbb{R}^{k\times k}$ represents the $k \times k$ identity matrix. We have thus completely specified $\|.\|_X$ in (3).

As can be seen from (7) the STLS problem is a constrained optimization problem with a quadratic objective function and nonlinear constraints. Therefore, taking into account the nature of the STLS problem, it should come as no surprise that all the algorithms for solving it will be iterative.

A straightforward algorithm for solving the STLS problem can be derived along the same lines as the heuristic algorithm developed in [21]: in each iteration the equality constraints of (7) are linearized around the current solution point (a solution point being determined by Δs and x).

Before describing the algorithm, we introduce some notation. Let us represent small perturbations on Δs and x by $\Delta\tilde{s} \in \mathbb{R}^{(m+n)\times 1}$ and $\Delta\tilde{x} \in \mathbb{R}^{n\times 1}$ respectively. Furthermore let

$$r(\Delta s, x) = (A + \Delta A)x - b - \Delta b$$

and $X \in \mathbb{R}^{m\times(m+n)}$ is defined by

$$X\Delta s = [\Delta A\ \Delta b]\begin{bmatrix} x \\ -1 \end{bmatrix}.$$

When e.g. $[\Delta A\ \Delta b]$ is Toeplitz this yields

$$X = \begin{bmatrix} -1 & x(n) & \cdots & x(1) & 0 & \cdots & \cdots & 0 \\ 0 & -1 & x(n) & \cdots & x(1) & 0 & & \vdots \\ \vdots & & \ddots & \ddots & & \ddots & & \vdots \\ 0 & & & \ddots & \ddots & & \ddots & 0 \\ 0 & \cdots & \cdots & 0 & -1 & x(n) & \cdots & x(1) \end{bmatrix}.$$

The iterative algorithm is obtained by replacing in (7) Δs and x by $\Delta s + \Delta\tilde{s}$ and $x + \Delta\tilde{x}$ respectively, followed by a linearization of the constraints around the current solution point $[\Delta s^T\ x^T]^T$ (simply omit second order terms that occur). We then obtain the following general algorithm outline:

Algorithm 1 *(STLS1).*

Input: $[A\ b] \in \mathbb{R}^{m\times(n+1)}$, $W \in \mathbb{R}^{(m+n)\times(m+n)}$

Output: the parameter vector $x \in \mathbb{R}^{n\times 1}$ and $\Delta s \in \mathbb{R}^{(m+n)\times 1}$ (i.e. the minimal representation of the matrix $[\Delta A\ \Delta b]$)

Step 1: Initialize Δs and x
Step 2: while *stopcriterion* not satisfied
Step 2.1: Solve the following equality constrained LS problem:
$\min_{\Delta\tilde{s},\Delta\tilde{x}} \|W^{1/2}(\Delta s + \Delta\tilde{s})\|_2^2$
such that $r(\Delta s, x) + J \begin{bmatrix} \Delta\tilde{s} \\ \Delta\tilde{x} \end{bmatrix} = 0$
Step 2.2: $\Delta s \leftarrow \Delta s + \Delta\tilde{s}$
$x \leftarrow x + \Delta\tilde{x}$
end

where $J = [X \; A + \Delta A]$ is the Jacobian of the constraints $r(\Delta s, x)$ w.r.t. $v \equiv [\Delta s^T \;\; x^T]^T$. A discussion on the initialization of Δs and x can be found in [16]. It should be noted however that problem (7) is a nonlinear optimization problem that suffers from many local minima, a problem that becomes more important with increasing dimensions of the problem. A straightforward choice of starting values, would be to take the LS solution of $Ax \approx b$ as the initial parameter vector and to set Δs 0, but better alternatives exist as explained in [16].

The choice of the stop criterion depends on the application at hand. In the remainder of the paper the following stop criterion is used:

$$\|\Delta s\|_2 < tol.$$

For solving the equality constrained LS problem in Step 2.1 the Generalized RQ (GRQ) factorization [4, 3] can be used. The latter GRQ factorization is a two-step procedure consisting of one RQ factorization followed by a QR factorization. For the RQ factorization a fast implementation exploiting the low displacement rank of the involved matrices could be developed. However, the consecutive QR factorization does not operate on one of the originally structured matrices but on one of the original matrices multiplied with an orthogonal matrix to the right. Since the latter matrix is typically unstructured, it is not possible to improve the computational performance of this second step. The problem with algorithm STLS1 is that through the elimination of the constraints, a lot of the original structure (present at the first level, i.e. the problem formulation level) is lost.

In order to be able to fully exploit the structure of an STLS problem we follow a different approach. Instead of eliminating the constraints of the STLS problem (7), we apply the Newton method for unconstrained optimization to the Lagrangian L of problem (7):

$$L(\Delta s, x, \lambda) = 0.5\Delta s^T W \Delta s - \gamma^T(b - Ax - X\Delta s),$$

where $\gamma \in \mathbb{R}^{m\times 1}$ is a vector of Lagrange multipliers. The straightforward application of the Newton method (see e.g. [10, 11]) on the Lagrangian L (for $W = I_{m+n}$) yields the following algorithm:

Algorithm 2 (STLS2).

Input: $[A\ b] \in \mathbb{R}^{m\times(n+1)}$

Output: the parameter vector $x \in \mathbb{R}^{n\times 1}$ and $\Delta s \in \mathbb{R}^{(m+n)\times 1}$ (i.e. the minimal representation of the matrix $[\Delta A\ \Delta b]$)

Step 1: Initialize Δs,x and γ

Step 2: while *stopcriterion* not satisfied

Step 2.1: Solve the following system of equations:

$$\begin{bmatrix} S & J^T \\ J & 0 \end{bmatrix} \begin{bmatrix} \Delta\tilde{s} \\ \Delta\tilde{x} \\ \Delta\tilde{\gamma} \end{bmatrix} = - \begin{bmatrix} g + J^T\gamma \\ r(\Delta s, x) \end{bmatrix}$$

Step 2.2: $\Delta s \leftarrow \Delta s + \Delta\tilde{s}$

$x \leftarrow x + \Delta\tilde{x}$

$\gamma \leftarrow \gamma + \Delta\tilde{\gamma}$

end

where $g = \begin{bmatrix} W & 0 \\ 0 & 0 \end{bmatrix} \begin{bmatrix} \Delta s \\ \Delta x \end{bmatrix} \in \mathbb{R}^{(m+2n)\times 1}$ is the gradient of the objective function in (7) and $J = [X\ \ A + \Delta A]$ is the Jacobian of the constraints $r(\Delta s, x)$ in (7), all w.r.t. $v \equiv [\Delta s^T\ \ x^T]^T$. Furthermore, for optimal convergence rate the matrix S should be set to $\nabla^2_{vv} L(\Delta s, x, \gamma)$. The latter means that in this case

$$S = \begin{bmatrix} I_{m+n} & 0 \\ 0 & 0 \end{bmatrix} - \sum_{i=1}^{m} \gamma(i) \nabla^2_{vv} r. \tag{8}$$

As shown in e.g. [10], S can also be chosen to be a positive definite approximation of $\nabla^2_{vv} L$, without changing the final solution of problem (7). Inclusion of the second term in (8) would render the structure of S rather complicated. We therefore only retain the first term in (8):

$$S = \begin{bmatrix} I_{m+n} & 0 \\ 0 & 0 \end{bmatrix}. \tag{9}$$

It is interesting to notice that Step 2.1 of algorithm STLS1 is basically the same as Step 2.1 of algorithm STLS2 (when the above approximation of S (see (9)) is used), although both algorithms are derived from a different starting point. The proof is easily obtained by applying the method of Lagrange multipliers to the subproblem described in Step 2.1 of algorithm STLS1. The system of equations that results from it is exactly the same as the system of equations that needs to be solved in Step

2.1 of algorithm STLS2 with the approximation of S described in (9). The latter observation also yields some additional insight in the convergence properties that can be expected for algorithms STLS1 and STLS2. By omitting the curvature of the constraints in (8) the convergence rate is no longer quadratic but super linear (see [10, 11]). However, this is largely compensated by the fact that one iteration can be implemented in a very fast way, exploiting the low displacement rank structure of the matrices involved in Step 2.1 of Algorithm STLS2 and if possible by taking advantage of the sparsity of the corresponding generators. This will be explained briefly in the following paragraph.

To see to which extent the efficiency of the kernel problem (i.e. Step 2.1 of algorithm STLS2) can be improved, the execution time for solving the following system of linear equations has to be considered:

$$Mz = b_1, \tag{10}$$

with

$$M = \begin{bmatrix} I_{(m+n)} & 0_{(m+n)\times n} & X^T \\ 0_{n\times(m+n)} & 0_{n\times n} & \Lambda^T \\ X & \Lambda & 0_{m\times m} \end{bmatrix}, \tag{11}$$

where $\Lambda = A + \Delta A$ and b_1 is the right hand side of the system of equations described in Step 2.1 of STLS2. The solution of (10) can be obtained by computing the LDL^T factorization of M, where L is lower triangular and D is a signature matrix. The solution is then found by solving the following linear systems:

$$\begin{aligned} Lz_2 &= b_1 \\ Dz_1 &= z_2 \\ L^T z &= z_1. \end{aligned} \tag{12}$$

The latter factorization can be computed in a fast way by an appropriate implementation of the generalized Schur algorithm [17] and a high computational efficiency is obtained by exploiting the low displacement rank of the matrix M and if possible by taking advantage of the sparsity of the corresponding generators.

Before applying the generalized Schur algorithm to M we observe that the matrix M is not strongly regular. In fact $\det(M(1:i,1:i)) = 0, i = m+n+1,\ldots,m+2n$. Hence a permutation matrix P is considered in order to transform M into the Toeplitz-block matrix $\tilde{K}$, i.e.,

$$\tilde{K} = PMP^T = \begin{bmatrix} I_{(m+n)} & X^T & 0_{(m+n)\times n} \\ X & 0_{m\times m} & \Lambda \\ 0_{n\times(m+n)} & \Lambda^T & 0_{n\times n} \end{bmatrix}.$$

It is easy to prove that $\tilde{K}$ is strongly regular if X and Λ are full rank. Considering the Schur complement of $I_{(m+n)}$ in $\tilde{K}$ we can obtain the following partial LDL^T decomposition of $\tilde{K}$ without any additional cost (of course the product XX^T is not explicitly computed),

$$\tilde{K}=\begin{bmatrix} I_{(m+n)} & & \\ X & I & \\ 0_{n\times(m+n)} & & I \end{bmatrix}\begin{bmatrix} I_{(m+n)} & \\ & \hat{K} \end{bmatrix}\begin{bmatrix} I_{(m+n)} & X^T & 0_{(m+n)\times n} \\ & I & \\ & & I \end{bmatrix},$$

where the matrix

$$\hat{K}=\begin{bmatrix} -XX^T & \Lambda \\ \Lambda^T & 0_{n\times n} \end{bmatrix}$$

of order $m+n$ is the Schur complement of $I_{(m+n)}$ in the matrix $\tilde{K}$. Then the problem is reduced to computing the LDL^T decomposition of $\hat{K}$. Further details on how this LDL^T decomposition can be computed in a very efficient way, can be found in [17].

4. Application

In this section we illustrate, by means of a real-life application, the benefit obtained by using STLS as an estimator instead of TLS. The STLS problem that has to be solved here is a deconvolution problem that occurs in renography [9]. The goal is to determine via deconvolution the so-called renal retention function of the kidney, which in system theoretic terms corresponds to the impulse response x_0 of the kidney as a system. This retention function visualizes the mean whole kidney transit time of one unit of a tracer injected into the patient, and it enables a physician to evaluate the renal function and renal dysfunction severity after transplantation. In order to obtain this impulse response, the following experiment is conducted. A radioactive tracer is injected in an artery of the patient. The measured input of the system is the by noise perturbed arterial concentration of the radioactive tracer as a function of time. This concentration is measured by means of a gamma camera, and thus in discretized time $u(k)$ represents the number of counts registered in the vascular region at the entrance of the kidney under study during the kth sampling interval. The measured output $y(k)$ (the so-called renogram) represents the (by noise perturbed) number of counts registered in the whole kidney region by the gamma camera in the kth sampling interval. Deconvolution analysis of the renogram is based on modeling the kidney as a linear time-invariant causal system with zero-initial state. Writing down the linear relationship between inputs and ouput for several time instances, we end up with an STLS problem (7) in

Table 1. This table shows the relative error $\|x - x_0\|_2/\|x_0\|_2$ for the TLS and STLS estimator, averaged over 100 runs, for different noise standard deviations σ_ν.

σ_ν	TLS	STLS
0.05	0.0010	0.0009
0.071	0.0015	0.0013
0.087	0.0018	0.0016
0.1	0.0021	0.0018
0.158	0.0032	0.0029
0.224	0.0046	0.0041
0.274	0.0056	0.0049
0.316	0.0064	0.0057

σ_ν	TLS	STLS
0.5	0.0104	0.0091
0.707	0.0138	0.0123
0.866	0.0177	0.0153
1	0.0204	0.0180
1.58	0.0313	0.0279
2.24	0.0454	0.0402
2.74	0.0576	0.0513
3.16	0.0660	0.0601

which the specific structure of A needs to be preserved (i.e. zero above the diagonal and Toeplitz). For details on the implementation, we refer to [18]. As described in the latter references it is quite straightforward to set up a simulation experiment for this deconvolution problem (again see [18]) for details). To compare the statistical accuracy of the STLS estimator with that of the TLS estimator, we perform for several noise standard deviations σ_ν a Monte-Carlo simulation consisting of 100 runs. In every run, we add a different realization of i.i.d. white Gaussian noise with standard deviation σ_ν to the noiseless input u_0 and the noiseless output y_0 of the previously described medical simulation example. The obtained noisy vectors u and y are the data used in the STLS problem. To compare the performance of the estimators at a noise level σ_ν, we average for both estimators the relative error $\|x - x_0\|_2/\|x_0\|_2$ over the different runs. Table 1 shows that in the case of the STLS estimator, the relative errors are 9% to 14% lower than in the case of the TLS estimator, confirming the statistical superior performance of the STLS estimator.

5. Conclusion

In this paper an overview is given of the different approaches for solving the STLS problems in which linear structures need to be preserved. Extensions towards nonlinearly structured matrices and general cost functions instead of L_2 cost functions are pointed out. Another extension, namely the misfit versus latency framework, is referred to.

It is shown how efficient algorithms can be developed for STLS problems involving linearly structured matrices. Using a simulation example that finds its origin in a medical application called renography, the superior statistical accuracy of STLS over TLS is illustrated.

Acknowledgments

This paper presents research results of the Belgian Programme on Interuniversity Poles of Attraction (IUAP P4-02 and P4-24), initiated by the Belgian State, Prime Minister's Office - Federal Office for Scientific, Technical and Cultural Affairs, of the Brite Euram Programme, Thematic Network BRRT-CT97-5040 'Niconet', of the Concerted Research Action (GOA) projects of the Flemish Government MEFISTO-666 (Mathematical Engineering for Information and Communication Systems Technology), of the IDO/99/03 project (K.U.Leuven) "Predictive computer models for medical classification problems using patient data and expert knowledge", and of the FWO (Fund for Scientific Research Flanders) projects G0200.00 and G078.01.

References

[1] T. J. Abatzoglou and J. M. Mendel. Constrained total least squares. In *Proceedings of IEEE International Conf. on Acoustics, Speech & Signal Processing*, pages 1485–1488, Dallas, 1987.

[2] T. J. Abatzoglou, J. M. Mendel, and G. A. Harada. The constrained total least squares technique and its applications to harmonic superresolution. *IEEE Transactions on Signal Processing*, 39:1070–1086, 1991.

[3] E. Anderson, Z. Bai, C. Bischof, J. Demmel, J. Dongarra, J. Du Croz, A. Greenbaum, S. Hammarling, A. McKenney, S. Ostrouchov, and D. Sorensen. *LAPACK Users' Guide*. SIAM, Philadelphia, 1995.

[4] E. Anderson, Z. Bai, and J. J. Dongarra. Generalized QR factorization and its applications. Technical Report CS-91-131 (LAPACK Working Note 31), Computer Science Dept., University of Tennessee, Knoxville, 1991.

[5] D.V. Bhaskar Rao and K.S. Arun. Model based processing of signals: A state space approach. *Proceedings IEEE*, 80:1277–1308, February 1992.

[6] B. De Moor. Structured total least squares and l_2 approximation problems. *Linear Algebra and its Applications, Special Issue on Numerical Linear Algebra Methods in Control, Signals and Systems, (Van Dooren, Ammar, Nichols, and Mehrmann, eds.)*, 188-189:163–207, July 1993.

[7] B. De Moor. Total least squares for affinely structured matrices and the noisy realization problem. *IEEE Transactions on Signal Processing*, 42:3004–3113, November 1994.

[8] P.D. Fiore and G.C. Verghese. Constrained maximum likelihood solution of linear equations. *IEEE Trans. on Signal Processing*, 48(3):671–679, 2000.

[9] J.S. Fleming and B.A. Goddard. A technique for the deconvolution of the renogram. *Phys. Med. Biol.*, 19:546–549, 1974.

[10] R. Fletcher. *Practical Methods of Optimization*. John Wiley & Sons, New York, 1987.

[11] P. E. Gill, W. Murray, and M. H. Wright. *Practical Optimization*. Academic Press, London, 1981.

[12] G. H. Golub and C.F. Van Loan. An analysis of the total least squares problem. *SIAM J. Numer. Anal.*, 17:883–893, 1980.

[13] P. Lemmerling, I. Dologlou, and S. Van Huffel. Speech compression based on exact modeling and structured total least norm optimization. In *Proceedings of ICASSP 98*, volume I, pages 353–356, Seattle, Washington, U.S.A., May 1998.

[14] P. Lemmerling, S. Van Huffel, and B. De Moor. The structured total least squares approach for nonlinearly structured matrices. *Numerical Linear Algebra with Applications*, 2001. to appear.

[15] P. Lemmerling and B. De Moor. Misfit versus latency. *Automatica*, 37(12):to appear, 2001.

[16] P. Lemmerling and S. Van Huffel. Analysis of the structured total least squares problem for Hankel/Toeplitz matrices. *Numerical Algorithms*, 27:89–114, 2001.

[17] N. Mastronardi, P. Lemmerling, and S. Van Huffel. Fast structured total least squares algorithms via exploitation of the displacement structure. in *this volume*.

[18] N. Mastronardi, P. Lemmerling, and S. Van Huffel. Fast structured total least squares algorithm for solving the basic deconvolution problem. *SIAM Journal on Matrix Anal. Appl.*, 22(2):533–553, 2000.

[19] A. Paulraj, R. Roy, and T. Kailath. A subspace rotation approach to signal parameter estimation. *Proceedings IEEE*, pages 1044–1045, July 1986.

[20] J. B. Rosen, H. Park, and J. Glick. Structured total least norm for nonlinear problems. *SIAM Journal on Matrix Analysis and Applications*, 20(1):14–30, 1999.

[21] J.B. Rosen, H. Park, and J. Glick. Total least norm formulation and solution for structured problems. *SIAM Journal on Matrix Anal. Appl.*, 17(1):110–128, 1996.

[22] R. Roy and T. Kailath. Esprit–estimation of signal parameters via rotational invariance techniques. *IEEE Transactions on Acoustics, Speech , Signal Processing*, 37(7):984–995, July 1989.

[23] A.J van der Veen, E. F. Deprettere, and A. L. Swindlehurst. Subspace based signal analysis using singular value decompostion. *Proceedings IEEE*, 81:1277–1308, September 1993.

[24] S. Van Huffel, H. Park, and J. B. Rosen. Formulation and solution of structured total least norm problems for parameter estimation. *IEEE Trans. on Signal Processing*, 44(10):2464–2474, 1996.

[25] S. Van Huffel and J. Vandewalle. *The Total Least Squares Problem: computational aspects and analysis*, volume 9. SIAM, Philadelphia, 1991.

FAST STRUCTURED TOTAL LEAST SQUARES ALGORITHMS VIA EXPLOITATION OF THE DISPLACEMENT STRUCTURE

Nicola Mastronardi
Department of Computer Science, K.U.Leuven,
Celestijnenlaan 200 A, 3001 Leuven-Heverlee, Belgium
nicola.mastronardi@cs.kuleuven.ac.be

Philippe Lemmerling and Sabine Van Huffel
ESAT-SISTA, K.U.Leuven, Kasteelpark Arenberg 10, B-3001 Leuven-Heverlee, Belgium
{philippe.lemmerling, sabine.vanhuffel}@esat.kuleuven.ac.be

Abstract We present fast algorithms and their stability properties for Toeplitz structured total least squares (STLS) problems.

The STLS problem can be formulated as the following constrained optimization problem

$$\min_{\Delta A, \Delta b, x} \|[\Delta A \; \Delta b]\|_F$$
$$\text{such that } (A + \Delta A)x = b + \Delta b$$
$$\text{and } [\Delta A \; \Delta b] \text{ has the same structure as } [A \; b].$$

This natural extension of the TLS problem is clearly more difficult to solve than the TLS problem, because of its highly nonlinear nature and the existence of many local minima. We focus here on the frequently occurring case where $[A \; b]$ is a Toeplitz matrix. The problem is solved in an iterative fashion, in which at each iteration the Karush-Kuhn-Tucker (KKT) equations of the locally linearized problem have to be solved. For this kernel routine we use a generalized Schur decomposition algorithm based on the low displacement rank of the KKT system matrix. By exploiting the sparsity of the associated generators, we obtain a fast algorithm that requires $O(mn + n^2)$ flops per iteration, where m and n are the number of rows and the number of columns of A, respectively. We also prove the stability of the latter kernel routine. The efficiency of the proposed fast implementation is compared to the efficiency of the straightforward implementation, which does not exploit the structure of

S. Van Huffel and P. Lemmerling (eds.), Total Least Squares and Errors-in-Variables Modeling, 93–106.

the involved matrices. The comparison is done on a recently introduced speech compression scheme in which the considered STLS problem constitutes one of the kernel problems. The numerical results confirm the high efficiency of the newly proposed fast implementation.

Keywords: Hankel/Toeplitz matrix, structured total least squares, displacement rank

1. Introduction

The iterative methods for solving Toeplitz Structured TLS problems proposed in [12] require to solve, at the k-th iteration, either a rectangular linear system (in the least squares sense)

$$A^{(k)}x = b^{(k)}, \quad A^{(i)} = \begin{bmatrix} X_{m,n-1}^{(k)T} & \Lambda_{m,n}^{(k)} \\ I_{m+n-1} & \end{bmatrix}, \; b^{(k)} \in \mathbb{R}^{2m+n-1}, \quad (1)$$

or a square linear system

$$C^{(k)}x = h^{(k)}, \; C^{(k)} = \begin{bmatrix} -X_{m,n+1}^{(k)}X_{m,n+1}^{(k)T} & \Lambda_{m,n}^{(k)T} \\ \Lambda_{m,n}^{(k)T} & \end{bmatrix}, \; h^{(k)} \in \mathbb{R}^{m+n}, \quad (2)$$

where $X_{m,l}^{(k)} \in \mathbb{R}^{m\times l}$ and $\Lambda_{m,n}^{(k)} \in \mathbb{R}^{m\times n}$ have the following Toeplitz structure

$$X_{m,l}^{(k)} = \begin{bmatrix} x_l^{(k)} \cdots & x_1^{(k)} & 0 & \cdots & 0 \\ 0 & x_l^{(k)} & \cdots & x_1^{(k)} & 0 & \vdots \\ \vdots & \ddots & \ddots & \ddots & \ddots & \vdots \\ 0 & \cdots & 0 & x_l^{(k)} & \cdots & x_1^{(k)} \end{bmatrix}, \Lambda_{m,n}^{(k)} = \begin{bmatrix} s_n^{(k)} & \cdots & s_2^{(k)} & s_1^{(k)} \\ s_{n+1}^{(k)} & \ddots & \ddots & s_2^{(k)} \\ \ddots & \ddots & \ddots & \vdots \\ s_{m+n-1}^{(k)} & \cdots & s_{m+1}^{(k)} & s_m^{(k)} \end{bmatrix},$$

and I_l is the identity matrix of order l. These matrices vary at each iteration and, in many applications, can be very large. In such cases, the computational burden required to solve these systems by means of classical algorithms (Gaussian elimination, QR algorithm, ...) can become prohibitively large. These algorithms can be speeded up by means of the *generalized Schur algorithm* (GSA), exploiting the so-called *displacement structure* of the involved matrices. The *displacement* of a Hermitian matrix was originally introduced in [6] and the *displacement theory* has been intensively studied in the last few decades (see [7] and the references therein).

The paper is organized as follows. In § 2 the basic concepts of displacement theory for symmetric strongly regular matrices are introduced. In STLS problems the involved matrices are of this kind [14, 10], indeed.

An extensive treatment of the displacement theory can be found in [7], including the non–symmetric and the complex case. In § 3 GSA is derived from the displacement theory. In § 4 a particular implementation of GSA for solving linear systems arising in Toeplitz structured TLS problems, exploiting the low displacement rank of the involved matrices and the structure of corresponding generators is considered. The stability property of this algorithm is discussed at the end of the section. In § 5 the increased efficiency of the proposed fast implementation is compared to the efficiency of the straightforward implementation, which does not exploit the structure of the involved matrices. The comparison is done on a recently introduced speech compression scheme in which the considered STLS problem constitutes one of the kernel problems. The conclusions can be found at the end of the paper.

2. Displacement theory for symmetric matrices

Let R be a symmetric strongly regular matrix and Φ be a strictly lower triangular matrix of order N, respectively. The displacement of R, with respect to Φ, is defined as

$$\nabla_\Phi R \triangleq R - \Phi R \Phi^T \ . \tag{3}$$

The matrix Φ, in the original definition (see [6]), is the lower shift matrix of order N,

$$Z = \begin{bmatrix} 0 & 0 & \cdots & 0 \\ 1 & 0 & \ddots & 0 \\ & \ddots & \ddots & \vdots \\ & & 1 & 0 \end{bmatrix}. \tag{4}$$

Indeed, the product ZRZ^T displaces R downwards along the main diagonal by one position, explaining the name *displacement* for $\nabla_Z R$. If $\nabla_\Phi R$ has low displacement rank, *independent* of N, then R is said to be *structured* with respect to the displacement ∇_Φ defined by (3), and α is called the *displacement rank* of R. We define the *displacement inertia* of R as the pair $\{p, q\}$, where p (resp. q) is the number of strictly positive (resp. negative) eigenvalues of $\nabla_\Phi R$. Of course, the displacement rank is $\alpha = p + q$. Therefore, we can write

$$\nabla_\Phi R = R - \Phi R \Phi^T = GJG^T, \ G \in \mathbb{R}^{N\times\alpha}, \ J = J^T = (I_p \oplus -I_q), \tag{5}$$

or, equivalently,

$$R = \Phi R \Phi^T + GJG^T. \tag{6}$$

The pair $\{G, J\}$ is called a ∇_Φ*−generator* of R. This representation is clearly not unique; for example, $\{G\Theta, J\}$ is also a generator for any

J-orthogonal matrix Θ (*i.e.*, for any Θ such that $\Theta J \Theta^T = J$). This is because

$$G \underbrace{\Theta J \Theta^T}_{J} G^T = GJG^T.$$

Moreover, the first p and the last q columns of G are called *positive* and *negative* ∇_Φ*–generators*, respectively, or, if there is no ambiguity, simply *positive* and *negative generators.* Let $\{G, J\}$ be a ∇_Φ–generator of R. Let f be the first nonzero row of G. $\{G, J\}$ is said to be in *proper form* if f has all the elements equal to zero with the exception of a single pivot element in position π. This pivot has to be in the first p positions when $fJf^T > 0$ (positive step) and in the last q positions when $fJf^T < 0$ (negative step) (note that the case $fJf^T = 0$ is ruled out by the strong regularity assumption on R [7]). Define $\Phi_i = \Phi^i$. It is straightforward to see that $\nabla_\Phi \Phi R \Phi^T = \Phi GJG^T \Phi^T$. Then

$$\Phi R \Phi^T = \Phi_2 R \Phi_2^T + \Phi GJG^T \Phi^T. \tag{7}$$

Replacing (7) in (6) it turns out $R = \Phi_2 R \Phi_2^T + \Phi GJG^T \Phi^T + GJG^T$. Proceeding recursively and using the fact that Φ is nilpotent, i.e., $\Phi_j = \mathbf{0}$, $j > N-1$, we can easily conclude that

$$R = \sum_{i=0}^{N-1} \Phi_i GJG^T \Phi_i^T. \tag{8}$$

3. Generalized Schur algorithm

The equivalence between the Schur reduction and the Gaussian elimination for computing the LDL^T factorization of a symmetric matrix R, where L is lower triangular and D diagonal, has been shown in [7]. However working directly on the representation (8) of R allows us to speed up the Gaussian elimination procedure. This leads to the so–called *Generalized Schur algorithm.*

Algorithm 1 *(Generalized Schur algorithm in proper form).* Let $R \in \mathbb{R}^{N \times N}$ be a symmetric matrix s.t. $\nabla_\Phi R = GJG^T$, for some full rank $G \in \mathbb{R}^{N \times \alpha}$. Set $G_1 = G$. Let l_i and d_i be the i–th column of L and the (i, i) entry of D, $i = 1, \ldots, N$.

for $i = 1, \ldots, N$
  transform G_i to proper form $\hat{G}_i$,
  compute l_i and d_i from the pivot column of $\hat{G}_i$,
  update $\hat{G}_i$ to G_{i+1} multiplying the pivot column of $\hat{G}_i$ by Φ.
end

As already mentioned, the computational burden of each step of GSA is given by

1 reduction of the generator matrix in proper form,

2 application of the matrix Φ to the pivot generator vector.

The reduction of a generator matrix in proper form can be accomplished in $O(\alpha(N-i))$ flops by means of Givens and hyperbolic transformations (see § 4). Hence, it turns out it is essential to consider a matrix Φ such that the displacement rank is $\alpha \ll N$, and the matrix–vector product

$$\Phi z, \quad z \in \mathbb{R}^N, \tag{9}$$

can be computed efficiently.

Example 1 *(Displacement of symmetric Toeplitz matrices).* Choosing Φ equal to the shift matrix (4), the displacement rank of symmetric Toeplitz matrices is 2,

$$T = \begin{bmatrix} c_1 & c_2 & \ddots & c_N \\ c_2 & c_1 & c_2 & \ddots \\ \ddots & \ddots & \ddots & \ddots \\ c_N & \ddots & c_2 & c_1 \end{bmatrix}, \quad \nabla_Z T = \begin{bmatrix} c_1 & c_2 & \cdots & c_N \\ c_2 & & & \\ \vdots & & & \\ c_N & & & \end{bmatrix}.$$

Assuming $c_1 = 1$, a generator for T is $\{u_1, v_1, (1 \oplus -1)\}$, where

$$u_1 = \begin{bmatrix} 1 \\ c_2 \\ \vdots \\ c_N \end{bmatrix}, v_1 = \begin{bmatrix} 0 \\ c_2 \\ \vdots \\ c_N \end{bmatrix} \text{ and } \nabla_Z T = \begin{bmatrix} 1 & 0 \\ c_2 & c_2 \\ \vdots & \vdots \\ c_N & c_N \end{bmatrix} \begin{bmatrix} 1 & \\ & -1 \end{bmatrix} \begin{bmatrix} 1 & 0 \\ c_2 & c_2 \\ \vdots & \vdots \\ c_N & c_N \end{bmatrix}^T.$$

$O(2(N-i))$ flops are required to reduce the generator G_i in proper form at the i–th iteration of GSA. Moreover (9) corresponds to shifting the entries of the vectors $\tilde{g}_i, i = 1, \ldots, N$, downward one position. Hence the complexity of GSA for symmetric Toeplitz matrices is $O(N^2)$.

4. The generalized Schur algorithm for linear systems arising in Toeplitz STLS problems

As already mentioned in § 1, at each iteration of Toeplitz STLS algorithms [12] we need to solve either a linear system of the form (1) or a linear system of the form (2). This section is focused on the solution of (2) by GSA. (1) can be solved in a similar way and details can be found in [14]. The MATLAB notation is used in the description of the algorithm. Let

$$Z = Z_m \oplus Z_n.$$

By inspection (see [7]), it can be seen that the displacement rank of $C^{(k)}$ with respect to Z is 4. Denote by $v_1 = X^{(k)}_{m,n+1}(1,:)^T$, and $v = v_1/\|v_1\|_2$. Let $y = -X^{(k)}_{m,n+1}v$ and $w = [\lambda(n), \lambda(n-1), \ldots, \lambda(1)]^T/\|v_1\|_2$. Then the generators of $C^{(k)}$ are defined in the following way.

$$\begin{aligned} g_1 &= [y^T, w^T]^T \\ g_2 &= [0, y(2:m)^T, w^T]^T \\ g_3 &= [0, \lambda(m+n-1), \lambda(m+n-2), \ldots, \lambda(n+1), 0.5, 0, \ldots, 0]^T \\ g_4 &= [0, \lambda(m+n-1), \lambda(m+n-2), \ldots, \lambda(n+1), -0.5, 0, \ldots, 0]^T, \end{aligned}$$

where g_2 and g_3 are positive, g_1 and g_4 are negative. We remark that the order of the columns in the generator matrix defined in § 2 (the positive and the negative generators are in the first and in the last columns of the matrix, respectively) can easily be retrieved by permuting the columns in (10). Similarly, the diagonal elements of the matrix J defined in (11) must be exchanged. Define

$$G_{Z,0} = \left[\begin{array}{cccc} g_1 & g_2 & g_3 & g_4 \end{array}\right] \tag{10}$$

and let

$$J = \mathrm{diag}(-1, 1, 1, -1) \tag{11}$$

Then $C^{(k)} = G_{Z,0} J G^T_{Z,0}$. Since the order of the matrix $C^{(k)}$ is $m+n$ the computational cost of the generalized Schur algorithm should be proportional to $(m+n)^2$. In § 4.1 we show that, exploiting the particular structure of the generators of $C^{(k)}$ the computational cost of the generalized Schur algorithm can be reduced to $O(mn+n^2)$. We observe that the matrix $C^{(k)}$ is indefinite. However, analyzing the generators and the Schur complement of $-X^{(k)}_{m,n+1}X^{(k)T}_{m,n+1}$ in $C^{(k)}$ we are able to say a priori that the steps of the algorithm for $i = 1, \ldots, m$ are negative, the steps for $i = m+1, \ldots, m+n$, are positive [10].

4.1 Description of the algorithm

At each iteration i of GSA, we look for a J–orthogonal matrix Θ_i in order to eliminate all elements of f_i,the i–th row of G_i with exception of one element. This can be done by choosing J–orthogonal matrices Θ such that

$$\left[\begin{array}{cc} * & 0 \end{array}\right] = \left[\begin{array}{cc} f_i(j) & f_i(k) \end{array}\right]\Theta.$$

Θ can be either a Givens rotation if $\{j,k\} \in \{1,4\}$ or $\{j,k\} \in \{2,3\}$, or a hyperbolic rotation elsewhere. Proceeding in this way we can eliminate all the entries of f_i with exception of a single pivot element. We also refer to an application of a Givens rotation to a pair of generators as an *updating* and an application of an hyperbolic rotation to a pair of generators as a *downdating* [3]. If $f_i J f_i^T < 0$ then we choose the index of the pivot element equal to one and set $D(i,i) = -1$ (negative step), else we choose the index equal to 2 and we set $D(i,i) = 1$ (positive step). The case $f_i J f_i^T = 0$ is ruled out by the strong regularity assumption on $C^{(k)}$ [7]. For stability reasons the downdating steps are performed by means of *mixed* hyperbolic rotations, also called *stabilized hyperbolic rotation.* [3].

To exploit the *structure* of the generators we divide the algorithm into 3 phases:

- 1st phase: iteration `for` $i = 1, \ldots, m-n$,
- 2th phase: iteration `for` $i = m-n+1, \ldots, m$,
- 3th phase: iteration `for` $i = m+1, \ldots, m+n$.

1st phase: iterations for $i = 1 : m - n$. Before describing this phase we observe that the vectors $g_3^{(i-1)}$ and $g_4^{(i-1)}$ differ only for the $(m+1)$th entry. This implies the updating of $g_1^{(i-1)}$ with $g_4^{(i-1)}$ and the downdating with $g_3^{(i-1)}$ modifies only the $(m+1)$th entry of $g_1^{(i-1)}$. Hence $g_1^{(i-1)}(1:m)$ and $g_2^{(i-1)}(1:m)$ continue to be the generator vectors at the beginning of the ith step for the LDL^T factorization of $-X_{m,n+1}^{(k)T} X_{m,n+1}^{(k)T}$. Thus each iteration of this phase is a negative step since $f_i J f_i^T < 0$. Now we describe how the generators are modified at each iteration of this phase. We have to update $g_1^{(i-1)}$ with $g_4^{(i-1)}$ and downdate with $g_3^{(i-1)}$. These vectors are

$$
\begin{aligned}
g_1^{(i-1)} &= [\underbrace{0,\ldots,0}_{i-1}, \xi_i, \ldots, \xi_{n+i}, \underbrace{0,\ldots,0}_{m-n-i}, \xi_{m+1}, \ldots, \xi_{m+n}]^T \\
g_4^{(i-1)} &= [\underbrace{0,\ldots,0}_{i-1}, \zeta_i, \ldots, \zeta_m, \zeta_{m+1}, \zeta_{m+2} \cdots, \zeta_{m+n}]^T \\
g_3^{(i-1)} &= [\underbrace{0,\ldots,0}_{i-1}, \zeta_i, \ldots, \zeta_m, \mu_{m+1}, \zeta_{m+2} \cdots, \zeta_{m+n}]^T.
\end{aligned}
\tag{12}
$$

The Givens rotation used to update $g_1^{(i-1)}$ with $g_4^{(i-1)}$ is

$$Q_{i-1} = \begin{bmatrix} c_G^{(i-1)} & -s_G^{(i-1)} \\ s_G^{(i-1)} & c_G^{(i-1)} \end{bmatrix}, c_G^{(i-1)} = \frac{\xi_i}{\sqrt{\xi_i^2 + \zeta_i^2}} \text{ and } s_G^{(i-1)} = \frac{\zeta_i}{\sqrt{\xi_i^2 + \zeta_i^2}}.$$

The updated vectors $\tilde{g}_1^{(i-1)}$ and $\tilde{g}_4^{(i-1)}$ are

$$\begin{aligned} \tilde{g}_1^{(i-1)} &= c_G^{(i-1)} g_1^{(i-1)} + s_G^{(i-1)} g_4^{(i-1)} \\ \tilde{g}_4^{(i-1)} &= -s_G^{(i-1)} g_1^{(i-1)} + c_G^{(i-1)} g_4^{(i-1)}, \end{aligned} \tag{13}$$

with $\tilde{g}_1^{(i-1)} = [\underbrace{0, \ldots, 0}_{i-1}, \tilde{\xi}_i, \ldots, \tilde{\xi}_{m+n}]^T \quad \left(\tilde{\xi}_i = \sqrt{\xi_i^2 + \zeta_i^2}\right)$. Moreover,

$$\tilde{g}_4^{(i-1)}(n+i+1:m) = c_G^{(i-1)} g_4^{(i-1)}(n+i+1:m), \tag{14}$$

since $g_1^{(i-1)}(j) = 0, j = n+i+1, \ldots, m$. The next step in the ith iteration is downdating $\tilde{g}_1^{(i-1)}$ with $g_3^{(i-1)}$ by means of a mixed hyperbolic rotation

$$H_{i-1} = \begin{bmatrix} 1 & 0 \\ \rho & \sqrt{1-\rho^2} \end{bmatrix} \begin{bmatrix} \frac{1}{\sqrt{1-\rho^2}} & 0 \\ 0 & 1 \end{bmatrix} \begin{bmatrix} 1 & \rho \\ 0 & 1 \end{bmatrix},$$

where ρ is such that $\begin{bmatrix} \hat{\xi}_i & 0 \end{bmatrix} = \begin{bmatrix} \tilde{\xi}_i & \zeta_i \end{bmatrix} H_{i-1}^T$. Taking (13) into account, it is straightforward to see that

$$\rho = -s_G^{(i-1)} \text{ and } \sqrt{1-\rho^2} = c_G^{(i-1)}.$$

The downdated vectors $\hat{g}_1^{(i-1)}$ and $\tilde{g}_3^{(i-1)}$ are

$$\hat{g}_1^{(i-1)} = \frac{\tilde{g}_1^{(i-1)} - s_G^{(i-1)} g_3^{(i-1)}}{c_G^{(i-1)}} = g_1^{(i-1)} - s_G^{(i-1)} \frac{g_3^{(i-1)} - g_4^{(i-1)}}{c_G^{(i-1)}}, \tag{15}$$

$$\begin{aligned} \tilde{g}_3^{(i-1)} &= -s_G^{(i-1)} \hat{g}_1^{(i-1)} + c_G^{(i-1)} g_3^{(i-1)} \\ &= -s_G^{(i-1)} g_1^{(i-1)} + \frac{g_3^{(i-1)} - (1 - c_G^{(i-1)2}) g_4^{(i-1)}}{c_G^{(i-1)}}. \end{aligned} \tag{16}$$

Hence, from (15) and (13), $\tilde{g}_3^{(1)}$ and $\tilde{g}_4^{(1)}$ continue to be equal, except for the $(m+1)$th entry. Furthermore, from (15), we observe that $g_1^{(i-1)}$ and $\hat{g}_1^{(i-1)}$ differ in their $(m+1)$th entry. Since $g_1^{(i-1)}(m+1) = 0$, then

$$\hat{g}_1^{(i-1)}(m+1) = -s_G^{(i-1)}(g_3^{(i-1)}(m+1) - g_4^{(i-1)}(m+1))/c_G^{(i-1)}.$$

We observe that it is not necessary to compute the whole vector in (14) since, at the next iteration, the corresponding entries of $g_1^{(i)}(n+i+2:m+1)$ are equal to 0. We need only to store the partial product

$$c_G^{(i-2)} \cdots c_G^{(2)} c_G^{(1)} \tag{17}$$

into a temporary variable, and multiply $g_4^{(i)}(n+i+1)$ with this variable at the beginning of the ith iteration. To finish the iteration $\hat{g}_1^{(i-1)}$ has to be downdated with $g_2^{(i-1)}$. This computation does not destroy the structure of the vectors since

$$\begin{aligned}
\hat{g}_1^{(i-1)} &= [\underbrace{0,\ldots,0}_{i-1},\underbrace{*,\cdots,*,*}_{n+1},\underbrace{0,\ldots,0}_{m-n-i},\underbrace{*,\cdots,*}_{n}]^T \\
g_2^{(i-1)} &= [\underbrace{0,\ldots,0}_{i-1},\underbrace{*,\cdots,*}_{n},\underbrace{0,\ldots,0}_{m-n-i+1},\underbrace{*,\cdots,*}_{n}]^T.
\end{aligned}$$

Let $\tilde{H}_{i-1}$ be the stabilized hyperbolic rotation such that

$$\begin{bmatrix} \breve{g}_1^{(i-1)} & \breve{g}_2^{(i-1)} \end{bmatrix} = \begin{bmatrix} \hat{g}_1^{(i-1)} & g_2^{(i-1)} \end{bmatrix} \tilde{H}_{i-1}^T$$

with $\breve{g}_2^{(i-1)}(i) = 0$. Then $\breve{g}_1^{(i-1)}$ becomes the ith column of L, $D(i,i) = -1$, and, for the next iteration, the updated vectors are

$$\begin{aligned}
g_1^{(i)} &= [0, \hat{g}_1^{(i-1)}(i+1:m+n)],\ g_1^{(i)}(m+1) = 0 \\
g_2^{(i)} &= \breve{g}_2^{(i-1)} \\
g_4^{(i)} &= \tilde{g}_4^{(i-1)} \\
g_3^{(i)} &= [\tilde{g}_4^{(i-1)}(1:m); \gamma; \tilde{g}_4^{(i-1)}(m+2:m+n)],
\end{aligned}$$

where $\gamma = c_G^{(i-1)} g_3^{(i-1)}(m+1) - s_G^{(i-1)} \hat{g}_3^{(i-1)}(m+1)$. The number of flops of this phase is $18mn - 18n^2$.

2nd phase: iterations for $i = m-n+1:m$. one. In this phase we do not need to store the product of the Givens coefficients $c_G^{(i-1)}$ into a temporary variable, since $g_1^{(i-1)}(k) \neq 0, k = i,\ldots,m+n$. We recall that $g_3^{(i-1)}$ and $g_4^{(i-1)}$ continue to be equal with exception of the $(m+1)$th entry. Thus $g_1^{(i-1)}(i:m)$ and $g_2^{(i-1)}(i:m)$ are the generator vectors at the ith iteration of the LDL^T factorization of $-X_{m,n+1}^{(k)T} X_{m,n+1}^{(k)T}$. Hence each iteration of this phase continues to be a negative step $(D(i,i) = -1)$. The number of flops of this phase is $13.5n^2$.

3rd phase: iterations for $i = m+1:m+n$. In this phase the vectors $g_3^{(i-1)}(m+1:m+n)$ and $g_4^{(i-1)}(m+1:m+n)$ are different. Each iteration of this phase is a positive step [10] $(D(i,i) = 1)$ and the ith column of L is $\breve{g}_2^{(i-1)}$. The number of flops of this phase is $9n^2$.

4.2 Stability of LDL^T factorization

The stability of GSA for STLS problems is studied in [15]. At each iteration of the latter algorithm, two hyperbolic rotations are performed. In [15] it is proved that the following stability results hold for the LDL^T factorization of $C^{(k)}$, provided the hyperbolic rotations are implemented in a stable way [3, 4].

Theorem 1. *[15] Let G be the generator matrix in (10) of $C^{(k)}$. Let L and D be the matrices of the LDL^T factorization of $C^{(k)}$ computed by means of the generalized Schur algorithm applying a sequence of Givens rotations and two mixed hyperbolic rotations per iteration. Then*

$$\|C^{(k)} - LDL^T\|_F \leq 74(m+n-1)(m+n)\varepsilon \left(2\sqrt{m+n}\|C^{(k)}\|_F + \|G\|_F^2\right).$$

Hence the proposed generalized Schur algorithm for STLS problems is backward stable.

5. Application

The STLS problem involving Toeplitz (or Hankel) matrices constitutes the kernel problem of many signal processing applications. In this section the fast implementation of GSA described in the previous section is applied to a STLS problem arising in a speech compression application [11, 9]. Section 5.1 gives a short description of the latter application. It is shown that the low-frequency part of the compression scheme is a Toeplitz STLS problem.

In Section 5.2 two different implementations of the STLS algorithm are applied to the latter speech compression application and compared. A first one, where linear systems (2), involved at each iteration of the algorithm, are solved by means of the GSA with exploitation of the structure of generators, and a second one, where linear systems (2) are solved by means of Gaussian elimination with partial pivoting [5]. Next, the computational efficiency of the two implementations is compared by means of the speech compression application.

5.1 Description

In this subsection we describe a subband based speech compression scheme. The incoming speech signal is processed on a frame by frame basis. In a first step a frame of the incoming signal is fed to a low-pass filter. This yields the low-frequency (LF) part. By subtracting the latter from the incoming signal, we obtain the high-frequency (HF), non-smoothed part. The motivation behind splitting the speech signal into two bands is that the LF part of a speech signal can efficiently be modeled by means of a linear model (Auto Regressive (AR) model). Instead of determining a classical AR model, leading to a vector of model parameters and a residual vector [13], we apply a minimal correction to the measured data (i.e. the LF part of the frame) such that the corrected measurements exactly satisfy an AR model. As shown in [11, 9], the latter procedure corresponds to solving a Toeplitz STLS problem that can be reformulated as follows:

$$\begin{array}{c} \min_{\Delta s, x} \frac{1}{2} \Delta s^T \Delta s \\ \text{such that } (A + \Delta A)x = b + \Delta b, \\ \text{and } [A\ b] \text{ has the same structure of } [\Delta A\ \Delta b], \end{array} \tag{18}$$

where $[A\ b]$ and $[\Delta A\ \Delta b]$ respectively are

$$\left[\begin{array}{cccc|c} s_{n+1} & s_n & \cdots & s_2 & s_1 \\ s_{n+2} & s_{n+1} & \ddots & \ddots & s_2 \\ \vdots & \ddots & \ddots & \ddots & \vdots \\ s_{m+n} & \ddots & \cdots & s_{m+1} & s_m \end{array}\right], \quad \left[\begin{array}{cccc|c} \Delta s_{n+1} & \Delta s_n & \cdots & \Delta s_2 & \Delta s_1 \\ \Delta s_{n+2} & \Delta s_{n+1} & \ddots & \ddots & \Delta s_2 \\ \vdots & \ddots & \ddots & \ddots & \vdots \\ \Delta s_{m+n} & \ddots & \cdots & \Delta s_{m+1} & \Delta s_m \end{array}\right]. \tag{19}$$

The vector s in (19) corresponds to the LF part of the incoming frame, the corrections we apply correspond to the vector Δs in (19) and the model for that particular frame is represented by the vector x. The compression comes about as follows. Due to the Toeplitz structure of $[A+\Delta A\ b+\Delta b]$ and the equality constraint in (18), it should be clear that we only need $s(1:n) + \Delta s(1:n)$ and $x \in \mathbb{R}^{n\times 1}$ in order to reconstruct the entire frame $s + \Delta s \in \mathbb{R}^{(m+n)\times 1}$. Thus, loosely speaking, we obtain a compression ratio of $2n/(m+n)$. For further implementation details we refer to [11, 9].

5.2 Numerical tests

As already mentioned in § 1, at each iteration of the iterative algorithm for solving STLS problems the solution of a structured linear system (2) is required. In this subsection we compare the efficiency of the STLS algorithm when linear systems (2) are solved in different ways. In particular we refer to the STLSf algorithm if linear systems (2) are solved by means of the GSA with exploitation of the structure of generators, the algorithm described in the previous section. We refer to the STLSs algorithm if linear systems (2) are solved by means of Gaussian elimination with partial pivoting. As shown in the previous section, the computational complexity of STLSf is $O(mn + n^2)$. The implementation STLSs has a computational complexity of $O((m + n)^3)$ per iteration.

The speech signal (see left hand side of figure 1) we compress is 14616 samples long. It is processed on a frame by frame basis, each frame being 504 samples long. As mentioned before each frame is split in a LF and a HF part. It is the LF part that is compressed using the STLS problem formulation. The LF part of the frame can be modelled sufficiently well using a 4th order model, thus yielding the following dimensions of the STLS problem (18): $m = 500$, $n = 4$. The right hand side of figure 1 shows the LF part of the original 6th frame (i.e. the signal s, full line) and the reconstructed LF part of the frame obtained after compression (i.e. the signal $s + \Delta s$, dashed line). The latter frame can be reconstructed using only the first 4 samples of $s + \Delta s$ and the parameter vector $x \in \mathbb{R}^{4\times 1}$.

Per frame, the two implementations STLSs and STLSf require the same (but varying) number of iterations, of course. The important number is thus the number of flops per iteration. For this particular problem these numbers are displayed in table 1. We clearly see the drastically increased computational performance obtained with STLSf.

To investigate the dependence of the computational cost of the different implementations on the size of the problem, we change the parameters of the compression scheme. The three implementations are applied using the following framelengths and estimated orders:

- framelength=254, estimated order: 4 (this means $m = 250$, $n = 4$)
- framelength=508, estimated order: 8 (this means $m = 500$, $n = 8$).

The results are shown on the second and third line of table 1.

Table 1. This table shows the increased performance of the implementation STLSf compared to the straightforward implementations STLSs, for different problem sizes. The numbers represent the flop count obtained with the MATLAB function `flops`. $flops_{STLSs}$ and $flops_{STLSf}$ represent the number of flops per iteration for respectively the implementations STLSs and STLSf.

frame length	estimated order	$m \times n$	$STLSs$	$STLSf$	$\frac{flops_{STLSs}}{flops_{STLSf}}$
504	4	500×4	$6.916e8$	117854	5868
254	4	250×4	$8.968e7$	59104	1517
508	8	500×8	$7.080e8$	$2.056e5$	3444

6. Conclusions

The iterative algorithms for solving Hankel/Toeplitz structured TLS problems described in [10, 14] require to solve, at each iteration, a structured linear system of order $m + n$. By exploiting the low displacement

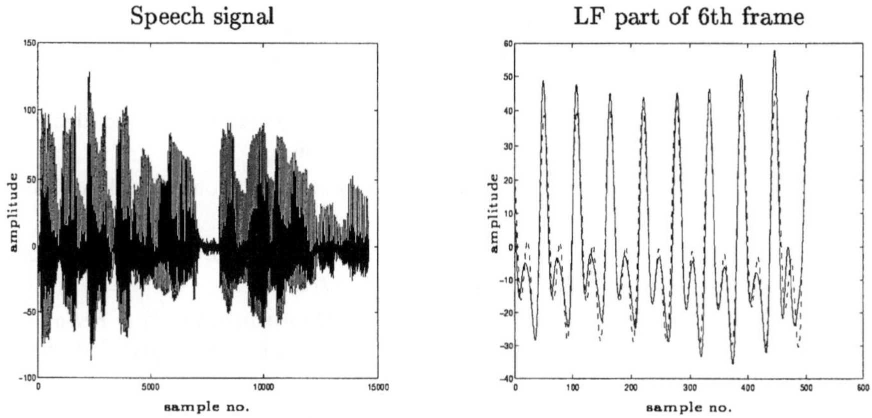

Figure 1. The left hand side of this figure shows the original speech signal. The right hand side shows the LF part of the 6th frame of the original signal (i.e. s, the full line) together with the reconstructed (after compression) LF part of the 6th frame (dashed line).

rank of the involved matrices GSA can be efficiently applied to compute the LDL^T of the involved matrices with $O((m+n)^2)$ computational complexity. By taking advantage of the sparsity of the corresponding generators we are able to even further improve the computational efficiency. The computational complexity for each iteration is $O(mn + n^2)$ whereas the straightforward implementation has a computational complexity of $O((m + n)^3)$. These numbers are confirmed by applying the different implementations in a recently proposed speech compression scheme that contains a Toeplitz STLS problem as kernel problem. Furthermore we have proved that this implementation of GSA for solving linear systems arising in STLS algorithms is backward stable.

Acknowledgments

This work is supported by IUAP P4-02 and P4-24, the Concerted Research Action projects MEFISTO-666 and FWO project G.0200.00.

References

[1] E. Anderson, Z. Bai, C. Bischof, J. Demmel, J. Dongarra, J. Du Croz, A. Greenbaum, S. Hammarling, A. McKenney, S. Ostrouchov, and D. Sorensen. *LAPACK Users' Guide*. SIAM, Philadelphia, 1995.

[2] E. Anderson, Z. Bai, and J. J. Dongarra. Generalized QR factorization and its applications. Technical Report CS-91-131 (LAPACK Working Note 31), Computer Science Dept., University of Tennessee, Knoxville, 1991.

[3] A.W. Bojanczyk, R.P. Brent, P. Van Dooren and F.R. De Hoog, A note on downdating the Cholesky factorization, *SIAM J. Sci. Stat. Comput.*, 1:210–220, 1980.

[4] S. Chandrasekaran, A.H. Sayed, Stabilizing the generalized Schur algorithm, *SIAM J. Matrix Anal. Appl.,* 17(4):950–983, 1996.

[5] G.H. Golub and C.F. Van Loan. *Matrix Computations.* Johns Hopkins University Press, Baltimore, 1989.

[6] T. Kailath, S. Kung and M. Morf, Displacement ranks of matrices and linear equations, *J. Math. Anal. Appl.*, 68:395–407, 1979.

[7] T. Kailath, Displacement structure and array algorithms, in *Fast Reliable Algorithms for Matrices with Structure*, T. Kailath and A. H. Sayed, Ed., SIAM, Philadelphia, 1999.

[8] T. Kailath, A.H. Sayed, Displacement structure: Theory and applications, *SIAM Review*, 37:297–386, 1995.

[9] P. Lemmerling, I. Dologlou, and S. Van Huffel, Speech compression based on exact modeling and structured total least norm optimization, in *Proceedings of the International Conference on Acoustics, Speech, and Signal Processing (ICASSP'98)*, Vol. 1, Seattle, WA, 353–356, 1998.

[10] P. Lemmerling, N. Mastronardi and S. Van Huffel, Fast algorithm for solving the Hankel/Toeplitz structured total least squares problem, *Numerical Algorithms*, 23:371–392, 2000.

[11] P. Lemmerling, N. Mastronardi and S. Van Huffel, Fast algorithms for structured total least squares in speech compression, Internal Report 01–55, ESAT-SISTA, K.U.Leuven (Leuven, Belgium), 2001.

[12] P. Lemmerling and S. Van Huffel, Structured total least squares: analysis, algorithms and applications, in *this volume.*

[13] L. Ljung, *System identification: theory for the user.* Prentice-Hall, Inc., 1987.

[14] N. Mastronardi, P. Lemmerling and S. Van Huffel, Fast structured total least squares algorithm for solving the basic deconvolution problem, *SIAM J. Matrix Anal. Appl.*, 22(2):533–553, 2000.

[15] N. Mastronardi, P. Van Dooren, and S. Van Huffel, On the stability of the generalized Schur algorithm, Proceedings of the conference *Numerical Analysis and Its Applications,* Rousse, Bulgaria, June 2000. Lecture Notes in Computer Science, 560–567, 1988, 2001.

THE EXTENDED STLS ALGORITHM FOR MINIMIZING THE EXTENDED LS CRITERION

Arie Yeredor

Dept. of Elect. Eng. - Systems, Tel-Aviv University

arie@eng.tau.ac.il

Abstract The recently-introduced Extended Least-Squares (XLS) parameters - estimation criterion is aimed at discriminating measurement errors from modelling errors (or "misfit" from "latency" errors). Using versatile weighting of "presumed errors", it encompasses the classical Least-Squares (LS) criterion on one hand, and the (Structured, or Constrained) Total LS [(S,C)TLS] criteria on the other hand. Thus, the STLS algorithm, originally aimed at solving TLS problems with structural constraints, can be modified, or "extended", to solve the XLS minimization problem. In this paper we introduce the Extended STLS algorithm, and demonstrate its use in the XLS context with estimating the parameters of a noisy Auto-Regressive (AR) process. We briefly compare the Extended STLS algorithm to other algorithms serving the same purpose.

Keywords: extended least squares (XLS), structured total least squares (STLS), extended STLS (ESTLS), system identification, latency error.

1. Introduction

The extended Least-Squares (XLS) estimation criterion has recently been introduced ([12, 15, 14]) as a tool for discriminating measurement errors from modelling errors. It has also been independently considered (with a slightly different terminology) ([9, 10]), as a tool for introducing a tradeoff of "misfit vs. latency". Specifically, assume that a vector of parameters $\boldsymbol{\theta}$ is to be estimated from a vector of observations $\boldsymbol{x}$, both related by the approximate set of equations

$$\boldsymbol{g}(\boldsymbol{x}, \boldsymbol{\theta}) \approx \mathbf{0}. \tag{1}$$

The inequality in (1) may often be attributed to two possibly distinct mechanisms: model mismatch (or "latency") and measurement

S. Van Huffel and P. Lemmerling (eds.), Total Least Squares and Errors-in-Variables Modeling, 107–117.

inaccuracies (or "misfit"). Model mismatch encompasses some "latent" inaccuracies that exist in (1) even when exact measurements are used. Measurement inaccuracies, on the other hand, account for possible deviations (or "misfit") of the measured data $\boldsymbol{x}$ from the true (unknown) data, say $\tilde{\boldsymbol{x}}$, with which (1) is exact (in the absence of model mismatch). In other words, (1) may be broken down into two distinct (in)equalities:

$$\boldsymbol{g}(\tilde{\boldsymbol{x}}, \boldsymbol{\theta}) \approx \mathbf{0} \qquad \boldsymbol{x} - \tilde{\boldsymbol{x}} \approx \mathbf{0} \tag{2}$$

where $\tilde{\boldsymbol{x}}$ is a vector of "presumed" ("accurate" or "latent") data, so that model mismatch and measurement inaccuracies are separately accounted for. While measurement inaccuracies can always be attributed to model errors, they are often caused by unrelated mechanisms. In such cases the distinction in (2) is justified.

The Least Squares (LS) estimation approach seeks parameters $\boldsymbol{\theta}$, which, together with the given measurements $\boldsymbol{x}$, minimize the (possibly weighted) Euclidean norm of $\boldsymbol{g}(\boldsymbol{x}, \boldsymbol{\theta})$:

$$\min_{\boldsymbol{\theta}}\{\boldsymbol{g}^T(\boldsymbol{x}, \boldsymbol{\theta})\boldsymbol{W}\boldsymbol{g}(\boldsymbol{x}, \boldsymbol{\theta})\} \Rightarrow \hat{\boldsymbol{\theta}}_{LS} \tag{3}$$

where $\boldsymbol{W}$ is some symmetric positive-definite weight matrix. $\hat{\boldsymbol{\theta}}_{LS}$ is actually the value that brings $\boldsymbol{g}(\boldsymbol{x}, \boldsymbol{\theta})$ as close as possible to its presumed value of $\mathbf{0}$. However, no substitution of the observed data is implied, thus $\boldsymbol{x} - \tilde{\boldsymbol{x}} = \mathbf{0}$.

On the other hand, the total LS (TLS) and its structurally-constrained variants (such as Constrained TLS (CTLS) ([1, 2, 5, 4]) and Structured TLS (STLS) ([6, 10])[1], can often be regarded as taking the opposite approach. Specifically, they attempt to find a vector of 'presumed' data $\tilde{\boldsymbol{x}}$ and an associated parameters vector $\boldsymbol{\theta}$, such that $\boldsymbol{g}(\tilde{\boldsymbol{x}}, \boldsymbol{\theta}) = \mathbf{0}$, while keeping to minimum the deviation of $\tilde{\boldsymbol{x}}$ from $\boldsymbol{x}$.

The more comprehensive XLS criterion accounts for both error sources simultaneously: Forming the concatenation of (2), we get

$$\tilde{\boldsymbol{g}}(\boldsymbol{x}, \tilde{\boldsymbol{x}}, \boldsymbol{\theta}) \triangleq \begin{bmatrix} \boldsymbol{g}(\tilde{\boldsymbol{x}}, \boldsymbol{\theta}) \\ \boldsymbol{x} - \tilde{\boldsymbol{x}} \end{bmatrix} \approx \mathbf{0}. \tag{4}$$

The XLS estimate of $\boldsymbol{\theta}$ is obtained by applying the LS criterion to $\tilde{\boldsymbol{g}}$. However, we now have to minimize with respect to $\tilde{\boldsymbol{x}}$ as well, obtaining

[1] In a somewhat confusing conflict of terms, Cadzow used the term CTLS to formulate a problem which slightly differs from Abatzoglou's original CTLS. Interestingly, STLS later turned out ([11]) to be equivalent to Abatzoglou's CTLS (differing only in the minimization approach, but not in the problem formulation).

as a by-product the XLS estimate of the presumed data:

$$\min_{\tilde{\boldsymbol{x}},\boldsymbol{\theta}}\{\tilde{\boldsymbol{g}}^T(\boldsymbol{x},\tilde{\boldsymbol{x}},\boldsymbol{\theta})\tilde{\boldsymbol{W}}\tilde{\boldsymbol{g}}(\boldsymbol{x},\tilde{\boldsymbol{x}},\boldsymbol{\theta})\} \Rightarrow \hat{\boldsymbol{\theta}}_{XLS} \quad (+\hat{\tilde{\boldsymbol{x}}}_{XLS}), \tag{5}$$

$\tilde{\boldsymbol{W}}$ being the extended weight, often chosen to be block-diagonal,

$$\tilde{\boldsymbol{W}} = \begin{bmatrix} \boldsymbol{W}_g & 0 \\ 0 & \boldsymbol{W}_x \end{bmatrix} \tag{6}$$

where $\boldsymbol{W}_g$ and $\boldsymbol{W}_x$ fit the dimensions of $\boldsymbol{g}(\boldsymbol{x},\boldsymbol{\theta})$ and $\boldsymbol{x}$, respectively, so that the XLS minimization (5) may assume the following form:

$$\min_{\tilde{\boldsymbol{x}},\boldsymbol{\theta}}\{\boldsymbol{g}^T(\tilde{\boldsymbol{x}},\boldsymbol{\theta})\boldsymbol{W}_g\boldsymbol{g}(\tilde{\boldsymbol{x}},\boldsymbol{\theta})+(\boldsymbol{x}-\tilde{\boldsymbol{x}})^T\boldsymbol{W}_x(\boldsymbol{x}-\tilde{\boldsymbol{x}})\} \Rightarrow \hat{\boldsymbol{\theta}}_{XLS} \quad (+\hat{\tilde{\boldsymbol{x}}}_{XLS}). \tag{7}$$

Proper choice of $\boldsymbol{W}_g$ and $\boldsymbol{W}_x$ reflects the desired balance between modelling error and measurement errors. LS and TLS emerge as limiting cases when $\boldsymbol{W}_g \to \boldsymbol{0}$ or $\boldsymbol{W}_x \to \boldsymbol{0}$, respectively. In addition, interesting statistical interpretations can be attributed to these weights in the context of statistical models ([12, 13]).

In order to pursue computationally appealing algorithms for the required minimization, we now restrict the context to pseudo-linear models, applicable in a variety of engineering problems such as estimating the parameters of a noisy Auto-Regressive (AR) stochastic process, or identifying a linear AR system with exogenous (ARX) noise.

By "pseudo-linear models" we refer to $\boldsymbol{g}(\tilde{\boldsymbol{x}},\boldsymbol{\theta})$ that are linear in $\tilde{\boldsymbol{x}}$ for any fixed value of $\boldsymbol{\theta}$, and vice versa. It is easy to observe, that $\boldsymbol{g}(\tilde{\boldsymbol{x}},\boldsymbol{\theta})$ complies with that definition iff it can be expressed in the form:

$$\boldsymbol{g}(\tilde{\boldsymbol{x}},\boldsymbol{\theta}) = \boldsymbol{A}(\tilde{\boldsymbol{x}})\boldsymbol{v}(\boldsymbol{\theta}) \tag{8}$$

where

$$\boldsymbol{A}(\tilde{\boldsymbol{x}}) = \boldsymbol{A}_0 + \sum_{n=1}^{N} \tilde{x}_n \boldsymbol{A}_n \tag{9}$$

with $\boldsymbol{A}_0, \boldsymbol{A}_1, \ldots \boldsymbol{A}_N$ a set of constant "model matrices", and $\boldsymbol{v}(\boldsymbol{\theta}) = \begin{bmatrix} v_0 & \boldsymbol{\theta}^T \end{bmatrix}^T$ where v_0 is some constant.

Despite the pseudo-linear restriction, no closed-form expression for minimization is currently known, but several iterative algorithms have been proposed ([12, 15, 14]). In this contribution we present the Extended STLS (ESTLS) algorithm, based on De Moor's STLS algorithm ([6]; see also chapters III-1 and III-2). Since STLS can be regarded as a limiting-case of XLS, the extension of the STLS solution into an XLS solution is intuitively appealing.

We begin by defining the general ESTLS problem and minimization approach. In section 3 we shall show how the XLS minimization can be viewed as an ESTLS problem, and outline the minimization algorithm. Section 4 contains a brief application example on estimating the parameters of a noisy speech-like AR process.

2. Extended STLS

Let $\boldsymbol{B}_0, \boldsymbol{B}_1, \dots \boldsymbol{B}_N$ be a set of $N+1$ "building blocks" matrices defining a "proper structure", such that every "properly structured" matrix $\boldsymbol{B}$ can be expressed via:

$$\boldsymbol{B}(\boldsymbol{b}) = \boldsymbol{B}_0 + \sum_{n=1}^{N} b_n \boldsymbol{B}_n \tag{10}$$

$\boldsymbol{b}$ being the vector of required coefficients. Assume also, that another vector of coefficients $\boldsymbol{a}$ is given, which, in turn, defines the properly structured "target matrix" $\boldsymbol{A}$:

$$\boldsymbol{A} = \boldsymbol{B}_0 + \sum_{n=1}^{N} a_n \boldsymbol{B}_n. \tag{11}$$

In STLS it is desired to find a vector $\boldsymbol{y}$ and another vector $\boldsymbol{b}$ which is nearest to $\boldsymbol{a}$ (in the sense of minimum norm of the difference $\|\boldsymbol{b} - \boldsymbol{a}\|_2^2$), with which $\boldsymbol{B}(\boldsymbol{b})$ satisfies $\boldsymbol{B}(\boldsymbol{b})\boldsymbol{y} = \boldsymbol{0}$.

ESTLS extends STLS by introducing additional linear constraints on $\boldsymbol{y}$, in the form of a fixed constraint matrix $\boldsymbol{T}$:

$$\min_{\boldsymbol{b},\boldsymbol{y}} \sum_{n=1}^{N} (b_n - a_n)^2 \quad s.t.: \quad \boldsymbol{B}(\boldsymbol{b})\boldsymbol{y} = \boldsymbol{0}, \boldsymbol{T}\boldsymbol{y} = \boldsymbol{0}, \boldsymbol{y}^T\boldsymbol{y} = 1. \tag{12}$$

This modification enables to use the ESTLS solution for XLS minimization by extending the row-dimension of the "model matrices" (9), such that a vector of "latent" errors (or "model errors") can be included in the pseudo-linear model. The added linear constraint enables to fix the desired ratio between allowable perturbations in this vector and the associated perturbation of data ("misfit", or "measurement errors"). We shall elaborate on this issue in section 3.

In fact, ESTLS can be cast into an STLS problem by properly transforming the matrix set, e.g. by augmenting all $\boldsymbol{B}_n$-s with $\boldsymbol{T}$, thereby incorporating the additional constraint $\boldsymbol{T}\boldsymbol{y} = \boldsymbol{0}$ into the original constraint $\boldsymbol{B}(\boldsymbol{b})\boldsymbol{y} = \boldsymbol{0}$. However, in order to derive a direct ESTLS solution, we construct the Lagrangian:

$$L(\boldsymbol{b},\boldsymbol{y},\boldsymbol{l},\boldsymbol{\xi},\lambda) = \frac{1}{2}\sum_{n=1}^{N}(b_n - a_n)^2 + \boldsymbol{l}^T\left[\boldsymbol{B}_0 + \sum_{n=1}^{N} b_n\boldsymbol{B}_n\right]\boldsymbol{y} - \boldsymbol{\xi}^T\boldsymbol{T}\boldsymbol{y} + \frac{1}{2}\lambda(\boldsymbol{y}^T\boldsymbol{y} - 1). \tag{13}$$

Differentiating with respect to $\boldsymbol{b}$, $\boldsymbol{y}$, $\boldsymbol{l}$, $\boldsymbol{\xi}$ and λ, equating zero and following some straightforward manipulations, we obtain the set:

$$\left[\boldsymbol{B}_0 + \sum_{n=1}^{N} a_n\boldsymbol{B}_n\right]\boldsymbol{y} = \left[\sum_{n=1}^{N}(\boldsymbol{l}^T\boldsymbol{B}_n\boldsymbol{y})\boldsymbol{B}_n\right]\boldsymbol{y} \tag{14a}$$

$$\left[\boldsymbol{B}_0 + \sum_{n=1}^{N} a_n\boldsymbol{B}_n\right]^T\boldsymbol{l} = \left[\sum_{n=1}^{N}(\boldsymbol{l}^T\boldsymbol{B}_n\boldsymbol{y})\boldsymbol{B}_n\right]^T\boldsymbol{l} + \boldsymbol{T}^T\boldsymbol{\xi} \tag{14b}$$

Defining matrices $\boldsymbol{D}_y$ and $\boldsymbol{D}_l$,

$$\boldsymbol{D}_y \triangleq \sum_{n=1}^{N}(\boldsymbol{B}_n\boldsymbol{y})(\boldsymbol{B}_n\boldsymbol{y})^T \qquad \boldsymbol{D}_l \triangleq \sum_{n=1}^{N}(\boldsymbol{B}_n^T\boldsymbol{l})(\boldsymbol{B}_n^T\boldsymbol{l})^T \tag{15}$$

and identifying the matrix $\boldsymbol{A}$ (11) on the left-hand sides of (14a) and (14b), we end up with:

$$\boldsymbol{A}\boldsymbol{y} = \boldsymbol{D}_y\boldsymbol{l} \qquad \boldsymbol{A}^T\boldsymbol{l} = \boldsymbol{D}_l\boldsymbol{y} + \boldsymbol{T}^T\boldsymbol{\xi} \tag{16}$$

to be solved subject to $\boldsymbol{y}^T\boldsymbol{y} = 1$ and $\boldsymbol{T}\boldsymbol{y} = \boldsymbol{0}$. It is useful to transform to a different couple of vectors, $\boldsymbol{u}$ and $\boldsymbol{v}$, and a scalar τ, which together solve the same set of equations:

$$\boldsymbol{A}\boldsymbol{v} = \boldsymbol{D}_v\boldsymbol{u}\tau \qquad \boldsymbol{A}^T\boldsymbol{u} = \boldsymbol{D}_u\boldsymbol{v}\tau + \boldsymbol{T}^T\boldsymbol{\xi} \tag{17}$$

(with $\boldsymbol{D}_v$ and $\boldsymbol{D}_u$ constructed similarly to $\boldsymbol{D}_y$ and $\boldsymbol{D}_l$), subject to slightly different constraints: $\boldsymbol{v}^T\boldsymbol{D}_u\boldsymbol{v}(= \boldsymbol{u}^T\boldsymbol{D}_v\boldsymbol{u}) = 1$ and $\boldsymbol{T}\boldsymbol{v} = \boldsymbol{0}$.

Apart from irrelevant rescaling of $\boldsymbol{T}$, the sets (16) and (17) are equivalent, since for every couple $(\boldsymbol{l},\boldsymbol{y})$ that solves (16), there exists a triplet $(\boldsymbol{u},\boldsymbol{v},\tau)$ (differing by scale only) which solves (17). The advantageous difference is, that with $(\boldsymbol{u},\boldsymbol{v},\tau)$, the target function $\|\boldsymbol{a} - \boldsymbol{b}\|^2$ can be verified to be given directly by τ^2. We therefore seek the triplet $(\boldsymbol{u},\boldsymbol{v},\tau)$ that solves (17) and has the minimal τ^2 among all such triplets.

If $\boldsymbol{D}_u$ and $\boldsymbol{D}_v$ were constant, (17) would be a non-homogeneous Restricted SVD (RSVD, see e.g. [8]) problem, which could be solved, e.g., by an iterative "Non-Homogeneous Inverse Iteration" algorithm (guaranteed to converge to the smallest τ and associated $\boldsymbol{u}$ and $\boldsymbol{v}$; see, e.g., App. A of [12]). However, since $\boldsymbol{D}_u$ and $\boldsymbol{D}_v$ depend on $\boldsymbol{u}$ and $\boldsymbol{v}$, we propose (as done in the STLS solution) to interlace updates of $\boldsymbol{D}_u$ and $\boldsymbol{D}_v$ between iterations (using "old" $\boldsymbol{u}$ and $\boldsymbol{v}$ to reconstruct "new" $\boldsymbol{D}_u$ and $\boldsymbol{D}_v$ (resp.)). It is stressed, however, that (as in STLS) such an algorithm is no longer guaranteed to converge. Yet, if it converges, the point of convergence is guaranteed to be a true (possibly local) minimum (since convergence implies simultaneous satisfaction of all the zero derivative equations).

3. XLS minimization as an ESTLS problem

We now turn to identify the XLS minimization problem in terms of an ESTLS problem. Recall that the pseudo-linear model is specified by a set of "building block" matrices $\boldsymbol{A}_0, \boldsymbol{A}_1, \ldots \boldsymbol{A}_N$, such that $\boldsymbol{A}(\tilde{\boldsymbol{x}})$ is a linear combination thereof, whose coefficients are the elements of $\tilde{\boldsymbol{x}}$. Similarly, we denote by $\boldsymbol{A}(\boldsymbol{x})$ the "measured" model matrix, which is a linear combination of the same "building block" matrices, using the measured data $x_1, x_2, \ldots x_N$ as coefficients. Let us further define an extended $p \times (q+2)$ matrix $\boldsymbol{Y}(\boldsymbol{x}) \stackrel{\triangle}{=} [\boldsymbol{0} \vdots \boldsymbol{A}(\boldsymbol{x})]$; a vector of scaled residual model errors, $\boldsymbol{\epsilon} \stackrel{\triangle}{=} \mu \boldsymbol{g}(\tilde{\boldsymbol{x}}, \boldsymbol{\theta})$, where μ is an arbitrary non-zero constant; and another extended matrix $\tilde{\boldsymbol{Y}}(\boldsymbol{\epsilon}, \tilde{\boldsymbol{x}}) \stackrel{\triangle}{=} [\boldsymbol{\epsilon} \vdots \boldsymbol{A}(\tilde{\boldsymbol{x}})]$.

In addition, we define $N+p$ new matrices: $\boldsymbol{B}_0, \boldsymbol{B}_1, \ldots \boldsymbol{B}_N$ are versions of $\boldsymbol{A}_0, \boldsymbol{A}_1, \ldots \boldsymbol{A}_N$, extended by a $p \times 1$ zeros column on the left:

$$\boldsymbol{B}_n = [\boldsymbol{0} \vdots \boldsymbol{A}_n] \quad n = 0, 1, \ldots N; \tag{18}$$

And $\boldsymbol{B}_{N+1}, \boldsymbol{B}_{N+2}, \ldots \boldsymbol{B}_{N+p}$ are additional "building-blocks", required to accommodate the p elements of $\boldsymbol{\epsilon}$ into the first column of $\boldsymbol{Y}(\boldsymbol{\epsilon}, \tilde{\boldsymbol{x}})$:

$$\boldsymbol{B}_{N+n} = [\boldsymbol{e}_n \vdots \boldsymbol{0}] \quad n = 1, 2, \ldots p \tag{19}$$

(here $\boldsymbol{0}$ denotes a $p \times (q+1)$ all-zeros matrix).

Consequently, we may now define the following coefficients vectors $\boldsymbol{a}$ and $\boldsymbol{b}$, used to express $\boldsymbol{Y}(\boldsymbol{x})$ and $\tilde{\boldsymbol{Y}}(\tilde{\boldsymbol{x}})$ (respectively) as linear combinations of $\boldsymbol{B}_0, \boldsymbol{B}_1, \ldots \boldsymbol{B}_{N+p}$:

$$\boldsymbol{a} = [\boldsymbol{x}^T \vdots \boldsymbol{0}^T]^T \qquad \boldsymbol{b} = [\tilde{\boldsymbol{x}}^T \vdots \boldsymbol{\epsilon}^T]^T \tag{20}$$

where $\boldsymbol{0}$ denotes a $p \times 1$ all-zeros vector. We therefore have:

$$\boldsymbol{Y}(\boldsymbol{a}) = \boldsymbol{B}_0 + \sum_{n=1}^{N+p} a_n \boldsymbol{B}_n \quad \tilde{\boldsymbol{Y}}(\boldsymbol{b}) = \boldsymbol{B}_0 + \sum_{n=1}^{N+p} b_n \boldsymbol{B}_n \tag{21}$$

The norm of the difference between $\boldsymbol{b}$ and $\boldsymbol{a}$ is given by:

$$\|\boldsymbol{b}-\boldsymbol{a}\|^2 = \|\tilde{\boldsymbol{x}}-\boldsymbol{x}\|^2 + \|\boldsymbol{\epsilon}\|^2 = \|\tilde{\boldsymbol{x}}-\boldsymbol{x}\|^2 + \mu^2\|\boldsymbol{g}(\tilde{\boldsymbol{x}},\boldsymbol{\theta})\|^2 \tag{22}$$

Thus, that difference may be expressed as:

$$\|\boldsymbol{b}-\boldsymbol{a}\|^2 = \boldsymbol{g}^T(\tilde{\boldsymbol{x}},\boldsymbol{\theta})\boldsymbol{W}_g\boldsymbol{g}(\tilde{\boldsymbol{x}},\boldsymbol{\theta}) + (\tilde{\boldsymbol{x}}-\boldsymbol{x})^T\boldsymbol{W}_x(\tilde{\boldsymbol{x}}-\boldsymbol{x}) \tag{23}$$

where $\boldsymbol{W}_g = \mu^2\boldsymbol{I}$ and $\boldsymbol{W}_x = \boldsymbol{I}$, yielding the XLS criterion (7). In order to maintain the relation between $\boldsymbol{\epsilon}$ and $\boldsymbol{g}(\tilde{\boldsymbol{x}},\boldsymbol{\theta})$, $\tilde{\boldsymbol{Y}}(\boldsymbol{b})$ has to satisfy $\tilde{\boldsymbol{Y}}(\boldsymbol{b})\boldsymbol{y} = \boldsymbol{0}$, where $\boldsymbol{y} = [-1/\mu\ 1\ \boldsymbol{\theta}^T]^T$ is the ESTLS minimization vector, which thus has to satisfy also $\boldsymbol{T}\boldsymbol{y} = \boldsymbol{0}$ with $\boldsymbol{T} = [\mu\ 1\ \boldsymbol{0}^T]$. If desired, additional linear constraints on $\boldsymbol{y}$, possibly reflecting some prior information on $\boldsymbol{\theta}$, may be incorporated by further augmenting $\boldsymbol{T}$.

Note that the matrices $\boldsymbol{B}_0, \boldsymbol{B}_1, \ldots \boldsymbol{B}_{N+p}$ are needed for notational purposes only, and have little computational effect. Their structure can be exploited to express the required matrices in terms of the model matrices $\boldsymbol{A}_1, \boldsymbol{A}_2, \ldots \boldsymbol{A}_N$, as evident in the algorithm summary:

Extended Structured Total Least Squares (ESTLS) Algorithm

Preliminaries:

The measurements: $\boldsymbol{x} = [x_1\ x_2\ \ldots\ x_N]^T$

$\boldsymbol{A}_0, \boldsymbol{A}_1, \ldots \boldsymbol{A}_N$ - the set of $p \times (q+1)$ model matrices.

A weight-balancing parameter: μ

Preparation / Initialization:

Construct the $1 \times (q+2)$ constraint matrix $\boldsymbol{T} = [\mu\ 1\ \boldsymbol{0}^T]$

(If desired, extend $\boldsymbol{T}$ to accommodate additional constraints on $\boldsymbol{\theta}$)

Denote by s the resulting row-dimension of $\boldsymbol{T}$ (normally $s = 1$).

Also, define the $s \times 1$ vector (normally a scalar) $\boldsymbol{t}_1$ and the $s \times (q+1)$ matrix (normally a vector) $\boldsymbol{T}_2$ via the partition: $\boldsymbol{T} = [\boldsymbol{t}_1 \vdots \boldsymbol{T}_2]$.

Construct the target matrix: $\boldsymbol{A}(\boldsymbol{x}) = \boldsymbol{A}_0 + \sum_{n=1}^{N} x_n \boldsymbol{A}_n$,

Find the QR decomposition of $\boldsymbol{A}(\boldsymbol{x})$: $\boldsymbol{A}(\boldsymbol{x}) = [\boldsymbol{Q}_1 \vdots \boldsymbol{Q}_2]\begin{bmatrix}\boldsymbol{R}\\ \boldsymbol{0}\end{bmatrix}$

where $\boldsymbol{Q} = [\boldsymbol{Q}_1 \vdots \boldsymbol{Q}_2]$ is unitary, $\boldsymbol{Q}_1$ is $p \times (q+1)$, $\boldsymbol{Q}_2$ is $p \times (p-q-1)$, $\boldsymbol{R}$ is $(q+1) \times (q+1)$ non-singular upper triangular. Calculate $\boldsymbol{R}^{-1}$.

Set initial guesses: $\boldsymbol{v}^{[0]}$ $((q+2) \times 1)$ and $\boldsymbol{u}^{[0]}$ $(p \times 1)$.

For $k = 1, 2, \ldots$ repeat until convergence:

Partition $\boldsymbol{v}^{[k-1]}$ into a scalar $v^{[k-1]}$ and a $(q+1)\times 1$ vector $\breve{\boldsymbol{v}}^{[k-1]}$:

$$\boldsymbol{v}^{[k-1]} = \left[v^{[k-1]} \vdots \breve{\boldsymbol{v}}^{[k-1]^T} \right]^T$$

Construct $\boldsymbol{D}_{v^{[k-1]}} = \sum_{n=1}^{N} (\boldsymbol{A}_n \breve{\boldsymbol{v}}^{[k-1]})(\boldsymbol{A}_n \breve{\boldsymbol{v}}^{[k-1]})^T + v^{[k-1]^2} \boldsymbol{I}$

and $\boldsymbol{D}_{u^{[k-1]}} = \begin{bmatrix} \boldsymbol{u}^{[k-1]^T} \boldsymbol{u}^{[k-1]} & \boldsymbol{0}^T \\ \boldsymbol{0} & \sum_{n=1}^{N} (\boldsymbol{A}_n^T \boldsymbol{u}^{[k-1]})(\boldsymbol{A}_n^T \boldsymbol{u}^{[k-1]})^T \end{bmatrix}$

Let $\gamma = \left(\boldsymbol{v}^{[k-1]^T} \boldsymbol{D}_{u^{[k-1]}} \boldsymbol{v}^{[k-1]} \right)^{1/4}$

Let $\tau^{[k-1]} = \boldsymbol{u}^{[k-1]^T} \boldsymbol{A}(\boldsymbol{x}) \breve{\boldsymbol{v}}^{[k-1]} / \gamma^2$

Construct: $\boldsymbol{P}_{11} = \boldsymbol{Q}_1^T \boldsymbol{D}_{v^{[k-1]}} \boldsymbol{Q}_1 / \gamma^2$ $\quad \boldsymbol{P}_{12} = \boldsymbol{Q}_1^T \boldsymbol{D}_{v^{[k-1]}} \boldsymbol{Q}_2 / \gamma^2$

$\boldsymbol{P}_{22} = \boldsymbol{Q}_2^T \boldsymbol{D}_{v^{[k-1]}} \boldsymbol{Q}_2 / \gamma^2$ $\quad \boldsymbol{P}_{21} = \boldsymbol{P}_{12}^T$

Calculate $\boldsymbol{P}_{22}^{-1}$ and $\boldsymbol{\Phi} = \boldsymbol{R}^{-1}(\boldsymbol{P}_{11} - \boldsymbol{P}_{12}\boldsymbol{P}_{22}^{-1}\boldsymbol{P}_{21})\boldsymbol{R}^{-T}$.

Calculate $\tilde{\boldsymbol{v}}^{[k-1]} = \boldsymbol{D}_{u^{[k-1]}} \boldsymbol{v}^{[k-1]} / \gamma^3$

and partition $\tilde{\boldsymbol{v}}^{[k-1]}$ into a scalar $\tilde{v}^{[k-1]}$ and a $q \times 1$ vector $\breve{\tilde{\boldsymbol{v}}}^{[k-1]}$:

$$\tilde{\boldsymbol{v}}^{[k-1]} = \left[\tilde{v}^{[k-1]} \vdots \breve{\tilde{\boldsymbol{v}}}^{[k-1]^T} \right]^T$$

Let $\begin{bmatrix} \boldsymbol{\xi}^{[k-1]} \\ v^{[k]} \end{bmatrix} = - \begin{bmatrix} \boldsymbol{T}_2 \boldsymbol{\Phi} \boldsymbol{T}_2^T & \boldsymbol{t}_1 / \tau^{[k-1]^2} \\ \boldsymbol{t}_1^T & 0 \end{bmatrix}^{-1} \begin{bmatrix} \boldsymbol{T}_2 \boldsymbol{\Phi} \breve{\tilde{\boldsymbol{v}}}^{[k-1]} \\ \tilde{v}^{[k-1]} \end{bmatrix}$

where $\boldsymbol{\xi}^{[k-1]}$ is $s \times 1$ and $v^{[k]}$ is a scalar.

Let $\boldsymbol{z}^{[k]} = \boldsymbol{R}^{-T} \left(\breve{\tilde{\boldsymbol{v}}}^{[k-1]} + \boldsymbol{T}_2^T \boldsymbol{\xi}^{[k-1]} \right) \tau^{[k-1]}$

Let $\boldsymbol{w}^{[k]} = -\boldsymbol{P}_{22}^{-1} \boldsymbol{P}_{21} \boldsymbol{z}^{[k]}$

Let $\boldsymbol{w}^{[k]} = -\boldsymbol{P}_{22}^{-1} \boldsymbol{P}_{21} \boldsymbol{z}^{[k]}$

Let $\boldsymbol{u}^{[k]} = \boldsymbol{Q}_1 \boldsymbol{z}^{[k]} + \boldsymbol{Q}_2 \boldsymbol{w}^{[k]}$

Let $\breve{\boldsymbol{v}}^{[k]} = \boldsymbol{\Phi} \left(\breve{\tilde{\boldsymbol{v}}}^{[k-1]} + \boldsymbol{T}_2^T \boldsymbol{\xi}^{[k-1]} \right) \tau^{[k-1]^2}$

Let $\boldsymbol{v}^{[k]} = [v^{[k]} \vdots \breve{\boldsymbol{v}}^{[k]^T}]^T$

Normalize: $\boldsymbol{u}^{[k]} = \boldsymbol{u}^{[k]} \cdot v_2^{[k]}$ $\quad \boldsymbol{v}^{[k]} = \boldsymbol{v}^{[k]} / v_2^{[k]}$

where $v_2^{[k]}$ denotes the second element of $\boldsymbol{v}^{[k]}$.

Upon convergence ($k = K$), the desired (readily scaled) vector is given by $\boldsymbol{v}^{[K]}$, so that $\hat{\boldsymbol{\theta}}_{XLS} = \boldsymbol{y}^{[K]}(3 : q+2)$ (i.e., the third through last elements of $\boldsymbol{v}^{[k]}$). As a by-product, $\hat{\hat{\boldsymbol{x}}}_{XLS} = \left[\hat{\hat{x}}_1 \; \hat{\hat{x}}_2 \cdots \hat{\hat{x}}_N \right]^T$ may be obtained from:

$$\hat{\hat{x}}_n = x_n - \tau^{[K]} \boldsymbol{u}^{[K]^T} \boldsymbol{A}_n \breve{\boldsymbol{v}}^{[K]} \quad n = 1, 2, \ldots N$$

(where $\breve{\boldsymbol{v}}^{[K]}$ is defined by the partition of $\boldsymbol{v}^{[K]}$. note that $\tau^{[K]}$ is unavailable from the last iteration. It may be substituted by $\tau^{[K-1]}$, which is available and almost identical due to convergence.)

Possible options for convergence criteria are to monitor the change in either $\boldsymbol{v}^{[k]}$, $\boldsymbol{u}^{[k]}$ or $\tau^{[k]}$ compared to a small threshold. A possible initialization for $\boldsymbol{v}^{[0]}$ is $\boldsymbol{v}^{[0]} = \left[-1/\mu \; 1 \; \hat{\boldsymbol{\theta}}^{[0]^T}\right]^T$, where $\hat{\boldsymbol{\theta}}^{[0]}$ is an intelligent guess of $\boldsymbol{\theta}$, obtained e.g. by ordinary LS. As for $\boldsymbol{u}^{[0]}$, the intuitive tendency is to set $\boldsymbol{u}^{[0]} = \mathbf{0}$, which is equivalent to assuming that $\tilde{\boldsymbol{x}} = \boldsymbol{x}$. However, this is unacceptable, since it would obviously disable the normalization operations involved. Therefore, either a non-zero random initial guess, or the more intelligent:

$$\boldsymbol{u}^{[0]} = \boldsymbol{D}_{v^{[0]}}^{-1} \boldsymbol{A}(\boldsymbol{x}) \breve{\boldsymbol{v}}^{[0]} \tag{24}$$

are recommended. Since convergence is not guaranteed and depends on the initial guess, wise selection of $\boldsymbol{v}^{[0]}$ and $\boldsymbol{u}^{[0]}$ remains an important open question.

4. Application example

As a brief application demonstration, we apply XLS to the problem of estimating the parameters of an AR process (such as a speech signal) from noisy observations. Let $\tilde{x}_n$, $n = 1, 2, \ldots N$ be an AR process of known order q (AR(q)). The underlying model may be written as:

$$\tilde{x}_n \approx -\sum_{k=1}^{q} \theta_k \tilde{x}_{n-k} \quad n = 1, 2, \ldots N \tag{25}$$

where $\boldsymbol{\theta} = [\theta_1 \; \theta_2 \; \cdots \; \theta_q]^T$ is the vector of unknown parameters. The initial conditions $\tilde{x}_0, \tilde{x}_{-1}, \ldots \tilde{x}_{-q+1}$ are assumed known.

In the case of strict equality in (25), $\tilde{x}_n$ is a deterministic process, which (for non-zero initial conditions) is a linear combination of exponential signals (as evident from the partial-fraction expansion of the Z-transform of $\tilde{x}_n$ if all the poles are of single multiplicity). The problem of estimating $\boldsymbol{\theta}$ from noisy observations of $\tilde{x}_n$ in that case has been treated extensively using STLS or equivalent techniques (e.g. [7], [3]). Generally, however, (25) randomly deviates from equality, rendering $\tilde{x}_n$ a stochastic AR process.

Given noisy measurements $x_n \approx \tilde{x}_n$, we wish to estimate $\boldsymbol{\theta}$ from $x_1, x_2, \ldots x_N$. Assuming, for simplicity, zero initial conditions, the pseudo-linear model (25) can be written in matrix form as in (8), with the $N \times (q+1)$ "model matrices" constructed as follows: $\boldsymbol{A}_0$ is all-zeros,

A_n $n = 1, 2, \ldots N$ are each all-zeros except for a diagonal of 1's, beginning at the $(n, 1)$ element. Following selection of a "weight balancing" parameter μ, the STLS algorithm outlined above can be readily applied.

In Fig. 1 we show a common convergence pattern obtained with an AR(10) process resembling a typical speech signal. The pattern is shown in terms of $\log_{10} ||\hat{\boldsymbol{\theta}}^{[k+1]} - \hat{\boldsymbol{\theta}}^{[k]}||$ vs. the iteration k, comparing to that of other XLS minimization algorithms, namely Alternating Coordinates Minimization (ACM) and the Extended CTLS (ECTLS) (see, e.g., [14]). ESTLS is seen to converge somewhat faster than ACM (which has a comparable computational load per iteration), and much faster than ECTLS (which has a significantly lower computational load per iteration). Interestingly, ESTLS has a somewhat "bumpy" convergence rate, alternating between very fast and very slow rates.

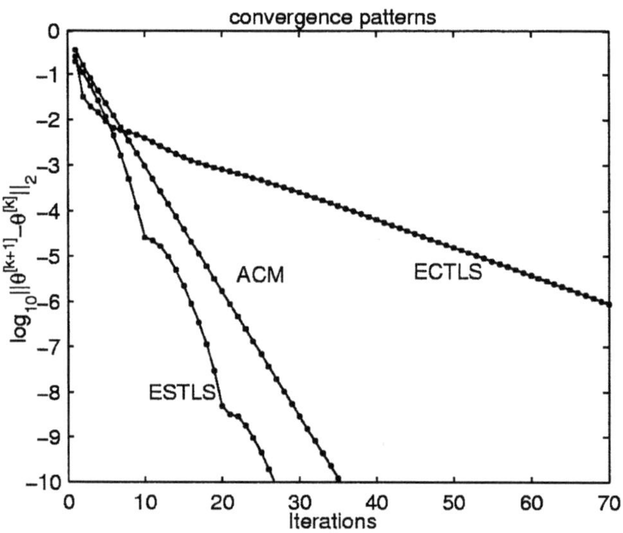

Figure 1. Convergence pattern for ESTLS, ACM and ECTLS algorithms for the XLS application example. The underlying process is AR(10) with poles at $0.9e^{\pm j2\pi f_k/f_s}$ $k = 1, 2, 3, 4, 5$ with $f_k = 400, 800, 1200, 2600, 3400[Hz]$ and $f_s = 8000[Hz]$. The signal to noise ratio was $18dB$ and the balancing parameter was $\mu = 0.2$. $N = 150$ samples were used (all algorithms used to the same data).

5. Conclusion

We presented the ESTLS problem and associated minimization algorithm, and showed how ESTLS can be used to minimize the XLS criterion. Since STLS is a limiting case of XLS (in terms of the "weighting balance"), this approach bears inherent heuristic justification. Comparing to other XLS minimization algorithms, its advantages or disadvantages (in terms of overall computational effort) are mainly application-dependent.

References

[1] T.J. Abatzoglou and J.M. Mendel. Constrained total least squares. *Proc. ICASSP-87*, pages 1485–1488, 1987.

[2] T.J. Abatzoglou, J.M. Mendel, and G.A. Harada. The constrained total least squares technique and its application to harmonic superresolution. *IEEE Trans. Signal Processing*, 39(5):1070–1087, 1991.

[3] Y. Bresler and A. Macovski. Exact maximum likelihood parameter estimation of superimposed exponential signals in noise. *IEEE Trans. Acoustics, Speech and Signal Processing*, 34(5):1081–1089, 1986.

[4] J.A. Cadzow. Total least squares, matrix enhancement, and signal processing. *Digital Signal Processing*, 4:21–39, 1994.

[5] J.A. Cadzow and D.M. Wilkes. Constrained total least squares: Signal and system modeling. *Proc. ICASSP-92*, 5:277–280, 1992.

[6] B. De Moor. Structured total least squares and L2 approximation problems. *Special Issue of Linear Algebra and Its Applications, on Numerical Linear Algebra Methods in Control, Signals, Systems*, 188/189:163–205, 1993.

[7] B. De Moor. Total least squares for affinely structured matrices and the noisy realization problem. *IEEE Trans. Signal Processing*, 42(11):3104–3113, 1994.

[8] B. De Moor and G. Golub. The restricted singular value decomposition: Properties and applications. *SIAM J. Matrix Anal. Appl.*, 12.:401–425, 1991.

[9] B. De Moor and P. Lemmerling. A new trade-off in linear system identification: Misfit versus latency. *Proceedings of the MTNS98, Padova, Italy*, 1998.

[10] P. Lemmerling. *Structured Total Least Squares: Analysis, Algorithms and Applications*. PhD thesis, Dept. of Electrical Engineering, K.U.Leuven, Belgium, 1999.

[11] P. Lemmerling, B. De Moor, and S. Van Huffel. On the equivalence of constrained total least squares and structured total least squares. *IEEE Trans. Signal Processing*, 44(11):2908–2911, 1996.

[12] A. Yeredor. *The Extended Least Squares Criterion for Discriminating Measurement Errors from Model Errors: Algorithms, Applications, Analysis*. PhD thesis, Tel-Aviv University, Dept. of Electrical Engineering - Systems, 1997.

[13] A. Yeredor. The joint MAP-ML estimation criterion and its relation to ML and to extended least squares. *IEEE Trans. Signal Proc.*, 48(12):3484–3492, 2000.

[14] A. Yeredor. The extended least squares criterion: Minimization algorithms and applications. *IEEE Trans. Signal Processing*, 49(1):74–86, 2001.

[15] A. Yeredor and E. Weinstein. The extended least squares and the joint maximum-a-posteriori - maximum likelihood estimation criteria. *Proceedings of ICASSP99*, 4:1813–1816, 1999. Phoenix, Arizona.

Part IV

NONLINEAR ERRORS-IN-VARIABLES MODELS AND STATISTICAL ESTIMATORS

BAYESIAN SMOOTHING FOR MEASUREMENT ERROR PROBLEMS

Scott M. Berry
Berry Consultants, 1039 Wellington Court Sycamore, IL 60178 USA
scott@berryconsultants.com

Raymond J. Carroll
Department of Statistics, Texas A&M University, College Station TX 77843-3143 USA
carroll@stat.tamu.edu

David Ruppert
School of Operations Research and Industrial Engineering,
Cornell University, Ithaca, NY 14853-3801 USA
davidr@orie.cornell.edu

Abstract In the presence of covariate measurement error, estimating a regression function nonparametrically is extremely difficult, the problem being related to deconvolution. In this paper we describe Bayesian approaches to modeling a flexible regression function when the predictor variable is measured with error. The regression function is modeled with smoothing splines.. We provide simulations with several nonlinear regression functions. Our simulations indicate that the *frequentist* mean squared error properties of the fully Bayesian method are better than those of previously proposed frequentist methods, at least in the examples we have studied.

Keywords: Bayesian methods, errors-in-variables model, functional relationship, generalized linear models, kernel regression, measurement error, nonparametric methods, SIMEX, splines, structural relationship.

1. Introduction

In this paper we review work of [1] on a fully Bayesian approach to the problem of nonparametric regression when the independent variables are

S. Van Huffel and P. Lemmerling (eds.), Total Least Squares and Errors-in-Variables Modeling, 121–130.

measured with error. This is known to be an extremely difficult problem in terms of global rates of convergence.

In [2] the assumption of global consistency is relaxed. They suggest two estimators: (a) a semiparametric one based on the SIMEX method of [3] which makes no assumptions about the unknown and unobserved covariates; and (b) a more parametric one that assumes that the unobserved covariates follow a mixture of normals distribution.

In [1], we proposed a new method for nonparametric function estimation when the covariate is measured with error. Our procedure can be looked at as the natural fully Bayesian extension of the techniques of [2]. It can also be looked upon as the extension to measurement error models of the MCMC technique of [7], or, viewed more broadly, the entire Bayesian formulation of smoothing splines (e.g., [10], [11]; [8], [9]).

The methodology in [1] and reviewed here is new in two respects. First, the adjustment for bias due to measurement error comes automatically from the Bayesian machinery. In contrast, other methods explicitly analyze the bias and devise a correction in a more ad hoc fashion. Second, and perhaps more importantly, the smoothing parameter selector, which also comes automatically from the Bayesian approach, is designed for the measurement error problem. Earlier work either did not propose a smoothing parameter selector or applied a smoothing parameter selector that ignores the effects of measurement error. However, measurement error has large effects on both bias and variance and a smoothing parameter that is optimal for correctly measured covariates may be far from optimal in the presence of measurement error.

In Section 2 we describe some background information on smoothing splines that is necessary for our development. In Section 3, we present our methodology. Section 4 presents simulations. The results indicate that even as a frequentist estimator our fully Bayesian method is at least competitive with that of [2], and sometimes much better. Section 5 presents a discussion of the results. Paper [1] should be consulted for more details.

2. Smoothing Splines

We present a brief introduction to smoothing splines. For additional information, see [10], [12], [6], [7], and [4].

Assume that $Y_i = m(X_i) + \epsilon_i$, where ϵ_i has mean zero and variance σ^2_ϵ. Let $[a, b]$ be the interval for which an estimate of m is sought. Let g be the best natural cubic spline (NCS) approximator of m, that is, the NCS that minimizes $\sum_{i=1}^n \{m(X_i) - g(X_i)\}^2$. If m is smooth, then the error in approximating m by g typically is negligible compared to the estimation error, so we assume that $m = g$.

A *smoothing spline* is defined as the minimizer over g of the penalized sum of squares

$$S(g) = \sum_{i=1}^{n} \{Y_i - g(X_i)\}^2 + \alpha \int_a^b \{g''(x)\}^2 \, dx, \tag{1}$$

for $\alpha > 0$. This minimizer is a NCS with knots at the distinct X_i values. The integral term of (1) is a roughness penalty and α, is the smoothing parameter. Let $\boldsymbol{g} = \{g(X_1), g(X_2), ..., g(X_n)\}^\mathsf{T}$. The penalty term can be written as $\alpha \int_a^b \{g''(x)\}^2 dx = \alpha \boldsymbol{g}^\mathsf{T} \boldsymbol{K} \boldsymbol{g}$, where $\boldsymbol{K}$ is an $n \times n$ dimensional matrix of rank $n - 2$, defined in [4]. The smoothing spline minimizing $S(g)$ is $\widehat{\boldsymbol{g}} = \boldsymbol{A}(\alpha)\boldsymbol{Y}$, where $\boldsymbol{A}(\alpha) = (\mathbf{I} + \alpha \boldsymbol{K})^{-1}$ and $\boldsymbol{Y} = (Y_1, \ldots, Y_n)^\mathsf{T}$. The vector $\widehat{\boldsymbol{g}}$ uniquely defines the smoothing spline.

The Bayesian approach to smoothing splines gives the vector $\boldsymbol{g}$ a prior density proportional to the "partially improper" Gaussian process:

$$(\alpha/2\sigma_\epsilon^2)^{M/2} \exp\left\{-(\alpha/\sigma_\epsilon^2) \boldsymbol{g}^\mathsf{T} \boldsymbol{K} \boldsymbol{g}\right\}, \tag{2}$$

where $M = n - 2$ and $\boldsymbol{K}$ is defined as above. Although both $\boldsymbol{K}$ and $\boldsymbol{g}$ depend on the knot locations, since $\boldsymbol{g}'\boldsymbol{K}\boldsymbol{g} = \int_a^b \{g''(x)\}^2 dx$ this prior is independent of the knot locations. If the observations Y_i are independent and normally distributed with mean $g(X_i)$ and variance σ_ϵ^2 then the posterior distribution for $\boldsymbol{g}$ is multivariate normal with mean $\widehat{\boldsymbol{g}} = \boldsymbol{A}(\alpha)\boldsymbol{y}$, and covariance matrix $\sigma_\epsilon^2 \boldsymbol{A}(\alpha)$.

3. General Model

We consider the following measurement error model,

$$Y_i = m(X_i) + \epsilon_i, \quad i = 1, ..., n, \tag{3}$$

where the ϵ_i are independent normal random variables with mean 0 and variance σ_ϵ^2. The X's are not observable, i.e., they are latent variables, but W that are surrogates for the Xs are observed;

$$W_{ij} = X_i + U_{ij}, \quad i = 1, ..., n, \quad j = 1, ..., m_i, \tag{4}$$

where the U_{ij} are independent normal errors with mean 0 and variance σ_u^2.

Model (4) is more general than it first looks. It can be interpreted as stating that a *known* function of the observed covariates (W) is the same function of the latent variables, plus independent, homoscedastic normally distributed measurement errors. We have written (4) as if the function were the identity function, but it could be anything, e.g., the logarithm. The reason for this generality is that in (3), the function $m(\cdot)$ is unknown, so that for example if $m_*(v) = m\{\exp(v)\}$, then $m_*\{\log(x)\} = m(x)$.

Denote $\boldsymbol{\theta} = (\boldsymbol{g}, \boldsymbol{X}, \sigma_\epsilon^2, \sigma_u^2, \alpha)$. The posterior density is

$$f(\boldsymbol{\theta}|\boldsymbol{Y}, \boldsymbol{W}) \propto f(\boldsymbol{Y}|\boldsymbol{g}, \boldsymbol{X}, \sigma_\epsilon^2)\, f(\boldsymbol{W}|\boldsymbol{X}, \sigma_u^2)\, f(\boldsymbol{g}|\alpha)\, f(\sigma_u^2)\, f(\sigma_\epsilon^2)\, f(\boldsymbol{X})\, f(\alpha). \tag{5}$$

3.1 Fully Bayesian Approach

In this section the fully Bayesian approach to this problem is developed. Prior distributions are placed on all parameters, including the structural parameters (μ_x, σ_x^2) and the variance components $(\sigma_\epsilon^2, \sigma_u^2)$, and the joint posterior distribution is calculated. One of the benefits of this approach is that observations of the smoothing spline are generated from the posterior and therefore we estimate the entire posterior distribution of g, not just its mode. Thus, various forms of "error bars" are straightforward to calculate. These credible sets take account of the measurement error of the independent variables and the use of a data-based smoothing parameter.

The method is as follows. Without loss of generality, we replace α/σ_ϵ^2 by γ. The prior distributions for σ_ϵ^2 and σ_u^2 are inverse-gamma distributions and the prior distribution for γ is a gamma distribution: $\sigma_\epsilon^2 \sim IG(A_\epsilon, B_\epsilon)$, $\sigma_u^2 \sim IG(A_U, B_U)$, and $\gamma \sim G(A_\gamma, B_\gamma)$. We use the following definitions of the inverse-gamma and gamma distributions:

$$f(x|A,B) = \frac{1}{\Gamma(A)B^A x^{A+1}} \exp\left(-\frac{1}{Bx}\right) I_{(0,\infty)}(x).$$

$$f(x|A,B) = \frac{1}{\Gamma(A)B^A} x^{A-1} \exp\left(-\frac{x}{B}\right) I_{(0,\infty)}(x).$$

For the simulations and examples in this paper we use a hierarchical Bayes approach. A normal distribution with mean μ_x and variance σ_x^2 is used, where $\mu_x \sim \text{Normal}(d_x, t_x^2)$ and $\sigma_x^2 \sim IG(A_x, B_x)$.

The hyperparameters that are fixed a priori and thus are "tuning constants" are denoted by Roman fonts. They are A_Y, B_Y, A_U, B_U, A_γ, B_γ, d_x, t_x^2, A_x, and B_x.

Assuming the hierarchical normal structure for $[X]$, the joint posterior is proportional to

$$\exp\Big\{-\frac{1}{2\sigma_\epsilon^2}\sum_{i=1}^{n}\{Y_i - g(X_i)\}^2 - \frac{1}{2\sigma_u^2}\sum_{i=1}^{n}\sum_{j=1}^{m_i}(W_{ij} - X_i)^2$$
$$-\frac{1}{2\sigma_x^2}\sum_{i=1}^{n}(X_i - \mu_x)^2 - \frac{1}{2t_x^2}(\mu_x - d_x)^2\Big\}$$
$$\times \exp\left\{-(\gamma/2)\boldsymbol{g}^{\mathsf T}\boldsymbol{K}\boldsymbol{g} - \frac{1}{B_\epsilon\sigma_\epsilon^2} - \frac{1}{B_U\sigma_u^2} - \frac{\gamma}{B_\gamma} - \frac{1}{B_x\sigma_x^2}\right\} \tag{6}$$
$$\times\sigma_\epsilon^{-2(n/2+A_\epsilon+1)}\sigma_u^{-2(\frac{1}{2}\sum_{i=1}^{n} m_i+A_U+1)}\sigma_x^{-2(n/2+A_X+1)}\gamma^{(A_\gamma+M/2-1)},$$

where $M = n - 2$ as in (2). The sampling is done using a successive substitution algorithm ([5]).

The complete conditional distributions for the parameters are

$$\boldsymbol{g}|\boldsymbol{X},\gamma,\sigma_\epsilon^2,\boldsymbol{Y},\mathbf{W} \sim \text{Normal}\left\{\boldsymbol{A}(\sigma_\epsilon^2\gamma)\boldsymbol{y},\ \sigma_\epsilon^2\boldsymbol{A}(\sigma_\epsilon^2\gamma)\right\};$$

$$X_i|\boldsymbol{W}_i,\boldsymbol{g},\sigma^2,\sigma_u^2,\boldsymbol{Y},\mathbf{W} \propto \exp\Big[-\frac{1}{2\sigma_u^2}\sum_{j=1}^{m_i}(W_{ij} - X_i)^2$$
$$-\frac{1}{2\sigma_\epsilon^2}\{Y_i - g(X_i)\}^2 - \frac{1}{2\sigma_x^2}(X_i - \mu_x)^2\Big]; \tag{7}$$

$$\sigma_\epsilon^2|\boldsymbol{g},\boldsymbol{X},\boldsymbol{Y},\mathbf{W} \sim IG\big(A_\epsilon + n/2, [1/B_\epsilon + (1/2)\sum_{i=1}^{n}\{Y_i - g(X_i)\}^2]^{-1}\big);$$

$$\gamma|\boldsymbol{g},\boldsymbol{X} \sim G\big(A_\gamma + \frac{M}{2}, [1/B_\gamma + \frac{1}{2}\boldsymbol{g}^{\mathsf T}\boldsymbol{K}\boldsymbol{g}]^{-1}\big);$$

$$\mu_x|\boldsymbol{X} \sim \text{Normal}\{(n\overline{X}t_x + d_x\sigma_x^2)/(nt_x^2 + \sigma_x^2), \sigma_x^2 t_x^2/(nt_x^2 + \sigma_x^2)\};$$

$$\sigma_x^2|\boldsymbol{X} \sim IG\big(A_x + n/2, \{B_x^{-1} + (1/2)\sum_{i=1}^{n}(X_i - \mu_x)^2\}^{-1}\big);$$
$$\sigma_u^2|\boldsymbol{X} \sim IG\big(A_U + (1/2)\sum_{i=1}^{n} m_i,$$
$$[1/B_U + (1/2)\sum_{i=1}^{n}\sum_{j=1}^{m_i}(W_{ij} - X_i)^2]^{-1}\big).$$

Observations from each of the complete conditionals are drawn iteratively in the order presented above. The generation of an observation of $\boldsymbol{g}$ is computationally difficult for smoothing splines because they have n knots. Because the values of $\boldsymbol{X}$, $\boldsymbol{K}$ (for smoothing splines) and γ are continually changing in the algorithm the matrix $\boldsymbol{A}(\sigma_\epsilon^2\gamma)$ (which is $n \times n$) and its inverse must be recomputed for each iteration of the MCMC algorithm. Reference [7] discusses an algorithm for generating observations of $\boldsymbol{g}$ in $O(n)$ operations.

The complete conditionals for the X_i's require a Metropolis-Hastings step. This is done by generating a candidate observation of X_i from a normal distribution with a mean of the current value of X_i and a standard deviation of $2\sigma_u^{(i)}/\sqrt{m_i}$, where $\sigma_u^{(i)}$ is the current value of σ_u in the MCMC algorithm. Using $\overline{W}_i$ as an estimate of X_i, without the information in the regression function, has a standard error of $\sigma_u^{(i)}/\sqrt{m_i}$. Using the rule-of-thumb of a candidate value having a standard deviation twice the standard deviation of the marginal posterior provides a conservative candidate distribution for X_i. In terms of efficiency of the Metropolis-Hastings step, in our experience it is better to overestimate this standard deviation than to underestimate it. The evaluation of the complete conditional for X_i is computationally straightforward.

Generating observations from each of the other complete conditionals is straightforward and fast. Because the position of $\boldsymbol{X}$ changes throughout the algorithm, when using smoothing splines, we keep track of the value of g at a uniformly distributed grid of points. For each realization of g in the sampler the value of g for each grid point is recorded. This enables us to keep track of pointwise moments and percentiles.

Having observations from the joint posterior distribution provides a powerful tool for inference. The pointwise mean curve is a natural estimate of the regression mean function m. Pointwise credible intervals can also be calculated very easily from the observations of $\boldsymbol{g}$. Functions (linear or non-linear) of the regression function can also be estimated, along with standard errors. This is the approach used by [11] in non-measurement error cases and [7] in non-measurement error semiparametric models. The work in [11] work predated the revolution in Bayesian computations and she treats the smoothing parameter as fixed. Reference [7] uses the Gibbs sampler to adjust the credible sets for uncertainty in the variance components that define the smoothing parameter. In the present work, use of the Gibbs sampler also adjusts the credible sets for measurement error.

For an example we consider simulated data. The data consist of 100 X's generated from a standard normal distribution. The responses, Y, were generated from a normal distribution with a standard deviation of

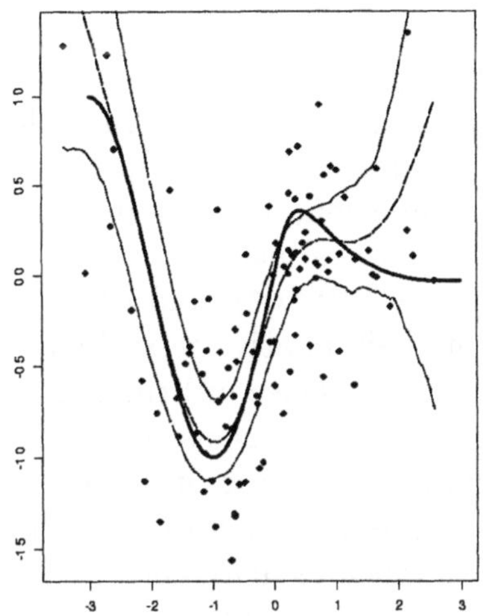

Figure 1. An example of the fully Bayesian Spline. The solid curve is the true regression function. The dashed middle curve is the mean of the posterior of the regression function. The dotted error bars represent the piecewise 90% credible intervals.

0.3 and a mean function of

$$m(x) = \frac{\sin(\pi x/2)}{1 + 2x^2\{\text{sign}(x) + 1\}}. \tag{8}$$

Each $m_i = 2$ and the W_{ij}, for $j = 1, 2$, are normally distributed with a mean of X_i and a standard deviation of 0.8. A pointwise posterior mean curve is used for the estimate of m. Credible curves are calculated by interpolating the pointwise $100(1 - \alpha)\%$ credible intervals. A burn-in time of 500 observations is used with 1000 observations from the posterior. Figure 1 shows the estimate of g and the 90% pointwise credible curves.

There are at least two possible methods for choosing the smoothing parameter for a smoothing spline. We place a prior distribution on γ. Reference [7] uses an identical procedure within semi-parametric models. It is worth noting that by placing a continuous density prior on γ, we have automatically given zero prior probability to the possibility of doing no smoothing at all. This is an automatic way of avoiding the possibility of gross undersmoothing that caused so much trouble for the methods of [2].

4. Simulations

We performed a series of simulations to compare our methods with those of [2]. In each case, 200 simulated data sets were generated, with X_i generated as independent normal random variables with mean μ_x and variance σ_x^2, with two replicates ($m_i = 2$), and with ϵ_i also normally distributed. In each simulation, for the fully Bayesian method, the following prior distributions are used: $\sigma_\epsilon^2 \sim IG(1,1)$, $\sigma_u^2 \sim IG(1,1)$, $\gamma \sim \text{Gamma}(3,1000)$, $\mu_x \sim N(0,10^2)$, and $\sigma_x^2 \sim IG(1,1)$. These priors were selected because of their relative flexibility. They are all proper, yet they are not strong, in the sense of bringing a lot of information to the problem. We found the results insensitive to moderate modifications of these priors. The flexibility of these priors are demonstrated by their success in the different regression functions used in the simulations.

For purposes of bias and mean squared error calculations, the smoothing spline estimates of g were computed on a grid of 101 points in the interval $[a,b]$, the interval chosen to contain most of the distribution for X. The mean squared biases and mean squared errors were computed over this grid.

The five cases considered were the following:

Case 1: The regression function, m is given in (8), with $n = 100$, $a = -2.0$, $b = 2.0$ $\sigma_\epsilon^2 = 0.3^2$, $\sigma_u^2 = 0.8^2$, $\mu_x = 0$ and $\sigma_x^2 = 1$.

Case 2: Same as Case 1 except $n = 200$.

Case 3: A modification of Case 1 above except that $n = 500$.

Case 4: Case 1 of [2], so that $m(x) = 1000x_+^3(1-x)_+^3$, $x_+ = xI(x>0)$, with $n = 200$, $a = 0.1$, $b = 0.9$ $\sigma_\epsilon^2 = 0.0015^2$, $\sigma_u^2 = (3/7)\sigma_x^2$, $\mu_x = 0.5$ and $\sigma_x^2 = 0.25^2$.

Case 5: A modification of Case 4 of [2], so that $m(x) = 10\sin(4\pi x)$, with $n = 500$, $a = 0.1$, $b = 0.9$ $\sigma_\epsilon^2 = 0.05^2$, $\sigma_u^2 = 0.141^2$, $\mu_x = 0.5$ and $\sigma_x^2 = 0.25^2$.

The methods compared were the following:

- naive smoothing spline fit ignoring measurement error;
- fully Bayesian approach;
- Structural method [2], 5 knots.
- Structural method, 15 knots.

Table 1 presents summary results for mean squared error. The striking feature of this table is that our Bayesian estimator has at least as good *frequentist* properties as the frequentist methods. In cases 1 and 2, it

Table 1. The mean squared error for the simulation. "Naive" is the naive smoothing spline, "Bayes" is the fully Bayesian method, and "Structural(m)" is the Structural regression spline of [2] with m knots. In each column, the smallest MSE values is in boldface.

Mean Squared Error $\times 10^2$								
Method	Case 1	2	3	4	5	6	7	8
Naive	6.91	5.57	5.38	11.55	3.79	5.77	5.84	16.48
Bayes	**2.84**	**1.56**	**1.47**	**1.95**	1/03	**2.69**	**2.49**	**7.41**
Structural(5)	8.17	3.82	1.73	2.17	2.03	7.27	7.91	16.84
Structural(15)	9.90	5.40	1.85	2.37	**0.80**	6.94	9.91	20.22

clearly dominates, having less than half the mean squared error of the other methods. In cases 3, its mean squared error efficiency is 20% greater than the structural spline with 15 knots, while it is only 25% less efficient in Case 5. The improvement of the fully Bayesian method over the frequentist methods is especially large for smaller sample sizes, e.g., Cases 1 and 6 (see below) where $n = 100$.

Clearly, even this limited simulation suggests that our Bayesian method is at least competitive with other methods proposed previously in the literature.

5. Discussion

The Bayesian approach to measurement error, modeling the mismeasured variables as latent random variables, and integrating them out, is a powerful one. In this paper we have developed a Bayesian method for nonparametric regression in the presence of measurement error. By modeling a smoothing spline from a Bayesian point of view we create algorithms to calculate the posterior distribution of the regression function. The resulting estimate accounts for the effects of measurement error both upon the estimator and upon the smoothing parameter. The resulting smoothing parameter selector appears to be the first that adjusts for the effects of measurement error.

The simulations demonstrate the flexibility of the fully Bayes approach, and even its efficiency in the frequentist sense, at least in the examples we have investigated. The fully Bayes approach also enable inference on more than just the regression function.

We believe that the reason that the fully Bayesian approach works better than previous proposals is this: often one can estimate the unknown X_i significantly more accurately using *all* information in the data about X_i rather than by using just the $W_{ij}, j = 1, \ldots, n_i$, see [1] for details. Clearly, this leaves open the possibility that with highly nonnormal errors, or highly heteroscedastic ones, the misspecified information from our simple model will lead to bias or other deleterious behavior in the Bayesian method.

Acknowledgments

Carroll's research was supported by a grant from the National Cancer Institute (CA–57030), and by the Texas A&M Center for Environmental and Rural Health via a grant from the National Institute of Environmental Health Sciences (P30–ES09106). Ruppert's research was supported by NSF Grant DMS–9804058.

References

[1] S. M. Berry, R. J. Carroll, and D. Ruppert. Bayesian smoothing and regression splines for measurement error problems. *Journal of the American Statistical Association*, 96:to appear, 2001.

[2] R. J. Carroll, J. D. Maca, and D. Ruppert. Nonparametric regression with errors in covariates. *Biometrika*, 86:541–554, 1999.

[3] J. R. Cook and L. A. Stefanski. Simulation–extrapolation estimation in parametric measurement error models. *Journal of the American Statistical Association*, 89:1314–1328, 1994.

[4] R. L. Eubank. *Nonparametric Regression and Spline Smoothing (Second Edition).* Marcel Dekker, Inc. New York, 1999.

[5] A. E. Gelfand and A. F. M. Smith. Sampling–based approaches to calculating marginal densities. *Journal of the American Statistical Association*, 85:398–409, 1990.

[6] P. J. Green and B. W. Silverman. *Nonparametric Regression and Generalized Linear Models: A Roughness Penalty Approach.* Chapman and Hall, London, 1994.

[7] T. Hastie and R. Tibshirani. Bayesian backfitting. *Statistical Science*, 15:193–223, 2000.

[8] D. Nychka. Bayesian confidence intervals for a smoothing spline. *Journal of the American Statistical Association*, 83:1134–1143, 1988.

[9] D. Nychka. The average posterior variance of a smoothing spline and a consistent estimate of the average squared error. *Annals of Statistics*, 18:415–428, 1990.

[10] G. Wahba. Improper priors, spline smoothing and the problem of guarding against model errors in regression. *Journal of the Royal Statistical Society, Series B*, 40:364–372, 1978.

[11] G. Wahba. Bayesian "confidence intervals" for the cross-validated smoothing spline. *Journal of the Royal Statistical Society, Series B*, 45:133–150, 1983.

[12] G. Wahba. *Spline models for observational data.* SIAM Press, Providence, 1990.

ON THE POLYNOMIAL MEASUREMENT ERROR MODEL

Chi-Lun Cheng
Institute of Statistical Science, Academia Sinica, Taipei, Taiwan, R.O.C
clcheng@stat.sinica.edu.tw

Hans Schneeweiss
Department of Statistics, University of Munich, Munich, Germany
schneew@stat.uni-muenchen.de

Abstract This paper discusses point estimation of the coefficients of polynomial measurement error (errors-in-variables) models. This includes functional and structural models. The connection between these models and total least squares (TLS) is also examined. A compendium of existing as well as new results is presented.

Keywords: polynomial errors-in-variables, total least squares, structural relationship, functional relationship, adjusted least squares, structural least squares.

1. Introduction: The model

Perhaps the most widely used models in applied statistics are regression models, linear or nonlinear. Measurement error (errors-in-variables) models are important alternatives to ordinary regression models because measurement error (ME) models assume both response and explanatory variables are measured with error, while ordinary regression models assume only the response variable to be measured with error. In many situations, ME models are more realistic than ordinary regression models.

The linear ME model has found extensive treatment in the literature, see, for example, the early review papers [32], [35], [27, Ch. 29], and, more recently, [12]; see also the monographs by Schneeweiss and Mittag [40], Fuller [17], and Cheng and Van Ness [13].

S. Van Huffel and P. Lemmerling (eds.), Total Least Squares and Errors-in-Variables Modeling, 131–143.

Since the mid-80's, non-linear measurement error models have been extensively studied. Most results are summarized in [7]. The present article reviews the polynomial ME model, which is the most natural extension of the linear model towards a nonlinear model.

The model we investigate can be described as follows. Assume that there are unobservable 'true' variables (ξ_i, η_i) that satisfy a polynomial relation,

$$\eta_i = \beta_0 + \beta_1 \xi_i + \cdots + \beta_k \xi_i^k, \qquad i = 1, \ldots, n, \tag{1}$$

for $k \geq 1$. One observes (x_i, y_i), which are the true variables plus additive errors $(\delta_i, \varepsilon_i)$; that is

$$x_i = \xi_i + \delta_i, \qquad y_i = \eta_i + \varepsilon_i, \qquad i = 1, \ldots, n, \tag{2}$$

where $(\delta_i, \varepsilon_i)$ are independent identically distributed with mean zero and covariance matrix

$$\mathbf{\Omega} = \begin{bmatrix} \sigma_\delta^2 & \sigma_{\delta\varepsilon} \\ \sigma_{\delta\varepsilon} & \sigma_\varepsilon^2 \end{bmatrix}.$$

If the ξ_i's are unknown constants, then the model is known as a *functional relationship*, while if the ξ_i's are independent identically distributed random variables and independent of the δ_i's and ε_i's, then the model is known as a *structural relationship*. The "structural parameters" are the intercept β_0, the slope coefficients $\beta_1, \ldots, \beta_k$, the error variances and covariance, σ_δ^2, σ_ε^2, and $\sigma_{\delta\varepsilon}$, and (in the structural model only) the parameters of the distribution of the ξ_i. In the functional model, the unknown constants ξ_i are known as "incidental parameters".

In addition to the measurement errors δ and ε, we may also have an "error in the equation", q say, which is to be added to the right hand side of (1). Although this error can be absorbed into the measurement error ε of the response variable, it is often worthwhile to distinguish between the cases with and without errors in the equation, see Section 3.1, see also [7, Sec. 2.3.2] for further comments.

The focus of this paper is the estimation of the structural parameters, especially the intercept and the slope coefficients, which are the parameters of major interest. It is well known that ordinary least squares (OLS) will produce inconsistent estimates, e.g., [20]. We are interested in constructing consistent estimators.

2. Identifiability and estimation without prior knowledge

We discuss the problem of identifiability only in the context of the structural relationship. For the functional relationship, identifiability is always satisfied, in a sense, but is of no use, as it does not guarantee the existence of consistent estimates, see [13, p. 162 and App. A].

We assume that $(\delta_i, \varepsilon_i)$ is jointly normal and, for simplicity, that $\sigma_{\delta\varepsilon} = 0$ unless specified otherwise.

Recall that in the simple linear ME model, Reiersøl [39] showed that the model is not identifiable if and only if ξ_i is normally distributed, see [4] for an extension of this result to the multivariate case. Therefore, we need extra information in order to find estimators that are consistent also when ξ is normal. The most commonly used additional assumptions are: $\sigma_\varepsilon^2/\sigma_\delta^2 = \lambda$ is known or σ_δ^2 is known.

As for the general nonlinear ME model, that is,

$$\eta_i = g(\xi_i, \boldsymbol{\beta}), \qquad i = 1, \ldots, n, \tag{3}$$

where $\boldsymbol{\beta} = [\beta_0, \beta_1, \ldots, \beta_k]^T$ is the unknown parameter vector and g is a real-valued known function, there is no identifiability characterization corresponding to that in the linear ME model. However, in the polynomial ME model, Kendall and Stuart [27, pp. 435–437] use a method of cumulants to find a consistent estimator in the structural model without additional assumptions. Their results imply that the structural polynomial model with $k > 1$ is identifiable if the errors $(\delta_i, \varepsilon_i)$ are normal.

Van Montfort [46, Ch. 2] uses higher moments instead of cumulants to estimate the parameters of a structural polynomial regression, see also [13, Ch. 6]. Recently, Huang and Huwang [23] constructed a consistent estimator by regressing both y and y^2 on the powers of x when ξ is normal. In the latter regression the parameters depend on those of the first regression. See also [25] for the corresponding Berkson model.

If one wishes to use ML estimation methods, the immediate problem encountered is that one needs to find the joint distribution of (x_i, y_i). This is not as straightforward as in the linear case. To the authors' best knowledge, there are no articles that deal with ML estimation in the polynomial ME model.

Although the above estimators are consistent, their efficiency may be rather low because they do not use knowledge of the error variances. In many practical situations such knowledge may well be available and can be used to construct more efficient estimators. We therefore now turn to such estimators.

3. Functional methods

We start with the functional variant of the ME model, i.e., the ξ_i's are taken to be fixed unknown quantities. The estimation methods developed on this basis can, however, also be used in the structural case, in particular, if no specific distribution has been assumed for the random

ξ_i's. We consider two major cases: when σ_δ^2 is known and when both error variances σ_δ^2 and σ_ε^2 (or their ratio) are known.

3.1 Regressor error variance known

If there is an error in the equation, q, it is very unlikely that its variance would be known *a priori*. On the other hand, the variance of the measurement error, δ, might well be known or at least estimable either from replicated data or from validation data. In this subsection we assume σ_δ^2 to be known (and possibly also higher moments of δ). In case $\sigma_{\delta\varepsilon} \neq 0$, this parameter should also be known.

If only σ_δ^2 is known, maximum likelihood estimation breaks down and does not yield sensible estimators, just as it happens in the linear case, see [33]. The problem with maximum likelihood estimation in the functional relationship is due to the presence of the incidental parameters ξ_i, whose number increases with sample size, see [37]. It is well known that in the presence of incidental parameters, the maximum likelihood estimator need not exist. Even if it exists, it need not be consistent.

Nevertheless consistent estimation of the model parameters is possible. One approach is based on a variant of the corrected score function approach, see [43], [36], and [6]. The following discussion is adapted from [10], see also [13, Ch. 6].

If we provisionally assume that the ξ_i's are known, then the model will return to the ordinary polynomial regression model, and ordinary least squares (OLS) induces the following unbiased estimating equation

$$\sum_{i=1}^{n}(y_i\xi_i^j - \beta_0\xi_i^j - \beta_1\xi_i^{j+1} - \cdots - \beta_k\xi_i^{j+k}) = 0, \tag{4}$$

for $j = 0, \ldots, k$. The problem with (4) is that the terms ξ_i^r, are not observable. If $\sigma_{\delta\varepsilon} = 0$ and if we can find unbiased estimates t_{ri} of ξ_i^r which do not involve any unknown parameters, then we can replace ξ_i^r by t_{ri} in (4) and solve the system for the estimators of $\beta_0, \beta_1, \ldots, \beta_k$.

In the following discussion, we omit the observation index i for ease of notation. Because $x = \xi + \delta$, we have, by the binomial theorem,

$$x^r = \sum_{j=0}^{r}\binom{r}{j}\xi^j\delta^{r-j}. \tag{5}$$

After replacing ξ^j in (5) by t_j and δ^{r-j} by its expectation, we can solve for t_j. The first five terms of t_j, for example, are

$$t_0 = 1, \quad t_1 = x, \quad t_2 = x^2 - \sigma_\delta^2, \quad t_3 = x^3 - 3x\sigma_\delta^2 - \mathbf{E}\delta^3,$$
$$t_4 = x^4 - 6x^2\sigma_\delta^2 - 4x\mathbf{E}\delta^3 - \mathbf{E}\delta^4 + 6\sigma_\delta^4,$$

Under the normality assumption of the error δ, we only need to know σ_δ^2 in order to find t_j. In this case we also have a recursion formula for obtaining t_j:

$$t_{j+1} = xt_j - \sigma_\delta^2 j t_{j-1},$$

with $t_0 = t_{-1} = 1$, see also [43]. Once the β's have been estimated, the unknown error variance σ_ε^2 can also be estimated consistently. If $\sigma_{\delta\varepsilon} \neq 0$, the term $y\xi^j$ has to be replaced in (4) by an expression h_j involving $\sigma_{\delta\varepsilon}$ such that $Eh_j = \eta\xi^j$. For $\sigma_{\delta\varepsilon} = 0$, $h_j = yt_j$ and for normal (δ, ε), $h_j = yt_j - \sigma_{\delta\varepsilon} j t_{j-1}$.

To sum up, let us introduce the matrices $\boldsymbol{H}_i$ with elements $(\boldsymbol{H}_i)_{jl} = t_{j+l,i}$, $j, l = 0, \ldots, k, i = 1, \ldots, n$, so that $\mathbf{E}\boldsymbol{H}_i = \left(\xi_i^{j+l}\right)_{j,l=0,\ldots,k}$, and the vectors $\boldsymbol{h}_i$, with $\mathbf{E}\boldsymbol{h}_i = \left(\eta_i \xi_i^j\right)_{j=0,\ldots,k}$. The "corrected" or "adjusted" estimating equation that replaces (4) can then be written as

$$\sum_{i=1}^n \boldsymbol{H}_i \hat{\boldsymbol{\beta}}_{ALS} = \sum_{i=1}^n \boldsymbol{h}_i.$$

Its solution, $\hat{\boldsymbol{\beta}}_{ALS}$, is the so-called *adjusted least squares* (ALS) estimate of $\boldsymbol{\beta}$. Chan and Mak [8] proposed a similar approach for the $\boldsymbol{\Omega}$ known case.

Under some regularity conditions, see [24], the resulting estimators are consistent and asymptotically normal. See also [7, p. 261 ff.], [16], and [3]. The asymptotic covariance matrix of $\hat{\boldsymbol{\beta}}_{ALS}$ is given by the formula

$$\boldsymbol{\Sigma}_{\hat{\boldsymbol{\beta}}_{ALS}} = \tfrac{1}{n} \bar{\boldsymbol{H}}^{-1} \bar{\boldsymbol{U}} \bar{\boldsymbol{H}}^{-1} \tag{6}$$

with $\bar{\boldsymbol{H}} = \frac{1}{n}\sum_{i=1}^n \boldsymbol{H}_i$, $\bar{\boldsymbol{U}} = \frac{1}{n}\sum_{i=1}^n (\boldsymbol{H}_i\hat{\boldsymbol{\beta}}_{ALS} - \boldsymbol{h}_i)(\boldsymbol{H}_i\hat{\boldsymbol{\beta}}_{ALS} - \boldsymbol{h}_i)^T$. Because of its form, (6) is often called a sandwich formula, see [7].

For small samples, $\hat{\boldsymbol{\beta}}_{ALS}$ turns out to be rather unstable, in particular if σ_δ^2 is large and if $k \geq 3$. In fact, the moments of $\hat{\boldsymbol{\beta}}_{ALS}$ do not exist, see [13], Chapter 2, for the linear case. Using an idea of Fuller's [17], the estimator can be improved for small samples by the following modification, see [11]. We use the concepts and notations of Subsection 3.2, see in particular (9). Then the modified ALS estimator of $\boldsymbol{\beta}$ is given by

$$\begin{aligned} \hat{\boldsymbol{\beta}}_{MALS} &= (\boldsymbol{M}_{tt} - a\bar{\bar{\boldsymbol{V}}})^{-1}(\boldsymbol{m}_{ty} - a\bar{\bar{\boldsymbol{v}}}), \\ a &= \begin{cases} (n-\alpha)/n & \text{if} \quad \hat{\rho} > 1 + \frac{1}{n} \\ \hat{\rho}(n-\alpha)/(n+1) & \text{if} \quad \hat{\rho} \leq 1 + \frac{1}{n}, \end{cases} \end{aligned}$$

where $\hat{\rho}$ is the smallest positive root of $\det(\boldsymbol{M} - \rho\tilde{\boldsymbol{W}}) = 0$ with

$$\tilde{\boldsymbol{W}} = \begin{bmatrix} \sigma_{\delta\varepsilon}^2/\sigma_\delta^2 & \bar{\hat{\boldsymbol{v}}}^T \\ \bar{\hat{\boldsymbol{v}}} & \bar{\hat{\boldsymbol{V}}} \end{bmatrix}$$

and some α to be chosen so that an estimator of good small sample properties results, e.g., $\alpha = k + 5$.

A multivariate extension of the ALS procedure including a small sample modification has recently been provided by Wall and Amemiya [47]. Actually the authors study a factor analysis (or "structural") model where the common factors obey a polynomial relation instead of the usual linear relation.

3.2 Ratio of error variances known

Consider now model (1) and (2), i.e., a polynomial without an error in the equation but with measurement errors in both variables. We can assume that in such a model $\boldsymbol{\Omega}$ is known.

Let us first consider maximum likelihood (ML), assuming the errors are normally distributed. For the general nonlinear model (3), with $\sigma_{\delta\varepsilon} = 0$, ML is equivalent to minimizing the function

$$Q(\boldsymbol{\beta}) = \sum_{i=1}^{n} \min_{\xi_i \in R} [(y_i - g(\xi_i, \boldsymbol{\beta}))^2 + \lambda(x_i - \xi_i)^2]$$

with respect to $\boldsymbol{\beta}$, where $\lambda = \sigma_\varepsilon^2/\sigma_\delta^2$. This method is called total least squares (TLS). Geometrically this means that in case $\lambda = 1$ one has to minimize the sum of squares of the perpendicular distances from the data points to the regression curve. TLS was first proposed by Adcock [1] and then re-discovered by many authors since. Golub and Van Loan [19] proposed numerical methods for TLS in a general linear framework. The more recent work is summarized in [45]. In the linear case, minimizing $Q(\boldsymbol{\beta})$ reduces to minimizing the function

$$q(\boldsymbol{\beta}) = \sum_{i=1}^{n} \frac{(y_i - \beta_0 - \beta_1 x_i)^2}{\sigma_e^2} \tag{7}$$

with $e_i = \varepsilon_i - \beta_1\delta_i$. This procedure, which corresponds to the algebraic side of TLS, is also called generalized or weighted least squares (WLS) and was introduced by Sprent [42].

Kendall and Stuart [27, p. 434] applied ML to the quadratic ($k = 2$) measurement error model. They did not find a closed form solution but had to resort to iterative methods just as in the general TLS approach.

For numerical procedures in the polynomial ME context see [38], see also [5].

In the linear model, the ML estimates of intercept and slope parameters are consistent. This is no more true for the nonlinear model, see [15].

Despite the inconsistency of the TLS estimate of $\boldsymbol{\beta}$, it might still be useful to use this method in practice, especially when σ_δ^2 is small. In such cases the bias, though still existent, might be almost negligible. In fact, there exists a small-σ_δ theory for TLS, see [48].

However, consistent estimation is possible. One can extend the WLS approach to the polynomial ME model. The idea comes from Wolter and Fuller [49], which they applied to the quadratic functional model with normal errors and known covariance matrix. Cheng and Schneeweiss [9] extended it to a polynomial of any degree, and the restriction of normally distributed errors was lifted. To be more specific, model (1)–(2) can be re-written as

$$y_i = \boldsymbol{\zeta}_i^T \boldsymbol{\beta} + \varepsilon_i, \qquad \boldsymbol{t}_i = \boldsymbol{\zeta}_i + \boldsymbol{\pi}_i, \tag{8}$$

where $\boldsymbol{\zeta}_i = [\xi_i^0, \ldots, \xi_i^k]^T$, $\boldsymbol{t}_i = [t_{0i}, \ldots, t_{ki}]^T$; and $\boldsymbol{\pi}_i = [\pi_{0i}, \ldots, \pi_{ki}]^T$ is implicitly defined by (8). Note that the new measurement errors π_{ri} have mean zero, covariance matrix $\boldsymbol{V}_i = \mathbf{E}(\boldsymbol{\pi}_i \boldsymbol{\pi}_i^T)$, and covariance vector $\boldsymbol{v}_i = \mathbf{E}(\boldsymbol{\pi}_i \varepsilon_i)$. Cheng and Schneeweiss [10], [9] gave unbiased estimates of $\boldsymbol{V}_i$ and $\boldsymbol{v}_i$, which were

$$\hat{\boldsymbol{V}}_i = \boldsymbol{t}_i \boldsymbol{t}_i^T - \boldsymbol{H}_i, \qquad \hat{\boldsymbol{v}}_i = \boldsymbol{t}_i y_i - \boldsymbol{h}_i.$$

Let $\boldsymbol{z}_i = (y_i, \boldsymbol{t}_i^T)^T$ and let $\boldsymbol{W}_i$ be the error covariance matrix of $\boldsymbol{z}_i$ and

$$\hat{\boldsymbol{W}}_i = \begin{bmatrix} \sigma_\varepsilon^2 & \hat{\boldsymbol{v}}_i{}^T \\ \hat{\boldsymbol{v}}_i & \hat{\boldsymbol{V}}_i \end{bmatrix}$$

its estimate. Moreover, let $\boldsymbol{M} = n^{-1}\Sigma \boldsymbol{z}_i \boldsymbol{z}_i^T$, $\boldsymbol{m}_{ty} = n^{-1}\Sigma \boldsymbol{t}_i y_i$, $\boldsymbol{M}_{tt} = n^{-1}\Sigma \boldsymbol{t}_i \boldsymbol{t}_i^T$ and $m_{yy} = n^{-1}\Sigma y_i^2$, then the estimate of $\boldsymbol{\beta}$ is given by

$$\hat{\boldsymbol{\beta}} = (\boldsymbol{M}_{tt} - \hat{\kappa} \bar{\hat{\boldsymbol{V}}})^{-1} (\boldsymbol{m}_{ty} - \hat{\kappa} \bar{\hat{\boldsymbol{v}}}), \tag{9}$$

where $\hat{\kappa}$ is the smallest positive root of

$$\det(\boldsymbol{M} - \kappa \bar{\hat{\boldsymbol{W}}}) = 0;$$

see also [34].

As pointed out in [49], the estimate (9) is the $\boldsymbol{\beta}$ that minimizes the function $w(\boldsymbol{\theta}) = \boldsymbol{\theta}^T \boldsymbol{M} \boldsymbol{\theta} / (\boldsymbol{\theta}^T \bar{\hat{\boldsymbol{W}}} \boldsymbol{\theta})^{-1}$, where $\boldsymbol{\theta}^T = (1, -\boldsymbol{\beta}^T)$. In other words, the estimate (9) can be viewed as a WLS estimate of the ("linearized") polynomial ME model (8). Note that if $k = 1$, then $w(\boldsymbol{\theta})$ reduces to (7).

The estimate (9) is consistent and asymptotically normal and its asymptotic covariance matrix is given in [9].

When the errors $(\delta_i, \varepsilon_i)$ are normal, then it is possible to relax the assumption of $\boldsymbol{\Omega}$ being known to knowing $\boldsymbol{\Omega}$ up to a proportionality factor. One can assume that, without loss of generality, this unknown factor is σ_ε^2 and

$$\boldsymbol{\Omega} = \begin{bmatrix} \sigma_\delta^2 & \sigma_{\delta\varepsilon} \\ \sigma_{\delta\varepsilon} & \sigma_\varepsilon^2 \end{bmatrix} = \sigma_\varepsilon^2 \begin{bmatrix} a & b \\ b & 1 \end{bmatrix},$$

where a and b are known constants. Note that the unknown factor σ_ε^2 can be estimated by

$$\hat{\sigma}_\varepsilon^2 = \hat{\boldsymbol{\theta}}^T \begin{bmatrix} m_{yy} & \boldsymbol{m}_{ty}^T \\ \boldsymbol{m}_{ty} & \bar{\boldsymbol{H}} \end{bmatrix} \hat{\boldsymbol{\theta}}.$$

Therefore, we need to estimate $\boldsymbol{\beta}$ and σ_ε^2 simultaneously. The exact numerical procedure has not been studied yet. It should be noted that we need the normality assumption of $(\delta_i, \varepsilon_i)$ because then all moments of $(\delta_i, \varepsilon_i)$ will be known up to some powers of σ_ε^2 once a and b are known.

4. Structural methods

In this section, we study the structural variant of the polynomial ME model. We assume that the ξ_i are iid $\mathrm{N}(\mu_\xi, \sigma_\xi^2)$. The errors are also assumed to be normal with $\sigma_{\delta\varepsilon} = 0$. σ_δ^2 is supposed to be known. $\mu_\xi (= \mu_x)$ and $\sigma_\xi^2 (= \sigma_x^2 - \sigma_\delta^2)$ can be estimated from the x_i alone without having resort to the model.

The method we present here is based on a quasi score function, which is derived from a conditional mean-variance model corresponding to the polynomial model. The following presentation is essentially due to Thamerus [44], see also [7].

Note that the error-free model (1) can be written as a mean-variance model

$$\begin{aligned} \mathbf{E}(y_i \mid \xi_i) &= \beta_0 + \beta_1 \xi_i + \cdots + \beta_k \xi_i^k \\ V(y_i \mid \xi_i) &= \sigma_\varepsilon^2. \end{aligned} \tag{10}$$

Now, the conditional distribution of ξ_i given x_i is also normal.

$$\begin{aligned} \xi_i \mid x_i &\sim \mathrm{N}(\mu_1(x_i), \tau^2) \\ \mu_1(x) &= \mu_x + (1 - \sigma_\delta^2/\sigma_x^2)(x - \mu_x) \\ \tau^2 &= \sigma_\delta^2(1 - \sigma_\delta^2/\sigma_x^2). \end{aligned}$$

Therefore we can easily derive from (10) a conditional mean-variance model of y given x:

$$\mathbf{E}(y_i \mid x_i) = \sum_{j=0}^{k} \beta_j \mu_j(x_i) =: m(x_i, \boldsymbol{\beta})$$
$$V(y_i \mid x_i) = \sigma_\varepsilon^2 + \sum_{j,l=0}^{k} \beta_j \beta_l \{\mu_{j+l}(x_i) - \mu_j(x_i)\mu_l(x_i)\} =: v(x_i, \boldsymbol{\beta}, \sigma_\varepsilon^2) \tag{11}$$

with

$$\mu_r(x) = \mathbf{E}(\xi^r \mid x) = \sum_{j=0}^{r} \binom{r}{j} \mu_j^* \mu_1(x)^{r-j}$$
$$\mu_j^* = \begin{cases} 0 & \text{if } \ j \text{ is odd} \\ 1 \cdot 3 \cdots (j-1)\tau^j & \text{if } \ j \text{ is even.} \end{cases}$$

The following estimating equations can then be set up:

$$\sum_{i=1}^{n} [\{y_i - m(x_i, \boldsymbol{\beta})\}/v(x_i, \boldsymbol{\beta}, \sigma_\varepsilon^2)] \boldsymbol{\mu}(x_i) = 0 \tag{12}$$
$$\sigma_\varepsilon^2 = \tfrac{1}{n} \sum_{i=1}^{n} [\{y_i - m(x_i, \boldsymbol{\beta})\}^2 - \sum_{j,l} \beta_j \beta_l \{\mu_{j+l}(x_i) - \mu_j(x_i)\mu_l(x_i)\}]$$

where $\boldsymbol{\mu}(x) = [\mu_0(x), \dots, \mu_k(x)]^T$. They can be solved by iteratively reweighted least squares and give rise to the so-called *structural least squares* (SLS) estimators of $\boldsymbol{\beta}$ and σ_ε^2. It should be noted that μ_x and σ_x^2 have to be replaced by their estimates in the estimating equations (12).

The estimator $\hat{\boldsymbol{\beta}}_{SLS}$ is consistent and asymptotically normal with asymptotic covariance matrix

$$\boldsymbol{\Sigma}_{\hat{\boldsymbol{\beta}}_{SLS}} = \tfrac{1}{n} [\mathbf{E}\{\boldsymbol{\mu}(x)\boldsymbol{\mu}^T(x)/v(x, \boldsymbol{\beta}, \sigma_\varepsilon^2)\}]^{-1} + O(\sigma_\delta^4). \tag{13}$$

The term $O(\sigma_\delta^4)$ is due to the estimation of μ_x and σ_x^2 and can be neglected in a first approximation, see [29].

The SLS method can be generalized to encompass models where the ξ_i follow a finite mixture of normal distributions and where the measurement errors could be heteroscedastic, see [44].

A very simple approximate method to SLS is the regression calibration (RCAL) method, see [7]. In this method, the ξ_i in equation (1) are simply replaced by their conditional means $\mu_1(x_i) = \mathbf{E}(\xi_i \mid x_i)$ and the resulting polynomial regression is estimated by OLS. The RCAL estimates are not consistent but their biases are typically very small.

5. Comparison of functional and structural methods

Since functional methods can also be used in a structural model, we can compare functional and structural methods within the context of a structural polynomial ME model. In particular Kukush et al [30] compare ALS and SLS in a structural model with normally distributed ξ_i. It might be expected that SLS is more efficient than ALS in this case, as it uses the information on the distribution of the ξ_i. However, it turns out that the asymptotic covariance matrices of ALS and SLS, see (6) and (13), are equal up to the order of σ_δ^2.

Comparisons with respect to OLS and to RCAL can also be carried out, see also [34].

When the ξ_i are not normally distributed, SLS breaks down, i.e., it leads to inconsistent estimates. Thus while SLS might be more efficient than ALS when the ξ_i are normal — but only in terms of order $O(\sigma_\delta^4)$ — the latter procedure is more robust with respect to the distribution of the ξ_i, see [28], [41].

6. Miscellaneous topics

We only studied point estimation. Another important issue is interval estimation. However, little is known in the polynomial ME model. In principle, one can use the asymptotic normality of the estimates of the β's to obtain (approximate) confidence regions on the parameters. But there is a serious problem regarding this approach due to the zero confidence level effect, see [18] and [13, Sec. 2.4].

The models we studied so far are all characterized by having additive measurement errors. Iturria et al [26] investigated a polynomial ME model with multiplicative measurement errors, that is, $x_i = \xi_i \delta_i$.

When we surveyed the structural polynomial ME model in Section 4, we restricted our attention to the case of normally distributed ξ_i. It is known that in the linear model a nonnormal distribution of ξ will produce a nonlinear regression function in the observables y and x, i.e., $\mathbf{E}(y|\,x)$ will be nonlinear, see [31]. Chesher [14] extends this result to the polynomial regression and gives explicit formulas for $\mathbf{E}(y|\,x)$ for any distribution of x and normal measurement errors. These relations might be exploited to find consistent estimates for the β_j's as long as they are identifiable from $\mathbf{E}(y|\,x)$ even if σ_δ^2 is not known.

In this paper we only studied polynomial ME models where either no further knowledge was available or where some knowledge about the error variances could be used. A further possibility to consistently estimate the β's is the use of instrumental variables and of indicator variables.

The usual instrumental variables approach for the linear model breaks down in nonlinear models, see [2]. Nevertheless consistent estimation using higher moments is possible, see [21] and [22] with an application to the estimation of Engel curves.

Acknowledgments

This work was supported by Deutsche Forschungsgemeinschaft (SFB-386), Deutscher Akademischer Austauschdienst, and National Science Council of Taiwan, Republic of China.

References

[1] R. J. Adcock. Note on the method of least squares. *Analyst*, 4:183–184, 1877.

[2] Y. Amemiya. Instrumental variable estimator for the nonlinear errors- in-variables model. *Journal of Econometrics*, 28:273–289, 1985.

[3] S. Baran. A consistent estimator in general functional errors-in-variables models. *Metrika*, 51:117 – 132, 2000.

[4] P. A. Bekker. Comment on identification in the linear errors in variables model. *Econometrica*, 54:215–217, 1986.

[5] H. I. Britt and R. H. Luecke. The estimation of parameters in nonlinear implicit models. *Technometrics*, 15:233–247, 1973.

[6] J. P. Buonaccorsi. A modified estimating equation approach to correcting for measurement error in regression. *Biometrika*, 83:433–440, 1996.

[7] R. J. Carroll, D. Ruppert, and L. A. Stefanski. *Nonlinear Measurement Error Models.* Chapman and Hall, New York, 1995.

[8] L. K. Chan and T. K. Mak. On the polynomial functional relationship. *Journal of the Royal Statistical Society, Series B*, 47:510–518, 1985.

[9] C.-L. Cheng and H. Schneeweiss. Note on two estimators for the polynomial regression with errors in the variables. In *Statistical Modeling: Proceedings of the 13th International Workshop on Statistical Modeling*, pages 141–147, New Orleans, Lousiana, 1998. B. Marx and H. Friedl.

[10] C.-L. Cheng and H. Schneeweiss. Polynomial regression with errors in the variables. *Journal of the Royal Statistical Society, Series B*, 60:189–199, 1998.

[11] C.-L. Cheng, H. Schneeweiss, and M. Thamerus. A small sample estimator for a polynomial regression with errors in the variables. *Journal of the Royal Statistical Society, Series B*, 62:699–709, 2000.

[12] C.-L. Cheng and J. W. Van Ness. On estimating linear relationships when both variables are subject to errors. *Journal of the Royal Statistical Society, Series B*, 57:167–183, 1994.

[13] C.-L. Cheng and Van Ness J. W. *Statistical Regression with Measurement Error.* Arnold, 1999.

[14] A. Chesher. Polynomial regression with covariate measurement error. *Discussion Papaer No. 98/448, University of Bristol, Department of Economics*, 1998.

[15] I. Fazekas, S. Baran, A. Kukush, and J. Lauridsen. On inconsistency of the least squares estimator in nonlinear functional relations. *Research Report N1, Odense University, Denmark, Department of Statistic and Demography*, 1998.

[16] I. Fazekas, S. Baran, A. Kukush, and J. Lauridsen. Asymptotic properties in space and time of an estimator in nonlinear functional errors-in-variables models. *Random Operators and Stochastic Equations*, 7:389–412, 1999.

[17] W. A. Fuller. *Measurement Error Models.* John Wiley and Sons, New York, 1987.

[18] L. J. Gleser and J. T. Hwang. The nonexistence of $100(1-\alpha)\%$ confidence sets of finite expected diameter in errors-in-variables and related models. *The Annals of Statistics*, 15:1351–1362, 1987.

[19] G. H. Golub and C. F. Van Loan. An analysis of total least squares problem. *SIAM J. Numer. Anal.*, 17:883–893, 1980.

[20] Z. Griliches and V. Ringstad. Error-in-the-variables bias in nonlinear contexts. *Econometrica*, 38:368–370, 1970.

[21] J. A. Hausman and W. K. Newey. Identification and estimation of polynomical errors-in-variables models. *Journal of Econometrics*, 50:273–295, 1991.

[22] J. A. Hausman, W. K. Newey, and J. L. Powell. Nonlinear errors in variables estimation of some engel curves. *Journal of Econometrics*, 50:205–233, 1995.

[23] Y. H. Huang and L. C. Huwang. On the polynomial structural relationship. *Canadian Journal of Statistics*, page to appear, 2001.

[24] P. J. Huber. The behavior of maximum likelihood estimates under nonstandard condition. In *Proceedings Fifth Berkeley Symposium on Mathematical Statistics and Probability*, volume 1, pages 221–233, University of California Press, Berkeley, 1967. L. LeCam and J. Neyman.

[25] Y. H. Huwang, L. C. amd Huang. On errors-in-variables in polynomial regression-berkson case. *Statistica Sinica*, 10:923–936, 2000.

[26] S. J. Iturria, R. J. Carroll, and D. Firth. Polynomial regression and estimating functions in the presence of multiplicative measurement error. *Journal of the Royal Statistical Society, Series B*, 61(3):547–561, 1999.

[27] M. G. Kendall and A. Stuart. *The Advanced Theory of Statistics*, volume 2. Hafner, New York, 4th edition, 1979.

[28] J. T. Kuha and J. Temple. Covariate measurement error in quadratic regression. *Discussion Paper 1999-W2, Nuffield College Oxford*, 1999.

[29] A. Kukush and H. Schneeweiss. A comparison of asymptotic covariance matrices of adjusted least squares and structural least squares in error ridden polynomial regression. *Discussion Paper 218, Sonderforschungsbereich 386, University of Munich*, 2000.

[30] A. Kukush, H. Schneeweiss, and R. Wolf. Comparison of three estimators in a polynomial regression with measurement errors. *Discussion Paper 233, Sonderforschungsbereich 386, University of Munich*, 2001.

[31] D. V. Lindley. Regression lines and the linear functional relationship. *Journal of the Royal Statistical Society, Supplement*, 9:218–244, 1947.

[32] A. Madansky. The fitting of straight lines when both variables are subject to error. *Journal of the American Statistical Association*, 54:173–205, 1959.

[33] L. Moberg and R. Sundberg. Maximum likelihood estimation of a linear functional relationship when one of the departure variances is known. *Scandinavian Journal of Statistics*, 5:61–64, 1978.

[34] M. S. Moon and R. F. Gunst. Polynomial measurement error modeling. *Computational Statistics and Data Analysis*, 19:1–21, 1995.

[35] P. A. P. Moran. Estimating structural and functional relationships. *Journal of Multivariate Analysis*, 1:232–255, 1971.

[36] T. Nakamura. Corrected score functions for errors-in-variables models: Methodology and application to generalized linear models. *Biometrika*, 77:127–137, 1990.

[37] J. Neyman and E. L. Scott. Consistent estimates based on partially consistent observations. *Econometrica*, 16:1–32, 1948.

[38] M. O'Neill, L. G. Sinclair, and F. J. Smith. Polynomial curve fitting when abscissas and ordinates are both subject to error. *Computer Journal*, 12:52–56, 1969.

[39] O. Reiersøl. Identifiability of a linear relation between variables which are subject to error. *Econometrica*, 18:375–389, 1950.

[40] H. Schneeweiß and H. J. Mittag. *Lineare Modelle mit fehlerbehafteten Daten.* Physica, Heidelberg, 1986.

[41] H. Schneeweiss and T. Nittner. Estimating a polynomial regression with measurement errors in the structural and in the functional case - a comparison. *Discussion Paper 197, Sonderforschungsbereich 386, University of Munich*, 2000.

[42] P. Sprent. A generalized least-squares approach to linear functional relationships. *Journal of the Royal Statistical Society, Series B*, 28:278–297, 1966.

[43] Leonard A. Stefanski. Unbiased estimation of a nonlinear function of a normal mean with application to measurement error models. *Communications in Statistics, Part A - Theory and Methods*, 18:4335–4358, 1989.

[44] M. Thamerus. Different nonlinear regression models with incorrectly observed covariates. In *Econometrics in Theory and Practice, Festschrift for Hans Schneeweiss.*, pages 31–44. Galata, R. and Kuechenhoff, H., 1998.

[45] S. Van Huffel and J. Vandewalle. *The Total Least Squares Problem: Computational Aspects and Analysis.* SIAM, Philadelphia, 1991.

[46] K. Van Montfort. *Estimating in Structural Models with Non-normal Distributed Variables: Some Alternative Approaches.* Leiden: DSWO, 1989.

[47] M. M. Wall and Y. Amemiya. Estimation for polynomial structural equation models. *Journal of the American Statistical Association*, 95:929–940, 2000.

[48] K. M. Wolter and W. A. Fuller. Estimation of nonlinear errors-in-variables models. *The Annals of Statistics*, 10:539–548, 1982.

[49] K. M. Wolter and W. A. Fuller. Estimation of the quadratic errors-in-variables model. *Biometrika*, 69:175–182, 1982.

ON CONSISTENT ESTIMATORS IN NONLINEAR FUNCTIONAL ERRORS-IN-VARIABLES MODELS

Alexander Kukush
ESAT-SISTA, K.U.Leuven, Kasteelpark 10, B-3001 Leuven-Heverlee, Belgium
akukush@esat.kuleuven.ac.be

Silvelyn Zwanzig
Uppsala University, Dep. of Mathematics, Box 480, S-75106 Uppsala, Sweden
20@math.uu.se

Abstract For the implicit nonlinear functional relation model a new contrast estimation procedure is proposed, where the deconvolution idea is used for eliminating the nuisance parameters in the usual minimum contrast function. Several examples are considered including L_1- and L_2-methods. Sufficient conditions for consistency are given.

Keywords: nonlinear errors-in-variables models, functional relationship, minimum contrast estimation, estimation function, total L_p norm approximation.

1. Introduction

The functional relation model is characterized by an increasing number of nuisance parameters and a fixed number of parameters of interest. It is an old fact that in models with an increasing number of nuisance parameters usual estimation procedures may fail, compare for instant the classical paper of Neyman, Scott [9]. The main problem is the elimination of the nuisance parameters in the estimating function, in such a way that the estimating procedure delivers consistent estimators. In linear models the least squares approach fulfills that and gives consistent and efficient estimators for the parameter of interest. The situation is different in nonlinear models.

In this paper we will present a method for construction of consistent estimators in nonlinear functional relation models. The main idea is a correction of a general minimum contrast estimation criterion. Under

S. Van Huffel and P. Lemmerling (eds.), Total Least Squares and Errors-in-Variables Modeling, 145–154.

the assumption that the minimal point is given, we are interested in the convergence of probability for increasing sample size. Our method is strongly related to Stefanski's corrected-score method, see [2, Ch. 6].

2. Corrected minimum contrast estimators

In order to emphasize the symmetric character of the observations we consider the implicit version of errors-in-variables models, which makes no difference between independent and dependent variables.

$$G(\eta_i, \beta) = 0, \quad Z_i = \eta_i + \varepsilon_i, \quad i = 1, .., n\,, \tag{1}$$

where ε_i are *i.i.d.* with expected value zero and covariance matrix Γ. The functional relation $G : \mathbb{R}^q \times \mathbb{R}^p \longrightarrow \mathbb{R}$ is a known smooth function. The expected values of the observations $\eta_i \in \mathcal{D} \subset \mathbb{R}^q$, form the nuisance parameter $\eta = (\eta_1, ..., \eta_n) \in \mathcal{D}^n$. The parameter of interest $\beta \in \mathcal{B} \subset \mathbb{R}^p$ has fixed dimension. The parameter consisting of the nuisance parameter and the parameter of interest is denoted by θ.

We apply the principle of *minimum contrast estimators* $\widetilde{\theta}$, which are defined as a measurable solution of

$$\widetilde{\theta} \in \arg\min_{\theta \in \Theta} \widetilde{C}_n(Z, \theta)\,. \tag{2}$$

Because of the model structure in (1) we will consider estimation criteria of the following structure.

$$\widetilde{C}_n(Z, \theta) = \frac{1}{n} \sum_{i=1}^{n} \widetilde{c}(Z_i, \eta_i, \beta)\,. \tag{3}$$

Examples for $\widetilde{C}_n(Z, \theta)$ are the total least squares estimator and L_1- type estimator with

$$\frac{1}{n} \sum_{i=1}^{n} \|Z_i - \eta_i\|^2, \quad \frac{1}{n} \sum_{i=1}^{n} |Z_i - \eta_i|\,. \tag{4}$$

A contrast function $\widetilde{C}_n(Z, \theta)$ is reasonable if for $C_n(\theta_0, \theta) = E_{\theta_0} \widetilde{C}_n(Z_i, \theta)$, with $\theta_0 = \left(\eta_1^0, ..., \eta_n^0, \beta_0\right)^T$, the following contrast condition is fulfilled:

(i) For all θ_0, θ

$$\Delta C_n(\theta) = C_n(\theta_0, \theta) - C_n(\theta_0, \theta_0) \geq \rho(d(\theta_0, \theta))\,, \tag{5}$$

where $d(\theta_0, \theta)$ denotes a seminorm in Θ and $\rho(.)$ a strong monotone function with $\rho(0) = 0$. In the least squares case we have $\Delta C_n(\theta) =$

$\left\|\eta - \eta^0\right\|^2$. The consistency of the minimum contrast estimators can be shown with help of the following inequality

$$\forall\epsilon > 0 \;\; P\left(d\left(\widetilde{\theta}, \theta_0\right) > \epsilon\right) \leq P\left(\sup_{\theta\in\Theta(\epsilon)} \frac{\left|\Delta C_n(\theta) - \Delta\widetilde{C}_n(\theta)\right|}{\rho\left(d\left(\theta, \theta_0\right)\right)} \geq 1\right) \quad (6)$$

where $\Theta(\epsilon) = \Theta \cap \{\theta : d(\theta_0, \theta) > \epsilon\}$ and $\Delta\widetilde{C}_n(\theta) = \widetilde{C}_n(\theta) - \widetilde{C}_n(\theta_0)$. The main problem of this approach is that the dimension of the parameter θ grows in the same order as the sample size. Under additional assumptions on the nuisance parameter set, which imply the consistent estimation of the nuisance parameters in [11] and [13] the consistency of the L_1- and L_2-norm estimators is shown.

But we are interested in an estimation of the parameter of interest only. We set $d(\theta_0, \theta) = \|\beta - \beta_0\|$. Thus we search for estimation criteria, where the nuisance parameter is eliminated and which fulfill (5). The naive approach is to use Z_i instead of η_i in (3) and to consider the estimator

$$\widetilde{\beta}_{naive} \in \arg\min_{\beta\in\mathcal{B}} \frac{1}{n}\sum_{i=1}^{n} \widetilde{c}(Z_i, Z_i, \beta).$$

But this estimator is inconsistent. The total least squares approach is

$$\widetilde{\beta}_{TLS} \in \arg\min_{\beta\in\mathcal{B}} \frac{1}{n}\sum_{i=1}^{n} \min_{\eta_i : G(\eta_i,\beta)=0} \|Z_i - \eta_i\|^2. \quad (7)$$

That works fine for linear models. But for nonlinear models (5) is not fulfilled. Our proposal is now to take instead of $\widetilde{C}(Z, \theta)$ another function $Q(Z, \beta)$, which is independent of the nuisance parameter η - but which has the same expectation as $\widetilde{C}(Z, \theta)$:

$$E_{\theta_0} Q(Z, \beta) = C_n(\theta_0, \theta). \quad (8)$$

Under $Q(Z, \beta) = \frac{1}{n}\sum_{i=1}^{n} q(Z_i, \beta)$, $C_n(\theta_0, \theta) = \frac{1}{n}\sum_{i=1}^{n} c(\eta_i, \beta)$ it is sufficient to solve (8) for each summand separately. Rewrite this as deconvolution problem: we want $q(Z_i, \beta)$ with

$$\forall\beta \in \mathcal{B} \quad \forall\eta_i \in \mathcal{D} \quad \int q(\eta_i + \varepsilon, \beta)\, p(\varepsilon)\, d\varepsilon = c(\eta_i, \beta), \quad (9)$$

where $p(\varepsilon)$ is the density of the i.i.d. errors in (1).

The new *corrected minimum contrast estimator* for the parameter of interest $\widetilde{\beta}$ is then defined as a measurable solution of

$$\widetilde{\beta} \in \arg\min_{\beta\in\mathcal{B}} \frac{1}{n}\sum_{i=1}^{n} q(Z_i, \beta). \quad (10)$$

Because the explicit solution of (9) is not given in every case, we require the existence of an approximate solution only. We call $q_\mu(Z_i, \beta)$ an approximate solution of (9) if for all μ greater then zero

$$\sup_{\beta \in \mathcal{B}} \sup_{\eta_i \in \mathcal{D}} \left| \int q_\mu(\eta_i + \varepsilon, \beta) \, p(\varepsilon) \, d\varepsilon - c(\eta_i, \beta) \right| \leq \mu. \tag{11}$$

Then the related *approximate corrected minimum contrast estimator* for the parameter of interest $\widetilde{\beta}_\mu$ is defined as a measurable solution of

$$\widetilde{\beta}_\mu \in \arg\min_{\beta \in \mathcal{B}} \frac{1}{n} \sum_{i=1}^{n} q_\mu(Z_i, \beta). \tag{12}$$

3. Special cases

We present examples, where (9) is solvable or where we can find a solution of (11). In both cases we have to consider integral equations. The solutions have to be given explicitly for each parameter in order to determine the minimal point. The key point is to find combinations of contrast functions $c(\eta_i, \beta)$ fulfilling (i) and model functions G and error densities $p(\varepsilon)$, which have explicit solutions at least of (11). For details see [7], and [8].

3.1 Corrected least squares estimator

Let us consider the explicit model in (1) with $\Gamma = \sigma^2 I$, $q = 2$ and

$$G(\eta_i, \beta) = \eta_{1i} - g(\eta_{2i}, \beta). \tag{13}$$

We set for the contrast $C_n(\theta_0, \theta)$ the sum of squares

$$\frac{1}{n} \sum_{i=1}^{n} \left(g(\eta_{2i}^0, \beta) - g(\eta_{2i}^0, \beta_0) \right)^2. \tag{14}$$

Then the contrast condition (i) is of Jennrich type: for some $c_0 > 0$

$$\frac{1}{n} \sum_{i=1}^{n} \left(g(\eta_{2i}^0, \beta) - g(\eta_{2i}^0, \beta_0) \right)^2 \geq c_0 \left\| \beta - \beta_0 \right\|^2. \tag{15}$$

The corrected least squares estimator $\widetilde{\beta}_{LSE}$ is measurable solution of

$$\widetilde{\beta}_{LSE} \in \arg\min_{\beta \in \mathcal{B}} \frac{1}{n} \sum_{i=1}^{n} \left((Z_{1i} - f(Z_{2i}, \beta))^2 + h(Z_{2i}, \beta) - f(Z_{2i}, \beta)^2 \right),$$

where f and h satisfy for all $\beta \in \mathcal{B}$ and all η_{2i}

$$E_{\eta_{2i}} f(Z_{2i}, \beta) = g(\eta_{2i}, \beta), \qquad E_{\eta_{2i}} h(Z_{2i}, \beta) = g^2(\eta_{2i}, \beta). \tag{16}$$

The estimator $\widetilde{\beta}_{LSE}$ is investigated in [7] and in [5], see also [13]. In a polynomial explicit model $\widetilde{\beta}_{LSE}$ coincides with the adjusted least squares estimator proposed by Cheng and Schneeweiss [3], this was mentioned in [1]. Astrometric applications of that estimator are considered in [12, 4].

3.2 An approximative corrected L_2-estimator

Consider the model (13) with $g(\eta_{2i}, \beta) = \beta\,|\eta_{2i}|\,, \beta \in [-d, d]$. Suppose that ε_{2i} is Laplace distributed, i.e. its density equals $\frac{1}{2}\exp(-|t|)$. We will apply the contrast (14). The first equation in (16), $Ef(Z_{2i}) = \beta\,|\eta_{2i}|$, has the formal solution involving the generalized δ–function: $f(z) = \beta(|z| - 2\delta(z))$. We approximate the δ–function by $\delta_\mu(z) = \mu^{-1}\varpi(\mu^{-1}z)$, where $\varpi(.)$ is a smooth probability density over $[-1, 1]$. Then the approximate corrected least squares estimator is defined by

$$\widetilde{\beta}_\mu \in \arg\min_{\beta \in \mathcal{B}} \frac{1}{n} \sum_{i=1}^{n} \left(Z_{1i}^2 - 2Z_{1i}\beta(|Z_{2i}| - 2\delta_\mu(Z_{2i})) + \beta^2(Z_{2i}^2 - 2)\right).$$

3.3 Regression functions of quotient form

Consider the explicit model (13) with a regression function of quotient form $g(\eta_{2i}, \beta) = g_1(\eta_{2i}, \beta)/g_2(\eta_{2i}, \beta)$ and transform it to an implicit model with $G(\eta_i, \beta) = g_2(\eta_{2i}, \beta)\eta_{1i} - g_1(\eta_{2i}, \beta)$. We set for the contrast $C_n(\theta_0, \theta)$:

$$\frac{1}{n}\sum_{i=1}^{n} G(\eta_i, \beta)^2.$$

The contrast condition (i) is fulfilled if $g_2(\eta_{2i}, \beta) > 0$ and if $g_1(\eta_{2i}, \beta)$ satisfies the Jennrich type condition (15). Models of quotient form occur in biometry for growth curves and also in chemistry. The following model is used for the determination of $\beta = (r_1, r_2)$, where r_1 and r_2 are the reaction rates in the copolymerization process

$$g(\eta_{2i}, \beta) = \frac{\eta_{2i}(r_1\eta_{2i} + 1)}{r_2 + \eta_{2i}}.$$

Assume that the moments $\mathbf{E}\varepsilon_{ji}^k = m_k$ of the error distribution in (1) are known. Then a solution $q(Z_i, \beta)$ of (9) is given by $Z_i = (y_i, x_i)$ and

$$r_1^2 f_4(x_i) - 2r_1r_2 f_2(x_i)\, y_i + r_2^2 f_2(y_i) + 2r_1 f_3(x_i)(1 - y_i) + 2r_2 x_i (f_2(y_i) - 1),$$

where $\mathbf{E}f_k(m+\varepsilon) = m^k$, especially $f_2(z) = z^2 - \sigma^2$, $f_3(z) = z^3 - 3z\sigma^2 - m_3$, $f_4(z) = z^4 - 6z^2\sigma^2 - 4zm_3 + 6\sigma^4 - m_4$. For details see [6], [7], and [14].

3.4 Corrected naive L_1-estimators

Consider the model (13), where ε_{2i} has canonical Laplace distribution and $g(.,\beta)$ has continuous first and second derivatives $g^{\xi}, g^{\xi\xi}$ and $\eta_{2i} \in [-a,a]$, for all i. The estimator is based on the L_1- estimation in ordinary nonlinear regression. We will apply the contrast $C_n(\theta_0,\theta)$:

$$\frac{1}{n}\sum_{i=1}^{n} \mathbf{E}_{\theta_0} \left| Z_{1i} - g(\eta_{2i}^0, \beta) \right|, \tag{17}$$

which satisfies the contrast condition (i), for details see [11]. Sufficient for the approximative corrected criteria function q_μ is the condition:

$$\sup_{\substack{Z_{1i}\in\mathbb{R} \\ \beta\in\mathcal{B} \\ \eta_{2i}\in[-a,a]}} \left| \int q_\mu \left(Z_{1i}, \eta_{2i} + t, \beta\right) \frac{1}{2} \exp(-|t|)dt - \left|Z_{1i} - g(\eta_{2i},\beta)\right| \right| \leq \mu. \tag{18}$$

The formal "exact" solution of (18) is

$$q_0\left(Z_i,\beta\right) = |R_i| - g^{\xi\xi}\ \ sign\left(R_i\right) - 2\left(g^{\xi}\right)^2 \delta\left(R_i\right),$$

where the derivatives are taken at (Z_{2i},β) and where $R_i = Z_{1i} - g(Z_{2i},\beta)$ is the naive residuum. We introduce approximations for the $\delta-$function, for the $sign(.)$ and for the $|.|$ function as well:

$$\delta_\mu(z) = \mu^{-1}\varpi(\mu^{-1}z),\ \ \Delta_\mu(s) = 2\int_0^s \delta_\mu(z)\,dz,\ \ A_\mu(t) = \int_0^t \Delta_\mu(s)\,ds,$$

where $\varpi(z)$ is a smooth even probability density over $[-1,1]$, fulfilling some regularity conditions. The approximate function satisfying (18) is

$$q_\mu\left(Z_i,\beta\right) = A_\mu(R_i) - g^{\xi\xi}\Delta_\mu\left(R_i\right) - 2\left(g^{\xi}\right)^2 \delta_\mu\left(R_i\right).$$

3.5 Corrected symmetric L_1 approach

Consider the model as above but we require additionally, that both errors are Laplace distributed. As contrast $C_n(\theta_0,\theta)$ function we choose now the empirical L_1 distance between regression functions

$$\frac{1}{n}\sum_{i=1}^{n} \left| g(\eta_{2i}^0,\beta) - g(\eta_{2i}^0,\beta_0) \right|.$$

The respected deconvolution problem is to find q_μ with

$$\sup_{\beta\in\mathcal{B}}\sup_{\eta_{2i}^0\in[-a,a]}\left|\mathbf{E}_{\theta_0}q_\mu\left(Z_i,\beta\right)-\left|g(\eta_{2i}^0,\beta)-g(\eta_{2i}^0,\beta_0)\right|\right|\leq\mu. \tag{19}$$

The formal "exact" solution of (19) can also be given by generalized functions. Using the approximations above we get

$$\begin{aligned} q_\mu\left(Z_i,\beta\right) &= A_\mu(R_i)+g^{\xi\xi}\Delta_\mu\left(R_i\right)-2\left(\left(g^\xi\right)^2+1\right)\delta_\mu\left(R_i\right) \\ &\quad -2g^{\xi\xi}\delta_\mu'\left(R_i\right)+2\left(g^\xi\right)^2\delta_\mu''\left(R_i\right). \end{aligned}$$

4. Consistency results

Our main theoretical results are the strong consistency of the estimators defined in (10) and (12).

Introduce two moment assumptions on $\overline{q}_i(\beta)=q\left(Z_i,\beta\right)-\mathbf{E}_{\theta_0}q\left(Z_i,\beta\right)$. They are related to the Whittle inequality, see [10].

(ii) There exist constants $c>0$, $k\geq1$, such that for all $\beta\in\mathcal{B}$

$$\frac{1}{n}\sum_{i=1}^{n}\left(\mathbf{E}_{\theta_0}\overline{q}_i\left(\beta\right)^{2k}\right)^{\frac{1}{k}}\leq c^{\frac{1}{k}}.$$

(iii) There exist a random variable $M_{(n)}$, a constant C and a real number $k\geq1$, such that for all n and for all $\beta,\beta'\in\mathcal{B}:\mathbf{E}_{\theta_0}{M_{(n)}}^k\leq C$ and

$$\frac{1}{n}\sum_{i=1}^{n}\left(\overline{q}_i(\beta)-\overline{q}_i(\beta')\right)^2\leq M_{(n)}\left\|\beta-\beta'\right\|^2.$$

Theorem 1. *Suppose Condition (i), (ii), (iii) with fixed* $k\geq1$.

1 Then for all $\tau>0$ *and for all* $n\geq1$

$$P_{\theta_0}\left(\left\|\widetilde{\beta}-\beta_0\right\|>\tau\right)\leq\text{const}\;\rho\left(\tau\right)^{-2k}n^{-\frac{2k^2}{2k+p}}$$

2 If $k>\frac{1+\sqrt{1+2p}}{2}$ *then* $\widetilde{\beta}\to\beta_0$ *a.s.*

Introduce the related versions of (ii) and (iii) for the normalized approximative corrected estimation function $\overline{q}_{i,\mu}(\beta)=q_\mu\left(Z_i,\beta\right)-\mathbf{E}_{\theta_0}q_\mu\left(Z_i,\beta\right)$.

(iv) There exist constants $c>0, 1<k<\infty$, $\gamma_1=\gamma_1\left(k\right)>0$, $\mu_0>0$ such that for all $\mu<\mu_0$ and for all β

$$\frac{1}{n}\sum_{i=1}^{n}\left(\mathbf{E}_{\theta_0}\left|\overline{q}_{i,\mu}(\beta)\right|^{2k}\right)^{\frac{1}{k}}\leq c^{\frac{1}{k}}\mu^{-\frac{\gamma_1(k)}{k}}.$$

(v) There exist a random variable $M_{(n)}$, a constant C and a real number $k \geq 1$ and $\gamma_2 = \gamma_2(k) > 0$, $\mu_0 > 0$, such that for all n , $\mu < \mu_0$ and for all $\beta, \beta' \in \mathcal{B}$: $\mathbf{E}_{\theta_0} M_{(n)}{}^k \leq C \mu^{-\gamma_2(k)}$ and

$$\frac{1}{n} \sum_{i=1}^{n} \left(\overline{q}_{i,\mu}(\beta) - \overline{q}_{i,\mu}(\beta')\right)^2 \leq M_{(n)} \left\| \beta - \beta' \right\|^2 .$$

Theorem 2. *Suppose (i), (iv), (v) with fixed $k \geq 1$ and $\gamma_2(k) \geq \gamma_1(k)$. Set in (12) for $a > 0$*

$$\mu = \mu(n) = a\, n^{-r}, \quad r = \left(\frac{2k^2}{2k+p} - \varkappa \right) \frac{2k}{2k\, \gamma_1(k) \;+\; p\, \gamma_2(k)} .$$

1 If $\varkappa > 0$ then for all $\tau > 0$ and for all $n \geq \left(\frac{2a}{\rho(\tau)}\right)^{\frac{1}{r}}$

$$P_{\theta_0} \left(\left\| \widetilde{\beta}_{\mu(n)} - \beta_0 \right\| > \tau \right) \leq \text{const} \; \left(\rho(\tau) - a n^{-r}\right)^{-2k} n^{-\varkappa}.$$

2 If $\varkappa > 1$ and $k > \frac{1+\sqrt{1+2p}}{2}$ then $\widetilde{\beta}_{\mu(n)} \to \beta_0$ a.s.

5. Proofs

The following lemma gives an inequality analoguos to (6) for the approximative criterion function.

Denote $S_n(\beta) = \frac{1}{n} \sum_{i=1}^{n} \overline{q}_{i,\mu}(\beta)$ and $\mathcal{B}(\tau) = \{\beta \in \mathcal{B} : \|\beta - \beta_0\| > \tau\}$.

Lemma 1. *Let suppose (i). Then for all $\tau > 0$*

$$P_{\theta_0} \left(\left\| \widetilde{\beta}_{\mu} - \beta_0 \right\| > \tau \right) \leq P_{\theta_0} \left(\sup_{\beta \in \mathcal{B}(\tau)} |S_n(\beta)| > \rho(\tau) - 2\mu \right).$$

Proof. Let $Q_\mu(\beta) = \frac{1}{n} \sum_{i=1}^{n} q_\mu(Z_i, \beta)$ and $\Delta Q_\mu(\beta) = Q_\mu(\beta) - Q_\mu(\beta_0)$ and $\Delta C_n(\beta) = C_n(\eta_0, \beta) - C_n(\eta_0, \beta_0)$.

Condition (i) implies $\inf_{\beta \in \mathcal{B}(\tau)} \Delta C_n(\beta) \geq \rho(\tau)$.

From (11) we have $|\Delta C_n(\beta) - \mathbf{E}_{\theta_0} \Delta Q_\mu(\beta)| \leq 2\mu$. Thus

$$\inf_{\beta \in \mathcal{B}(\tau)} \mathbf{E}_{\theta_0} \Delta Q_\mu(\beta) \geq \rho(\tau) - 2\mu. \tag{20}$$

From the definition of $\widetilde{\beta}_\mu$ we have $\Delta Q_\mu\left(\widetilde{\beta}_\mu\right) \leq 0$. Under $\widetilde{\beta}_\mu \in \mathcal{B}(\tau)$, (20) implies that

$$\begin{aligned} \rho(\tau) - 2\mu &\leq \inf_{\beta \in \mathcal{B}(\tau)} \mathbf{E}_{\theta_0} \Delta Q_\mu(\beta) - \inf_{\beta \in \mathcal{B}(\tau)} \Delta Q_\mu(\beta) \\ &\leq \sup_{\beta \in \mathcal{B}(\tau)} \left| \Delta Q_\mu(\beta) - \mathbf{E}_{\theta_0} \Delta Q_\mu(\beta) \right|, \end{aligned}$$

which proves the lemma, for $S_n(\beta) = \Delta Q_\mu(\beta) - \mathbf{E}_{\theta_0}\Delta Q_\mu(\beta)$. □

The next lemma delivers the uniform convergence of the approximative criterion function.

Lemma 2. *Assume (iv) and (v) with $\gamma_2(k) \geq \gamma_1(k)$. Then there exist a constant* const*, such that for all $\epsilon > 0, n \geq 1$*

$$P_{\theta_0}\left(\sup_{\beta\in\mathcal{B}} |S_n(\beta)| > \epsilon\right) \leq \text{const}\ \epsilon^{-2k}\ \mu^{-\frac{\gamma_1(k)\,2k\,+\,\gamma_2(k)\,p}{2k+p}}\ n^{-\frac{2k^2}{2k\,+p}}.$$

Proof. Imbed $\mathcal{B}$ into a ball $K(\beta_0, R) = \{\beta : \|\beta - \beta_0\| \leq R\}$. For $0 < d < C_1$ the compact set $\mathcal{B}$ can be covered by a finite number of balls $K(\beta^l, d)$, $l = 1, ..., N(d)$, with $N(d) \leq C_2\ d^{-p}$, where C_1 and C_2 are constants. Then

$$\sup_{\beta\in\mathcal{B}} |S_n(\beta)| \leq \max_{l=1,..,N(d)} \left|S_n\left(\beta^l\right)\right| + \sup_{\|\beta''-\beta'\|\leq d} \left|S_n(\beta'') - S_n(\beta')\right| = S_1 + S_2.$$

$$\begin{aligned} P = P_{\theta_0}&\left(\sup_{\beta\in\mathcal{B}} |S_n(\beta)| \geq \epsilon\right) \\ &\leq \sum_{l=1}^{N(d)} P_{\theta_0}\left(\left|S_n\left(\beta^l\right)\right| \geq \frac{\epsilon}{2}\right) + P_{\theta_0}\left(S_2 \geq \frac{\epsilon}{2}\right) = P_1 + P_2. \end{aligned} \tag{21}$$

We apply the Whittle inequality for a real number $s, s \geq 2$, then

$$P_1 \leq \text{const}\, d^{-p} n^{-k} \mu^{-\gamma_1(k)} \epsilon^{-2k}.$$

From condition (v) we have $S_2^2 \leq M_{(n)} d^2$ and applying the Chebychev inequality we obtain

$$P_2 \leq \epsilon^{-2k} 4^k d^{2k} \mathbf{E} M_{(n)}^k \leq \text{const}\ \epsilon^{-2k} \mu^{-\gamma_2(k)} d^{2k}.$$

Summarizing we obtain $P \leq \text{const}\,\epsilon^{-2k}\left[\mu^{-\gamma_1(k)} d^{-p}\, n^{-k} + \mu^{-\gamma_2(k)} d^{2k}\right]$. Choosing d in an optimal way by $d_{opt}^{2k+p} = \text{const}_1\ n^{-k} \mu^{(\gamma_2(k)-\gamma_1(k))} \leq \text{const}_2$, we obtain the statement. □

Proof of Theorem 2:

For $\mu = an^{-r}$, $n > \left(\frac{2a}{\rho(\tau)}\right)^{\frac{1}{r}}$ we have $\rho(\tau) - 2\mu > 0$. Lemma 1 implies

$$P_{\theta_0}\left(\left\|\widetilde{\beta}_\mu - \beta_0\right\| > \tau\right) \leq P_{\theta_0}\left(\sup_{\beta\in\mathcal{B}} |S_n(\beta)| \geq \frac{1}{2}(\rho(\tau) - 2\mu)\right).$$

We apply Lemma 2 with $2\epsilon = \rho(\tau) - 2\mu$ and get the first statement of Theorem 2. The second statement follows by the Borel Cantelli Lemma. □

Proof of Theorem 1: The proof goes along the line above. We can interpret condition (ii) as condition (iv) with $\gamma_1(k) = 0$ and condition (iii) as condition (v) with $\gamma_1(k) = 0$. Therefore the following analogue to Lemma 2 holds

$$P_{\theta_0}\Big(\sup_{\beta\in\mathcal{B}} |S_n(\beta)| > \epsilon\Big) \leq \text{const}\, \epsilon^{-2k}\, n^{-\frac{2k^2}{2k+p}}.$$

By Lemma 1 with $\mu = 0$ we get Theorem 1. □

References

[1] S. Baran. A consistent estimator in general functional errors-in-variables model. *Metrika* 51(2):117–132, 2000.

[2] R. J. Carroll, D. Ruppert and L. A. Stefanski. *Measurement Errors in Nonlinear Models.* Chapman and Hall, London, 1995.

[3] C.-L. Cheng and H. Schneeweiss. The polynomial regression with errors in the variables. *J. R. Statist. Soc. B*, 60(1):189–199, 1998.

[4] Ch. De Vegt and S. Zwanzig. Comparison of reference frame linking methods, Preprint-No. 2001-03, April 2001, University of Hamburg, Schwerpunkt Mathematische Statistik und Stochastische Prozesse, 23, 2001.

[5] I. Fazekas and A. Kukush. Asymptotic properties of an estimator in nonlinear functional errors-in-variables models. *Computers Math. Appl.* 34(10):23–39, 1997.

[6] S. E. Keeler and P. M. Reilly. The design of experiments when there are errors in all the variables. *Can. J. Chem. Eng.* 70:774–778, 1992.

[7] A. Kukush, and S. Zwanzig. On a corrected contrast estimator in the implicit nonlinear functional relation model. Preprint No. 97–11, Institute of Mathematical Stochastics. University of Hamburg, 1997.

[8] A. Kukush, and S. Zwanzig. On adaptive minimum contrast estimators in the implicit nonlinear functional regression models. *Ukr. Math. Journal.* 53(9):1204–1209, 2001.

[9] J. Neyman and E. L. Scott. Consistent estimates based on partially consistent observations. *Econometrica* 16(1):1–32, 1948.

[10] P. Whittle. Bounds for the moments of linear and quadratic forms in independent variables. *Theory Prob. and Appl.* 5:302–305, 1960.

[11] S. Zwanzig. On L1-norm estimators in nonlinear regression and in nonlinear error-in-variables models. In *IMS Lecture Notes-Monograph Series*, 31:101–118, 1997.

[12] S. Zwanzig. Application of Hipparcos dates: A new statistical method for star reduction. Journees 1997, Academy of Sciences of the Czech Republic, Astronomical Institute, 142–144, 1997.

[13] S. Zwanzig. Estimation in nonlinear functional error-in-variables models. Habilitationsschrift. University of Hamburg. Department of Mathematics, 1998.

[14] S. Zwanzig. Experimental design for nonlinear error-in-variables models, in *Advances in Stochastic Simulation Methods* Ed. Balakrishnan, Melas, Ermakov. 153–163, 2000.

ON CONSISTENT ESTIMATORS IN LINEAR AND BILINEAR MULTIVARIATE ERRORS-IN-VARIABLES MODELS

Alexander Kukush, Ivan Markovsky and Sabine Van Huffel

ESAT-SISTA, K.U.Leuven

Kasteelpark 10, B-3001 Leuven-Heverlee, Belgium

{Alexander.Kukush, Ivan.Markovsky, Sabine.VanHuffel}@esat.kuleuven.ac.be

Abstract We consider three multivariate regression models related to the TLS problem. The errors are allowed to have unequal variances.

For the model $AX = B$, the elementwise-weighted TLS estimator is considered. The matrix $[A\ B]$ is observed with errors and has independent rows, but the errors in a row are correlated. In addition, the corresponding error covariance matrices may differ from row to row and some of the columns are allowed to be error-free. We give mild conditions for weak consistency of the estimator, when the number of rows in A increases. We derive the objective function for the estimator and propose an iterative procedure to compute the solution.

In a bilinear model $AXB = C$, where the data A, B, C are perturbed by errors, an adjusted least squares estimator is considered, which is consistent, i.e. converges to X, as the number m of rows in A and the number q of columns in B increase.

A similar approach is applied in a related model, arising in motion analysis. The model is $v^T F u = 0$, where the vectors u and v are homogeneous coordinates of the projections of the same rigid object point in two images, and F is a rank deficient matrix. Each pair (u, v) is observed with measurement errors. We construct a consistent estimator of F in three steps: a) estimate the measurement error variance, b) construct a preliminary matrix estimate, and c) project that estimate on the subspace of singular matrices.

A simulation study illustrates the theoretical results.

Keywords: linear and bilinear errors-in-variables models, elementwise-weighted total least squares, adjusted least squares, essential/fundamental matrix.

S. Van Huffel and P. Lemmerling (eds.), Total Least Squares and Errors-in-Variables Modeling, 155–164.

1. Introduction

We study three multivariate functional relation models. The errors are allowed to have unequal variances, which makes the asymptotic analysis of the estimators more complicated. Our goal is to derive consistent estimators, which admit a numerically reliable computation. *Consistent estimator* is an estimator that converges to the true value of the regression parameters, as the sample size tends to infinity.

2. Model $AX = B$

Let $A \in \mathbb{R}^{m \times n}$ and $B \in \mathbb{R}^{m \times d}$ be *observations*, and $X \in \mathbb{R}^{n \times d}$ be a *parameter* of interest. We suppose that

$$A = A_0 + \tilde{A}, \quad B = B_0 + \tilde{B}, \tag{1}$$

and that there exists $X_0 \in \mathbb{R}^{n \times d}$ such that

$$A_0 X_0 = B_0. \tag{2}$$

Here A_0 and B_0 are non-stochastic true values, and $\tilde{A}$ and $\tilde{B}$ are *errors*, and the matrix X_0 yields the true values of the parameters. The measurements are represented by the rows of A and B. We fix the dimension of X_0, and let the number of measurements m increase. For $i \geq 1$ let

$$\delta_{ij} := \tilde{a}_{ij}, \text{ if } j = 1, \ldots, n, \quad \text{and} \quad \delta_{i,n+k} := \tilde{b}_{ik}, \text{ if } k = 1, \ldots, d.$$

If all errors δ_{ij} are non-correlated and equally sized, the TLS technique is appropriate to solve an overdetermined set $AX \approx B$, see, e.g. [18]. In [17] the generalized TLS formulation is considered, where some of the columns of A are assumed to be known exactly, and the errors in the remaining data are allowed to be correlated and not equally sized, but they should be row-wise i.i.d. for consistency. In the present paper we confirm the claim from [17, p.313] about the consistency of the estimator in that model and further relax the conditions.

2.1 The elementwise-weighted TLS problem

Our main assumptions are as follows.

1 The rows of $(\delta_{ij},\ i \geq 1,\ j = 1, \ldots, n + d)$ are independent with zero mean and finite second moments.

Let $\sigma_{ij}^2 := \mathbf{E}\delta_{ij}^2$. We allow some of σ_{ij}^2 to be equal to zero.

2 For a fixed $J \subset \{1, \ldots, n + d\}$, every $j \notin J$ and $i \geq 1$ satisfy $\sigma_{ij}^2 = 0$. For $i \geq 1$ denote $\Sigma_i := \text{cov}(\delta_{ij}, j \in J)$.

3 There exists $\kappa > 0$ such that for every $i \geq 1$, $\lambda_{\min}(\Sigma_i) \geq \kappa^2$.

For instance we have ordinary regression if $J = \{n+1, \ldots, n+d\}$, i.e. if only B is observed with errors and A is error-free. We allow the Data Least Squares Problem, see [5], where $J \subset \{1, \ldots, n\}$, i.e. A is observed with errors and B is error-free.

Introduce two matrices by

$$X_{\text{ext}}^0 := \begin{bmatrix} X_0 \\ -I_d \end{bmatrix} = (x_{jk}^0,\ j = 1, \ldots, n+d,\ k = 1, \ldots, d)$$

with identity matrix $I_d \in \mathbb{R}^{d \times d}$, and $X_J^0 := (x_{jk}^0,\ j \in J,\ k = 1, \ldots, d)$.

We use similar notation X_{ext} and X_J for every $X \in \mathbb{R}^{n \times d}$.

4 $\operatorname{rank}\left(X_J^0\right) = d$.

Now, let $[\Delta A\ \Delta B] = (\Delta a_{ij},\ i = 1, \ldots, m,\ j = 1, \ldots, n+d)$, and column vectors $\Delta a_i^J := (\Delta a_{ij},\ j \in J) \in \mathbb{R}^{|J| \times 1}$.

The elementwise-weighted total least squares (EW-TLS) problem consists in finding the value $\hat{X}$ of unknown matrix X and the values $\Delta\hat{A}, \Delta\hat{B}$ of unknown error matrices $\tilde{A}, \tilde{B}$ minimizing the weighted sum of squared corrections

$$\min_{X, \Delta A, \Delta B} \sum_{i=1}^{m} ||\Sigma_i^{-\frac{1}{2}} \Delta a_i^J||^2 \tag{3}$$

subject to the constraints

$$\begin{aligned} &\operatorname{rank}(X_J) = d, \quad (A - \Delta A)X = B - \Delta B, \\ &\text{and} \quad \Delta a_{ij} = 0, \text{ for } 1 = i, \ldots, m,\ j \notin J. \end{aligned} \tag{4}$$

The EW-TLS estimator $\hat{X}$ of X_0 in the model (1),(2) is a measurable value of X, which solves the optimization problem (3),(4).

Now, we derive an objective function for $\hat{X}$. Denote

$$(AX - B)^T := [r_1(X), \ldots, r_m(X)],$$

and

$$S_i := \operatorname{cov}(\delta_{ij},\ j = 1, \ldots, n+d), \quad i = 1, \ldots, m.$$

Theorem 1. *Let the conditions 1 to 4 hold. The EW-TLS estimator $\hat{X}$ of X_0 in the model (1),(2) exists iff there exists an unconstrained minimum of the function*

$$Q(X) := \sum_{i=1}^{m} r_i(X)^T \left(X_J^T \Sigma_i X_J\right)^{-1} r_i(X), \quad \operatorname{rank}(X_J) = d. \tag{5}$$

Then $\hat{X}$ is a minimum point of $Q(X)$, and the values $\Delta\hat{A}$, $\Delta\hat{B}$ in the corresponding problem (3),(4) are given by

$$[\Delta\hat{A}\ \Delta\hat{B}] = \begin{bmatrix} r_1^T(\hat{X})\left(\hat{X}_J^T\Sigma_i\hat{X}_J\right)^{-1}\hat{X}_{\text{ext}}^T S_1 \\ \vdots \\ r_m^T(\hat{X})\left(\hat{X}_J^T\Sigma_i\hat{X}_J\right)^{-1}\hat{X}_{\text{ext}}^T S_m \end{bmatrix}, \text{ where } \hat{X}_{\text{ext}} := \begin{bmatrix} \hat{X} \\ -I_d \end{bmatrix}.$$

In the particular case $J = \{1, \ldots, n+d\}$ and $\Sigma_i = I_{n+d}$, $i = 1, \ldots, m$, the EW-TLS estimator coincides with the TLS-estimator. Then the function (5) equals

$$Q(X) = \operatorname{tr}\left((AX - B)\left(X_{\text{ext}}^T X_{\text{ext}}\right)^{-1}(AX - B)^T\right), \tag{6}$$

and the TLS estimator is known to minimize the function (6), see [16]. Thus (5) can be considered as a generalization of the objective function (6) for the case of unequally sized errors.

2.2 Consistency of the EW-TLS estimator

The next conditions guarantee existence of the estimator and its consistency. We denote $q := d(|J| - (d+1)/2)$, and let $\underline{\lambda}$ and $\bar{\lambda}$ be the minimal and maximal eigenvalues of $A_0^T A_0$, respectively.

5 For fixed $r \geq 2$, $r > \frac{1}{2}\left(q + \max\{(n-1), q\}\right)$,

$$\sup_{i\geq 1,\ j\in J} \mathbf{E}|\delta_{ij}|^{2r} < \infty. \tag{7}$$

6 $\underline{\lambda}/\sqrt{m} \to \infty$, as $m \to \infty$.

7 $\underline{\lambda}^2/\bar{\lambda} \to \infty$, as $m \to \infty$.

The next two conditions are stronger versions of 6, 7.

8 For some $m \geq 1$, $\sum_{m=m_0}^{\infty} \left(\sqrt{m}/\underline{\lambda}\right)^r < \infty$.

9 For some $m \geq 1$, $\sum_{m=m_0}^{\infty} \left(\bar{\lambda}/\underline{\lambda}^2\right)^r < \infty$.

Theorem 2. *Suppose that the conditions 1 to 7 hold. In the model (1),(2) we have:*

a) the EW-TLS estimator $\hat{X}$ exists with probability tending to one, as $m \to \infty$,

b) $\hat{X} \xrightarrow{\text{P}} X_0$, as $m \to \infty$,

c) instead of 6,7, assume 8,9, where r comes from 5. Then $\hat{X} \to X_0$ a.s., as $m \to \infty$.

Theorem 2b) extends the weak consistency result of Gallo [7], who considered a univariate homoscedastic model, i.e. B is a vector and the rows in $[\tilde{A}\ \tilde{B}]$ are i.i.d. The eigenvalue asymptotics 6,7 are exactly his conditions, and he demanded 5, with $2r = 4$. In a homoscedastic case, we can weaken 5 and set $2r = 4$ in (7), thus obtaining a direct generalization of Gallo's result to the multivariate homoscedastic model.

In a homoscedastic multivariate model (1),(2) Gleser [8] required that $A_0^T A_0/m \to A_\infty > 0$, which implies that both $\underline{\lambda}$ and $\bar{\lambda}$ have the order m. He showed the strong consistency assuming that the fourth moments of the errors are limited. Theorem 2c) covers Gleser's result except for the condition that the moments of order $4+\delta$ be bounded (but we do not require that the errors are row-wise identically distributed).

In a homoscedastic model the asymptotic properties of $\hat{X}$ are derived from the explicit formula of the estimator. It is impossible to obtain such a formula in our case where all errors may be differently sized. Instead we show that the estimator is a solution of the unconstrained problem $\min_X Q(X)$, and analyze the asymptotic behavior of the objective function (5).

2.3 Computation of the EW-TLS estimator

We describe a numerical scheme to compute the estimator. Theorem 2a) proves that the minimum point $\hat{X}$ exists with probability tending to one, and that it is a root of the equation $Q'(X) := \mathcal{F}(X,Z) = 0$.

A direct computation gives

$$\mathcal{F}(X,Z) = \sum_{i=1}^{m} \left(a_i r_i^T (Z^T S_i Z)^{-1} - P S_i^{\frac{1}{2}} Z (Z^T S_i Z)^{-1} r_i r_i^T (Z^T S_i Z)^{-1} \right).$$

Here $r_i = r_i(X)$ and $Z = Z(X) := X_{\text{ext}}$ are functions of X, $P := [I_n\ 0_{n\times d}]$ and $A^T = [a_1 \cdots a_m]$. The proposed iterative algorithm for solving $\mathcal{F}(X,Z) = 0$ is as follows.

1) Compute an approximation $X_{(0)}$, using, e.g. weighted least squares.

2) Given an approximation $X_{(j)}$, we define $Z_{(j)} := \left[\begin{smallmatrix} X_{(j)} \\ -I_d \end{smallmatrix}\right]$ and $X_{(j+1)}$ is found from the optimization problem

$$\min_X ||\mathcal{F}(X, Z_{(j)})||_F^2,$$

which can be solved by Newton-type algorithms.

We mention the paper [15]. The authors consider a univariate problem (1),(2), where X and B are vectors, derive an objective function like (5) and propose an algorithm, which is related to the one presented above. They also solve the equation $\mathcal{F}(X, Z) = 0$, using a linear approximation with respect to X, while our approximation $\mathcal{F}(X, Z_{(j)})$ is quadratic with respect to X.

2.4 Simulation examples

In order to demonstrate the asymptotic properties of the EW-TLS estimator, we set up a simulation experiment. The true data $[A_0\ b_0]$ is a random matrix with independent elements, uniformly distributed in the interval $[0, 1]$. The true value X_0 of the parameters is known. The errors are selected in two different ways:

1) $\operatorname{cov}(\tilde{a}_i) = \sigma_i^2 I$ and $\operatorname{var}(\tilde{b}_i) = \sigma_i^2$, where σ_i, $i = 1, \ldots, m$ are uniformly distributed random variables in the interval $[0.02, 0.1]$, and

2) $\operatorname{cov}(\operatorname{vec}[\tilde{A}\ \tilde{B}]) = \sigma^2 I$, where $\sigma = 0.1$.

The plots on figures 1a and 1b show the relative error of estimation (averaged over 1000 noise realizations) as a function of the sample size m for the ordinary least squares (OLS) estimator, the weighted least-squares (WLS) estimator, the TLS estimator and the EW-TLS estimator. The OLS estimator uses a wrong assumption that A is noise free and the errors in b are equally distributed. The WLS uses the covariance information for $\tilde{b}$ but also assumes that A is noise free. The TLS estimator allows errors in both A and b but assumes that they are equally distributed. The EW-TLS estimator uses the full covariance information.

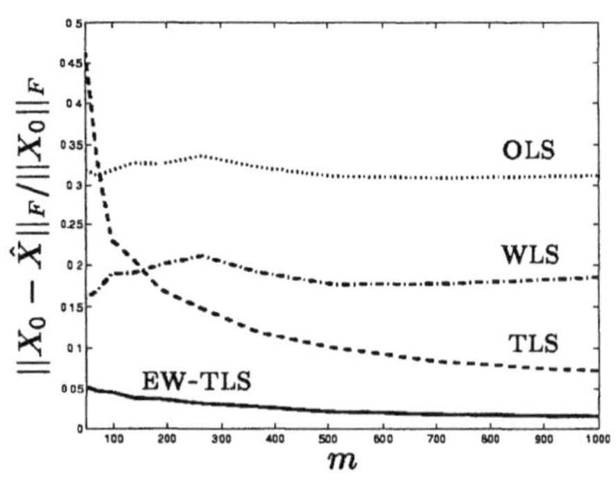

Figure 1a. Differently sized errors

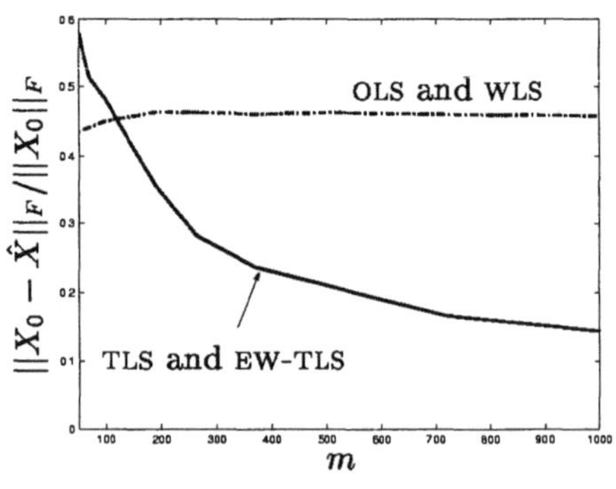

Figure 1b. Equally sized errors

3. Model $AXB = C$

Let $A \in \mathbb{R}^{m\times n}$, $B \in \mathbb{R}^{p\times q}$, $C \in \mathbb{R}^{m\times q}$ be *observations* and $X \in \mathbb{R}^{n\times p}$ be a *parameter* of interest. We suppose that

$$A = A_0 + \tilde{A}, \quad B = B_0 + \tilde{B}, \quad C = C_0 + \tilde{C}, \tag{8}$$

and that there exists $X_0 \in \mathbb{R}^{n\times p}$ such that

$$A_0 X_0 B_0 = C_0. \tag{9}$$

Here A_0, B_0, and C_0 are non-stochastic *true values* and $\tilde{A}$, $\tilde{B}$, and $\tilde{C}$ are *errors*. The matrix X_0 is the nominal or true value of the parameter. The measurements are represented by the rows of A, the columns of B and the elements of C. We fix the dimension of X_0, and let the numbers m and q increase.

We suppose that the errors $\{\tilde{a}_{ij},\ i \geq 1,\ 1 \leq j \leq n\}$, $\{\tilde{b}_{kl},\ 1 \leq k \leq p,\ l \geq 1\}$ and $\{\tilde{c}_{il},\ i \geq 1,\ l \geq 1\}$ form three independent arrays of random variables, which are centered and possess finite second order moments.

It is known that orthogonal regression is inconsistent for nonlinear measurement error models, [6, 13]. The orthogonal regression estimator in the bilinear model (8),(9) is a (weighted) TLS-estimator, see [18], therefore due to nonlinearity of the model, the TLS estimator is inconsistent in this case.

Let

$$V_{\tilde{A}} := \mathbf{E}\tilde{A}^T\tilde{A},\ V_{\tilde{B}} := \mathbf{E}\tilde{B}\tilde{B}^T.$$

The adjusted least squares (ALS) estimator is given by

$$\hat{X} := (A^T A - V_{\tilde{A}})^{\dagger}(A^T C B^T)(BB^T - V_{\tilde{B}})^{\dagger}. \tag{10}$$

It is similar to the ALS estimator proposed for a polynomial errors-in-variables model in [2]. The estimator (10) is consistent and asymptotically normal. Using the idea of [3] it is possible to construct a small sample modification of the estimator, which is more stable for small m and q and is asymptotically equivalent to the ALS estimator. Simulation results show that it reduces the error of estimation for small sample sizes [9].

The optimality of the estimator (10) is a topic of further research. For the model (1),(2) the ALS estimator is asymptotically efficient provided $V_{\tilde{A}}$ is known exactly and $\mathbf{E}\tilde{b}_{kl}^2$ are known up to a constant factor, see [11]. We conjecture that the estimator (10) is asymptotically efficient if $V_{\tilde{A}}$ and $V_{\tilde{B}}$ are known exactly and $\mathbf{E}\tilde{c}_{il}^2$ are known up to a constant factor.

4. Model $v^T F u = 0$

The next functional relation deals with the exploitation of the epipolar constraint information for the construction of the fundamental matrix for uncalibrated images, which once decomposed, solves the structure from motion problem, see [14, 4].

Let $u_i = [u_i(1)\ u_i(2)\ 1]^T \in \mathbb{R}^{3\times 1}$ and $v_i = [v_i(1)\ v_i(2)\ 1]^T \in \mathbb{R}^{3\times 1}$, $i = 1, \ldots, m$, represent the homogeneous pixel coordinates in the first and second image respectively. The model is

$$v_i^T F u_i = 0, \quad i = 1, \ldots, m,$$

where $F \in \mathbb{R}^{3\times 3}$ is the *fundamental matrix* with rank $(F) = 2$, and F is a parameter of interest. We suppose that the vectors u_i, v_i are observed with errors,

$$u_i = u_{0,i} + \tilde{u}_i, \quad \text{and} \quad v_i = v_{0,i} + \tilde{v}_i, \quad i = 1, \ldots, m,$$

and that there exists $F_0 \in \mathbb{R}^{3\times 3}$, with rank $(F_0) = 2$ and $||F_0||_F = 1$ such that

$$v_{0,i}^T F_0 u_{0,i} = 0, \quad i = 1, \ldots, m.$$

The matrix $F_0 \in \mathbb{R}^{3\times 3}$ is the true fundamental matrix F. The vectors $u_{0,i}$ and $v_{0,i}$ are the true values of the measurements u_i and v_i, respectively, and $\tilde{u}_i$, $\tilde{v}_i$ represent the measurement errors.

We derive a strongly consistent estimator for F_0 by taking more realistic assumptions than those in [14]. We suppose that the error terms $\{\tilde{u}_i,\ \tilde{v}_i,\ i \geq 1\}$ are independent and centered, and

$$\operatorname{cov}(\tilde{u}_i) = \operatorname{cov}(\tilde{v}_i) = \sigma_0^2 T, \quad i \geq 1$$

with $T := \operatorname{diag}(1, 1, 0)$ and fixed but unknown $\sigma_0^2 > 0$, $\sigma_0^2 \in (0, d^2]$.

The estimator of F_0 proposed below is the corrected minimum contrast estimator, considered in [12] in a more general context. It is related to the method of corrected score functions [1].

Let

$$S_m(\sigma^2) := \sum_{i=1}^{m} \left(u_i u_i^T - \sigma^2 T\right) \otimes \left(v_i v_i^T - \sigma^2 T\right),$$

$$Q_m(\sigma^2) := \left|\lambda_{\min}\left(S_m(\sigma^2)\right)\right|, \quad 0 \leq \sigma^2 \leq d^2.$$

A consistent estimator for σ_0^2 is given by

$$\hat{\sigma}^2 \in \arg \min_{0 \leq \sigma^2 \leq d^2} Q_m(\sigma^2).$$

Let $\hat{f}_1$ be a normalized eigenvector belonging to the minimal eigenvalue of $S_m(\hat{\sigma}^2)$. A preliminary estimator $\hat{F}_1$ is a matrix with $\mathrm{vec}(\hat{F}_1) = \hat{f}_1$. The final estimator $\hat{F}$ is obtained from $\hat{F}_1$ by projecting it to the set of rank deficient matrices, e.g. by computing the SVD of $\hat{F}_1$. Under appropriate conditions, $\mathrm{dist}(\hat{F}, \{\pm F_0\}) \to 0$ a.s., as $m \to \infty$. For details see [10].

We show the performance of the derived estimators $\hat{F}$ and $\hat{\sigma}^2$ by generating artificial data for which the true value F_0 is known. The fundamental matrix F_0 is a randomly chosen rank two matrix with unit Frobenius norm. The true coordinates $u_{0,i}$ and $v_{0,i}$ have third components equal to one, and the first two components are randomly chosen vectors in $\mathbb{R}^{2\times 1}$ with unit norm and random direction. The perturbations $\tilde{u}_i$ and $\tilde{v}_i$ are selected according to the assumptions stated in the paper. The estimation is repeated $N = 1000$ times with the same true data and different noise realizations. The presented results are the averaged ones.

We compare three estimators: the TLS estimator, the estimator $\hat{F}$ using the true noise variance σ_0^2, and the estimator $\hat{F}$ using the estimated noise variance $\hat{\sigma}^2$. Figure 2a shows the relative error of estimation as a function of the sample size m, and Figure 2b shows the convergence of the estimate $\hat{\sigma}^2$.

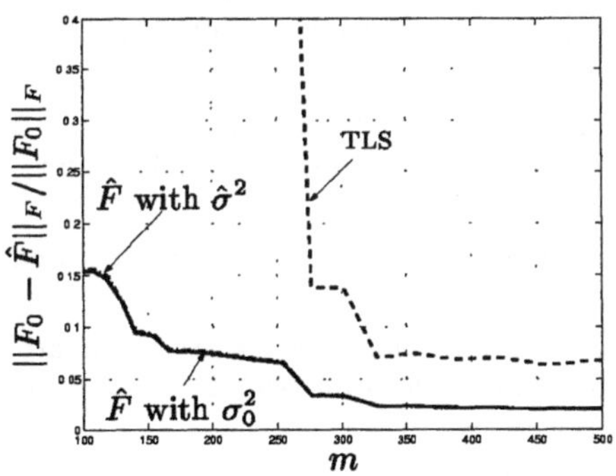

Figure 2a. Estimation of F_0

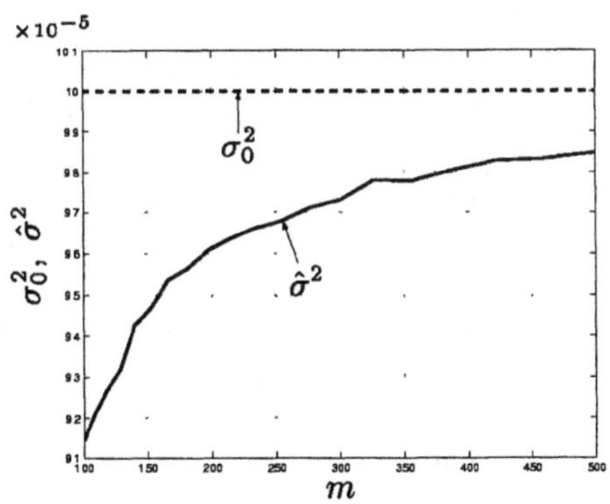

Figure 2b. Estimation of σ_0^2

Acknowledgments

A. Kukush is supported by a postdoctoral research fellowship of the Belgian office for Scientific, Technical and Cultural Affairs, promoting Scientific and Technical Collaboration with Central and Eastern Europe.

This work is supported by IUAP P4-02 and P4-24, the Concerted Research Action projects MEFISTO-666 and FWO project G.0200.00.

References

[1] R. J. Carroll, D. Ruppert, and L. A. Stefanski. *Measurement Error in Nonlinear Models.* Number 63 in Monographs on Statistics and Applied Probability. Chapman & Hall/CRC, 1995.

[2] C. Cheng and H. Schneeweiss. Polynomial regression with errors in the variables. *J. R. Statistical Society B*, (60):189–199, 1998.

[3] C. Cheng, H. Schneeweiss, and Markus Thamerus. A small sample estimator for a polynomial regression with errors in the variables. *J. R. Statistical Society*, (62):699–709, 2000.

[4] G. Cirrincione. TLS and constrained TLS neural networks for computer vision. In *this volume.*

[5] E. Dowling and R. DeGroat. The equivalence of the TLS and minimum norm methods. *IEEE Trans. on Signal Proc.*, 39:1891–1892, 1991.

[6] W. A. Fuller. *Measurement Error Models.* New York: Wiley, 1987.

[7] P. P. Gallo. Consistency of regression estimates when some variables are subject to error. *Comm. Stat. B-Theory Methods*, (11):973–893, 1982.

[8] L. J. Gleser. Estimation in a multivariate "errors in variables" regression model: Large sample results. *Ann. Stat.*, (9):24–44, 1981.

[9] A. Kukush, I. Markovsky, and S. Van Huffel. Consistent adjusted least squares estimator for errors-in-variables model $AXB = C$. Technical Report 01-72, K.U.Leuven, Dept. EE, K.U.Leuven, 2001. Submitted to Metrika.

[10] A. Kukush, I. Markovsky, and S. Van Huffel. Consistent fundamental matrix estimation in a quadratic measurement error model arising in motion analysis. Technical Report 01–64, ESAT-SISTA, Dept. EE, K.U.Leuven, 2001.

[11] A. Kukush and E.-O. Maschke. The efficiency of adjusted least squares in the linear functional relationship. Discussion Paper 208, SFB 386, Ludwig-Maximilians-University of Munich, 2000.

[12] A. Kukush and S. Zwanzig. On consistent estimators in nonlinear functional errors-in-variables models. In *this volume.*

[13] A. Kukush and S. Zwanzig. On inconsistency of the least squares estimator in nonlinear functional error-in-variables models. Preprint N96-12, Institut für Mathematische Stochastik, Universität Hamburg, 1996.

[14] M. Mühlich and R. Mester. The role of total least squares in motion analysis. In H. Burkhardt, editor, *Proc. European Conference on Computer Vision (ECCV'98)*, Springer Lecture Notes on Computer Science, pages 305–321. Springer-Verlag, June 1998.

[15] A. Premoli and M. L. Rastello. The parametric quadratic form method for solving total least squares problems with elementwise weighting. In *this volume.*

[16] P. Sprent. A generalized least-squares approach to linear functional relationships. *J. R. Statistical Society B*, (28):278–297, 1966.

[17] S. Van Huffel and J. Vandewalle. Analysis and properties of the generalized total least squares problem $AX \approx B$ when some or all columns in A are subject to error. *SIAM J. Matrix Anal.*, 10(3):294–315, 1989.

[18] S. Van Huffel and J. Vandewalle. *The Total Least Squares Problem: Computational Aspects and Analysis.* SIAM, Philadelphia, 1991.

IDENTIFICATION OF SEMI-LINEAR MODELS WITHIN AN ERRORS-IN-VARIABLES FRAMEWORK

Rik Pintelon and Johan Schoukens
Vrije Universiteit Brussel, department ELEC, Pleinlaan 2, 1050 Brussel, Belgium
{Rik.Pintelon, Johan.Schoukens}@vub.ac.be

Abstract Semi-linear models are models which are linear-in-the-observations and (non)linear in the model parameters. Assuming that all observations are noisy (errors-in-variables framework), the Cramér-Rao lower bound of the model parameters is calculated, and the stochastic properties (strong convergence, convergence rate, strong consistency, asymptotic normality, asymptotic efficiency) of the Markov estimator are analyzed. It follows that in general the Markov estimator is strongly consistent, and asymptotically inefficient (in case of Gaussian errors the estimator does not reach the Cramér-Rao lower bound). Sufficient conditions for the asymptotic efficiency of the Markov estimator are given. The theory is applicable to, for example, signal modelling and multivariable system identification, both in time and frequency domains.

Keywords: Markov estimator, system identification, signal processing, semi-linear errors-in-variables model, estimating function.

1. Introduction

Consider the following general model based on N observations

$$M_0(\theta) + M_1(\theta)z = 0 \tag{1}$$

which is linear-in-the-measurements $z \in \mathbb{R}^{sN}$ and (non-)linear-in-the-model-parameters $\theta \in \mathbb{R}^{n_\theta}$. $M_0(\theta) \in \mathbb{R}^{rN}$, $M_1(\theta) \in \mathbb{R}^{rN \times sN}$ with $s \geq r$ has rank rN and n_θ, s and r fixed integers, independent of the number of observations N. Each time a new observation is added, the number of model equations and the number of measurements increase with, respectively, r and s. In frequency domain applications, (1) often has a block diagonal structure

$$M_{0k}(\theta) + M_{1k}(\theta)z_k = 0 \text{ for } k = 1, 2, \ldots, N \tag{2}$$

with $M_{0k}(\theta) \in \mathbb{R}^r$, $M_{1k}(\theta) \in \mathbb{R}^{r \times s}$, and $z_k \in \mathbb{R}^s$. The relationship with (1) is given by

S. Van Huffel and P. Lemmerling (eds.), Total Least Squares and Errors-in-Variables Modeling, 165–177.

$$M_0^T(\theta) = [M_{01}^T(\theta)\ldots M_{0N}^T(\theta)],\ M_1(\theta) = \mathrm{diag}(M_{10}(\theta), \ldots, M_{1N}(\theta)), \text{ and}$$
$$z^T = [z_1^T \ldots z_N^T] \tag{3}$$

Two special cases of model (1) are worth mentioning.

1.1 Signal Model

Putting $s = r$ and $M_1(\theta) = -I_{sN}$ in (1) gives the signal model
$$z = M_0(\theta) \tag{4}$$

This is typically the form encountered when estimating a linear combination of basis signals such as sinewaves, cisoids, exponential functions, ...

1.2 Transfer Function Model

Putting $M_0(\theta) = 0$, $M_1(\theta) = [A(\theta)\ \ -B(\theta)]$ and $z^T = [y^T\ \ u^T]$ gives the transfer function model
$$A(\theta)y = B(\theta)u \tag{5}$$
where $y \in \mathbb{R}^{rN}$, and $u \in \mathbb{R}^{qN}$ with $r + q = s$. $A(\theta) \in \mathbb{R}^{rN \times rN}$ is a regular matrix. The model equations of a linear time-invariant multivariable system can be written under this form [1], [2].

2. The Markov Estimator

An estimate $\hat{\theta}$ of the model parameters θ of the semi-linear model (1) is calculated using noisy observations $z = z_0 + n_z$ of the true (unknown) values z_0. For transfer function models (5) it boils down to a classical errors-in-variables problem. Since z_0 is unknown, it should also be estimated and is parametrized as z_p. The Markov estimator minimizes the squared residuals $(z - z_p)$ weighted with the noise covariance matrix $C_{n_z} = \mathrm{Cov}(n_z)$, taking into account the model equations (1). This constrained minimization problem, with parameters θ and z_p, can be solved using Lagrange multipliers $\lambda \in \mathbb{R}^{rN}$ [3]
$$\frac{1}{2}(z - z_p)^T C_{n_z}^+ (z - z_p) + \lambda^T (M_0(\theta) + M_1(\theta) z_p) \tag{6}$$
with $+$ the Moore-Penrose pseudo-inverse [4]. Singular noise covariance matrices are allowed to cover the case where parts of the measurements may be known exactly. This is, for example, the case in transfer function modeling with known inputs. Since (6) is quadratic in z_p and linear in λ, z_p and λ can be explicitly eliminated. This gives the following expression for z_p (see Appendix A)
$$C_{n_z} C_{n_z}^+ z_p = C_{n_z} C_{n_z}^+ z - C_{n_z} M_1^T(\theta)(M_1(\theta) C_{n_z} M_1^T(\theta))^{-1}(M_0(\theta) + M_1(\theta) z) \tag{7}$$

It makes it possible to eliminate the parameters z_p in (6), which results in a significant reduction of the size of the minimization problem. The following

Markov cost function is found

$$V_{\text{Markov}}(\theta, z) = \frac{1}{2}e^T(\theta, z)C_e^{-1}(\theta)e(\theta, z) = \frac{1}{2}\varepsilon^T(\theta, z)\varepsilon(\theta, z) \tag{8}$$

with

$$e(\theta, z) = M_0(\theta) + M_1(\theta)z, \quad C_e(\theta) = M_1(\theta)C_{n_z}M_1^T(\theta), \quad \varepsilon(\theta, z) = \Lambda(\theta)e(\theta, z) \tag{9}$$

and where $\Lambda(\theta) \in \mathbb{R}^{rN \times rN}$ satisfies $\Lambda^T(\theta)\Lambda(\theta) = C_e^{-1}(\theta)$ (see Appendix A). Note that $\text{Cov}(e(\theta, n_z)) = C_e(\theta)$ and $\text{Cov}(\varepsilon(\theta, n_z)) = I_{rN}$. Minimizing the cost function (8) w.r.t. θ gives the Markov estimates of the model parameters

$$\hat{\theta}(z) = \arg\min_{\theta} V_{\text{Markov}}(\theta, z) \tag{10}$$

The Markov estimates $\hat{z}$ of the true observations z_0 are found by evaluating (7) at $\theta = \hat{\theta}(z)$

$$C_{n_z}C_{n_z}^+\hat{z} = C_{n_z}C_{n_z}^+z - C_{n_z}M_1^T(\hat{\theta}(z))C_e^{-1}(\hat{\theta}(z))e(\hat{\theta}(z), z) \tag{11}$$

Note that this formula estimates the observations lying in the regular space of C_{n_z}. Those lying in the null space of C_{n_z} are known exactly.

The Markov estimates require the knowledge of the noise covariance matrix C_{n_z}. It can be estimated from independent repeated experiments (see Chapter 8 of [2]). This consumes a lot of computer time and memory space if N is large. In frequency domain identification only a (block) diagonal version of C_{n_z} is required. Using the block diagonal structure (2) of the model equations and replacing C_{n_z} in (8) by $\text{diag}(C_{n_{z1}}, \ldots, C_{n_{zN}})$, with $C_{n_{zk}} = \text{Cov}(n_{zk})$, gives the simplified Markov cost function

$$V_{\text{Markov}}(\theta, z) = \frac{1}{2}\sum_{k=1}^{N} e_k^T(\theta, z_k)C_{e_k}^{-1}(\theta)e_k(\theta, z_k) \tag{12}$$

with $e_k(\theta, z_k) = M_{0k}(\theta) + M_{1k}(\theta)z_k$, $C_{e_k}(\theta) = \text{Cov}(e_k(\theta, n_{zk})) = M_{1k}(\theta)C_{n_{zk}}M_{1k}^T(\theta)$ and $z_k = z_{0k} + n_{zk}$. Neglecting the non-diagonal terms of $C_{e_k}(\theta)$ in (12), the Markov cost function can even be simplified further to

$$V_{\text{Markov}}(\theta, z) = \frac{1}{2}\sum_{k=1}^{N}\sum_{i=1}^{r}\frac{e_{k[i]}^2(\theta, z_k)}{\text{var}(e_{k[i]}(\theta, n_{zk}))} \tag{13}$$

The stochastic properties of the minimizers of (8), (12) and (13) are analyzed in Section 4.

3. The Cramér-Rao lower bound

The concept of Cramér-Rao lower bound requires the existence of a true model

$$M_0(\theta_0) + M_1(\theta_0)z_0 = 0 \tag{14}$$

with z_0 the true observations and θ_0 the true model parameters, and the knowledge of the probability density function of the measurements $z = z_0 + n_z$.

The Cramér-Rao lower bound for unbiased estimators is constructed under the following assumption.

Assumption 3.1. *(Gaussian errors): The observations z_0 are deterministic and the errors n_z are normally distributed with known covariance matrix C_{n_z}.*

The log-likelihood function becomes

$$\ln f_z(z, z_p, \theta) = -\frac{1}{2}(z - z_p)^T C_{n_z}^{+}(z - z_p) + c \tag{15}$$

with c a constant independent of z_p and θ, and where the parameters z_p and θ satisfy the constraint (1)

$$M_0(\theta) + M_1(\theta) z_p = 0 \tag{16}$$

Straightforward calculation of the Fisher information matrix $Fi(z_0, \theta_0)$ from (15) is impossible since z_p contains too many unknowns. Indeed, the parameters of z_p lying in the singular space of C_{n_z} are known exactly and should not appear in the CR-bound. Next, (16) puts rN linear constraints on the entries of z_p. The resulting rN linear dependent variables should not also appear in the CR-bound.

Hence, the first step in calculating the CR-bound consists of reducing the sN parameters z_p to the $\text{rank}(C_{n_z}) - rN$ parameters x: $z_p = z_p(x, \theta)$ (see Appendix B). It facilitates writing the log-likelihood function as

$$\ln f_z(z, x, \theta) = -\frac{1}{2}(z - z_p(x, \theta))^T C_{n_z}^{+}(z - z_p(x, \theta)) + c \tag{17}$$

The corresponding Fisher information matrix $Fi(x_0, \theta_0)$ is

$$Fi(x_0, \theta_0) = \begin{bmatrix} F_{xx} & F_{x\theta} \\ F_{x\theta}^T & F_{\theta\theta} \end{bmatrix} \tag{18}$$

Applying the inverse of block matrices to $CR(x_0, \theta_0) = Fi^{-1}(x_0, \theta_0)$ gives the Cramér-Rao lower bound on the model parameters

$$CR(\theta_0) = Fi^{-1}(\theta_0) = (F_{\theta\theta} - F_{x\theta}^T F_{xx}^{-1} F_{x\theta})^{-1} \tag{19}$$

Filling out the explicit expressions of $F_{\theta\theta}$, F_{xx} and $F_{x\theta}$ in (19) gives, after some calculations (see Appendix B),

$$\begin{aligned} Fi(\theta_0) &= V_{\text{Markov}}''(\theta_0, z_0) \\ &= \left(\frac{\partial e(\theta, z_0)}{\partial \theta_0}\right)^T C_e^{-1}(\theta_0)\left(\frac{\partial e(\theta, z_0)}{\partial \theta_0}\right) \\ &= \left(\frac{\partial \varepsilon(\theta, z_0)}{\partial \theta_0}\right)^T \left(\frac{\partial \varepsilon(\theta, z_0)}{\partial \theta_0}\right) \end{aligned} \tag{20}$$

This shows that the Fisher information matrix of the model parameters $Fi(\theta_0)$ equals the Hessian of the Markov cost function (8), evaluated at the true observations and the true model parameters. Hence, the larger the eigenvalues of the Hessian matrix, the smaller the CR-bound of the model parameters.

4. Properties Markov Estimator

In this section we analyze the stochastic properties of the Markov estimator $\hat{\theta}(z)$ for $N \to \infty$. The following result is instrumental in the analysis: under some suitable noise assumptions (mixing condition on n_z) $\hat{\theta}(z)$, the minimizing argument of $V_{\text{Markov}}(\theta, z)$ converges strongly to $\tilde{\theta}(z_0)$, the minimizer of $V_{\text{Markov}}(\theta) = \mathcal{E}\{V_{\text{Markov}}(\theta, z)\}$ (proof: see [2]).

4.1 Consistency

The concept of consistency requires that the true model (see eq. (14)) belongs to the considered model set. Besides the model parameters, (a part of) the measurements are also estimated. First, the consistency of the estimates $\hat{\theta}(z)$ (10) of the model parameters is analyzed. Next, the consistency of the estimates $\hat{z}$ (11) of the measurements is discussed.

4.1.1 Model Parameters

Assuming that the noise n_z is independent of the true observations z_0 and using

$$e(\theta, z) = e(\theta, z_0) + M_1(\theta)n_z \text{ and } \operatorname{trace}(e^T C_e^{-1} e) = \operatorname{trace}(C_e^{-1} e e^T),$$

we find the expected value of the cost function (8)

$$V_{\text{Markov}}(\theta) = \mathcal{E}\{V_{\text{Markov}}(\theta, z)\} = \mathcal{E}\{V_{\text{Markov}}(\theta, z_0)\} + rN/2 \tag{21}$$

Since $\mathcal{E}\{V_{\text{Markov}}(\theta_0, z_0)\} = 0$ and $rN/2$ is θ-independent, it follows from (21) that θ_0 minimizes $V_{\text{Markov}}(\theta)$. Since we have shown that $\tilde{\theta}(z_0) = \theta_0$, it follows that the Markov estimate (10) is strongly consistent.

Note that the expected values of the simplified cost functions (12) and (13) are also given by (21). We conclude that the Markov estimates of the (block) diagonal model (2) based on the simplified cost functions (12), (13) are still strongly consistent. Removing some parts of the non-diagonal elements of C_{n_z} does not influence the consistency property: the minimal requirement is that each residual is weighted with its variance. For frequency domain system identification, this means that the correlation of the errors n_z over the frequencies and between the different outputs can be neglected. However, the correlation of the errors n_z between an output and all the inputs may not be removed since it influences $\operatorname{var}(e_{k[i]}(\theta, z_k))$; otherwise consistency is lost.

4.1.2 Observations

In general, the estimates $\hat{z}$ (11) of the observations are inconsistent. This is due to the fact that the uncertainty of z in (11) does not decrease by making more observations (no averaging effect occurs in $C_{n_z} C_{n_z}^{+} z$). z cancels in (11) for signal models (4) and transfer function models (5) with known input (excitation) signals u_0. In these cases strongly consistent estimates of the observations are obtained through, respectively, $\hat{z} = M_0(\hat{\theta}(z))$ and $\hat{y} = A^{-1}(\hat{\theta}(z))B(\hat{\theta}(z))u_0$ (proof: see Appendix C).

4.2 Strong Convergence

If model errors are present ($e(\tilde{\theta}(z_0), z_0) \neq 0$) one can wonder why the Markov estimator should be preferred over, for example, the nonlinear least squares estimator

$$V_{\mathrm{NLS}}(\theta, z) = \frac{1}{2}e^T(\theta, z)e(\theta, z) \tag{22}$$

The reason for this is that the Markov estimate $\hat{\theta}(z)$ converges to a value $\tilde{\theta}(z_0)$ which is independent of the signal-to-noise ratio. Indeed, replacing n_z by υn_z (and hence C_{n_z} by $\upsilon^2 C_{n_z}$) in the expected value of the cost function (21), gives

$$V_{\mathrm{Markov}}(\theta) = \upsilon^2 \mathcal{E}\{V_{\mathrm{Markov}}(\theta, z_0)\} + rN/2 \tag{23}$$

It shows that $\tilde{\theta}(z_0)$, the minimizing argument of (23) is independent of υ, and that $\tilde{\theta}(z_0)$ can be interpreted as the solution of the noiseless problem. The same is true for the simplified Markov estimators (12) and (13). This is not the case for the nonlinear least squares estimator (22). Indeed, the expected value of (22) equals

$$V_{\mathrm{NLS}}(\theta) = \mathcal{E}\{V_{\mathrm{NLS}}(\theta, z)\} = \mathcal{E}\{V_{\mathrm{NLS}}(\theta, z_0)\} + \mathrm{trace}(C_e(\theta)) \tag{24}$$

Replacing n_z by υn_z (and hence C_{n_z} by $\upsilon^2 C_{n_z}$) in (24), gives

$$V_{\mathrm{NLS}}(\theta) = \mathcal{E}\{V_{\mathrm{NLS}}(\theta, z)\} = \mathcal{E}\{V_{\mathrm{NLS}}(\theta, z_0)\} + \upsilon^2\mathrm{trace}(C_e(\theta)) \tag{25}$$

Clearly, the minimizer of (25) depends on υ.

4.3 Asymptotic Normality

It makes no sense to estimate parameters if no quality stamp on the result can be given. Otherwise, any random guess is a valuable estimate. For example, the quality stamp can be a region around the estimated value where the limit (true) value lies within some confidence level. Therefore, we would like to know the distribution function of $\hat{\theta}(z)$ for finite N. In most cases it is impossible to calculate, and we can only make statements about the asymptotic distribution function of $\hat{\theta}(z)$. Under some suitable noise assumptions (n_z is mixing of order infinity) it can be shown that $\sqrt{N}(\hat{\theta}(z) - \tilde{\theta}(z_0))$ converges to a Gaussian random variable $\sqrt{N}\delta_\theta(z)$ with zero mean and covariance matrix $\mathrm{Cov}(\sqrt{N}\delta_\theta(z))$

$$\begin{aligned} \mathrm{Cov}(\sqrt{N}\delta_\theta(z)) &= V_N''^{-1}(\tilde{\theta}(z_0))Q_N(\tilde{\theta}(z_0))V_N''^{-1}(\tilde{\theta}(z_0)) \\ Q_N(\tilde{\theta}(z_0)) &= N\mathcal{E}\{V_N'^T(\tilde{\theta}(z_0), z)V_N'(\tilde{\theta}(z_0), z)\} \end{aligned} \tag{26}$$

with $V_N(\theta, z) = V_{\mathrm{Markov}}(\theta, z)/N$, $V_N(\theta) = \mathcal{E}\{V_{\mathrm{Markov}}(\theta, z)\}/N$ and $'$ the derivative w.r.t. θ (proof: see [2]). It follows that the uncertainty on the estimated parameters is small if the eigenvalues of the Hessian matrix $V_N''(\tilde{\theta}(z_0))$ are large. Although the asymptotic normality neither guarantees the convergence of $\mathrm{Cov}(\sqrt{N}\hat{\theta}(z))$ to $\mathrm{Cov}(\sqrt{N}\delta_\theta(z))$, nor the existence of $\mathrm{Cov}(\sqrt{N}\hat{\theta}(z))$, it makes it possible to construct uncertainty regions on $\hat{\theta}(z)$ with a given confidence level. This is sufficient for our purposes.

4.4 Asymptotic Efficiency

The concept of asymptotic efficiency requires that the true model (see eq. (14)) belongs to the model set. Under this condition, expression (26) of the covariance matrix becomes

$$\begin{aligned}
\mathrm{Cov}(\sqrt{N}\delta_\theta(z)) &= V_N''^{-1}(\theta_0) + V_N''^{-1}(\theta_0) q_N(\theta_0) V_N''^{-1}(\theta_0) \\
q_N(\theta_0) &= N\mathcal{E}\{ v_N'^T(\theta_0, n_z) v_N'(\theta_0, n_z)\} + \\
&\qquad 2\mathrm{herm}\left(\mathcal{E}\{\left(\frac{\partial \varepsilon(\theta, z_0)}{\partial \theta_0}\right)^T\}\mathcal{E}\{\Delta(\theta_0, n_z) v_N'(\theta_0, n_z)\}\right)
\end{aligned} \tag{27}$$

with $\mathrm{herm}(A) = (A + A^H)/2$ (proof: see Appendix D). Comparing the Cramér-Rao lower bound (20) for normally distributed errors n_z and deterministic z_0 to the asymptotic covariance matrix (27), shows that the Markov estimates are, in general, asymptotically inefficient ($V_N''(\theta_0) = V_N''(\theta_0, z_0)$ for deterministic z_0). The inefficiency term $V_N''^{-1}(\theta_0) q_N(\theta_0) V_N''^{-1}(\theta_0)$ is, however, small w.r.t. $V_N''^{-1}(\theta_0)$ for sufficiently large signal-to-noise ratios.

For some noise covariance matrices the inefficiency term is zero. Indeed, the Markov estimates (10) are asymptotically efficient if $\mathrm{rank}(C_{n_z}) = rN$ for any $N \geq N_0$ (proof: see Appendix E). Note that the condition $\mathrm{rank}(C_{n_z}) = rN$ is automatically satisfied for signal models. In frequency domain identification of multivariable systems, it implies that the number of non-coherent noise sources must be equal to the number of outputs. Note that the efficiency result is not valid for the estimates based on the simplified Markov cost functions (12) and (13).

4.5 Robustness

The consistency, asymptotic normality, convergence rate and asymptotic bias properties of the Markov estimator (10) for (block) diagonal models (2) are robust w.r.t. to the knowledge of some non-diagonal parts of C_{n_z} (compare the simplified cost functions (12) and (13) to (8)). This is not the case for the asymptotic efficiency: removing the non-diagonal elements of C_{n_z} increases the uncertainty of the estimates.

4.6 Practical Calculation of Uncertainty Bounds

The expression (26) for $\mathrm{Cov}(\delta_\theta(Z))$ is not tractable and will be approximated.

Replacing $e(\tilde{\theta}(z), z_0)$ by $\mu e(\tilde{\theta}(z), z_0)$ and n_z by υn_z, an approximation for "small" model errors ($\mu \to 0$) and "large" signal-to-noise ratios ($\upsilon \to 0$) is given by

$$\mathrm{Cov}(\delta_\theta(z)) = C_\theta(I_{n_\theta} + O(\upsilon) + O(\mu) + O(\mu^2\upsilon^{-2})\lambda(z_0))$$

$$C_\theta = \left[\mathcal{E}\left\{\left(\frac{\partial\varepsilon(\theta, z_0)}{\partial\tilde{\theta}(z_0)}\right)^T\left(\frac{\partial\varepsilon(\theta, z_0)}{\partial\tilde{\theta}(z_0)}\right)\right\}\right]^{-1} = \upsilon^2 O(N^{-1}) \tag{28}$$

where $\lambda(z_0) = 1$ for random z_0 and $\lambda(z_0) = 0$ for deterministic z_0 (see [2]). Formula (28) shows that in the presence of model errors ($\mu \neq 0$), the uncertainty of the estimated model parameters does not decrease to zero for random z_0 as the noise level υ tends to zero. Intuitively, this can be understood as follows: in the absence of observation noise, $n_z = 0$, the model errors still depend on the particular realization of z_0. Hence, $\hat{\theta}(z)$ depends on z_0, and $\mathrm{Cov}(\delta_\theta(z)) = \mu^2 O(N^{-1})$.

Expression (28) requires the knowledge of the true observations z_0 and the (true) model parameters θ_0 or $\tilde{\theta}(z)$, which are not available. Approximations of the asymptotic covariance matrix are obtained by replacing z_0 by z and θ_0 or $\tilde{\theta}(z_0)$ by $\hat{\theta}(z)$. Mostly the following approximation is used

$$\mathrm{Cov}(\hat{\theta}(z)) \approx \left[\left(\frac{\partial\varepsilon(\theta, z)}{\partial\hat{\theta}(z)}\right)^T\left(\frac{\partial\varepsilon(\theta, z)}{\partial\hat{\theta}(z)}\right)\right]^{-1} \tag{29}$$

Note that the right hand side of (29) is calculated in Newton based minimization methods of cost function (8). Formula (29) allows for the calculation of uncertainty bounds of any model related quantity $f(z, \hat{\theta}(z))$ via linearizing $f(z, \hat{\theta}(z))$ at z_0, $\tilde{\theta}(z_0)$.

5. Conclusions

Remarkably enough the Markov estimator is in general not asymptotically efficient for Gaussian errors n_z. This is not in contradiction with the well known asymptotic efficiency property of maximum likelihood estimators [4]. Indeed, this property has been shown assuming that the number of estimated parameters does not increase with the amount of data, while the number of free parameters in the errors-in-variables problem is $\mathrm{rank}(C_{n_z}) - rN + n_\theta$ and increases with N. The Markov estimator is asymptotically efficient if $\mathrm{rank}(C_{n_z}) - rN + n_\theta$ is independent of N. This is true for the two following cases: (i) the number of non-coherent noise sources $\mathrm{rank}(C_{n_z})$ equals the number of model equations rN (see section 4.4); (ii) the noiseless observations z_0 can be modeled with a number of parameters that is independent of N. For example, case (ii) is satisfied in system identification problems if the noiseless input can be modeled as filtered white noise or as a periodic signal with a FIXED (N-independent) number of frequencies.

Note that the results presented in this paper also apply to complex observations $z \in \mathbb{C}^{sN}$ and/or complex model parameters $\theta \in \mathbb{C}^{n_\theta}$. To see this it is sufficient to stack the real and imaginary parts of z and/or θ on top of each other in (1). Simplified expressions can be obtained if the errors $n_z \in \mathbb{C}^{sN}$ are circular complex distributed: $\mathcal{E}\{ n_z n_z^T \} = 0$ (see [2]).

Acknowledgement

This research is sponsored by the Fund for Scientific Research Flanders, the Flemish Community (Concerted Action IMMI) and the Belgian Government (IUAP IV/2).

Appendix A: Constrained Minimization (6)

Expressing the stationarity of the cost function (6) w.r.t. z_p and λ gives

$$-C_{n_z}^{+}(z - z_\mathrm{p}) + M_1^T(\theta)\lambda = 0 \tag{30}$$

$$M_0(\theta) + M_1(\theta) z_\mathrm{p} = 0 \tag{31}$$

λ is solved from (30) by left multiplication with $M_1(\theta) C_{n_z}$

$$\lambda = (M_1(\theta) C_{n_z} M_1^T(\theta))^{-1} M_1(\theta) C_{n_z} C_{n_z}^{+}(z - z_\mathrm{p}) \tag{32}$$

Since C_{n_z} is a symmetric positive semi-definite matrix, it can be decomposed into singular values as $C_{n_z} = U \Sigma U^T$, where U is an orthogonal matrix ($U^T U = U U^T = I_{sN}$) and Σ a diagonal matrix ($\mathrm{rank}(\Sigma) = \mathrm{rank}(C_{n_z})$) containing the sorted singular values. Defining

$$U^T z_\mathrm{p} = \begin{bmatrix} W_{1\mathrm{p}} \\ W_{20} \end{bmatrix} \tag{33}$$

where $\dim(W_{1\mathrm{p}}) = \mathrm{rank}(C_{n_z})$, it follows that

$$z_\mathrm{p} = U \begin{bmatrix} W_{1\mathrm{p}} \\ W_{20} \end{bmatrix} \text{ and } C_{n_z} C_{n_z}^{+} z_\mathrm{p} = U \begin{bmatrix} I_{r_c} & 0 \\ 0 & 0 \end{bmatrix} U^T z_\mathrm{p} = U \begin{bmatrix} W_{1\mathrm{p}} \\ 0 \end{bmatrix} \tag{34}$$

with $r_c = \mathrm{rank}(C_{n_z})$. $W_{1\mathrm{p}}$ stands for the (linear combination of) measurements lying in the regular space of C_{n_z} and is estimated, while W_{20} represents the (linear combination of) measurements lying in the null space of C_{n_z} and is known exactly. Hence, we have

$$z = U \begin{bmatrix} W_1 \\ W_{20} \end{bmatrix} \text{ and } C_{n_z} C_{n_z}^{+} z = U \begin{bmatrix} W_1 \\ 0 \end{bmatrix} \tag{35}$$

Rewriting (31) and (32) using (34) and (35) allows to eliminate $W_{1\mathrm{p}}$ in (32)

$$\lambda = (M_1(\theta) C_{n_z} M_1^T(\theta))^{-1} (M_0(\theta) + M_1(\theta) z) \tag{36}$$

Substituting (36) into (30) and left multiplication of (30) by C_{n_z} gives (7). Using (34) it can be seen that (7) is independent of W_{20}, which is known exactly, and, hence, makes it possible to estimate the measurements W_{1p} lying in the regular space of C_{n_z} only. Since $C_{n_z}^+ = (C_{n_z}^+ C_{n_z})C_{n_z}^+(C_{n_z}C_{n_z}^+)$ and $C_{n_z}^{+T} = C_{n_z}^+$, the cost function (6), where z_p satisfies (31), can be written as

$$\frac{1}{2}[C_{n_z}C_{n_z}^+(z-z_p)]^T C_{n_z}^+ [C_{n_z}C_{n_z}^+(z-z_p)] \tag{37}$$

Substituting (7) into (37) using $C_{n_z}C_{n_z}^+C_{n_z} = C_{n_z}$ gives (8) directly.

Appendix B: Cramér-Rao Lower Bound for Semi-Linear Models

The proof consists of two parts. In the first part we reduce the sN parameters z_p to the $r_c - rN$ parameters x_0, with $r_c = \mathrm{rank}(C_{n_z})$. In the second part the Fisher information matrix $Fi(x_0, \theta_0)$ of the observations x and the model parameters θ is reduced to the Fisher information matrix $Fi(\theta_0)$.

(i) The known observations lie in the null space of C_{n_z} ($sN - r_c$ parameters) and are separated from the unknown observations by decomposing the parameter vector z_p as in (34). Using (34), with $U = [U_1 U_2]$ and $U_1 \in \mathbb{R}^{sN \times r_c}$, the constraint (31) can be written as

$$M_1(\theta)U_1W_{1p} = -M_0(\theta) - M_1(\theta)U_2W_{20} \tag{38}$$

$C_e(\theta) = M_1(\theta)C_{n_z}M_1^T(\theta)$ has, by assumption, full rank rN and, therefore, $r_c \geq rN$ and $\mathrm{rank}(M_1(\theta)U_1) = rN$. Since $r_c \geq rN$, (38) has, in general, infinitely many solutions for W_{1p}. They can be found by adding one particular solution of (38) to the solution of the homogenous part of (38). The solution W_{1h} of the homogenous part lies in the null space of $M_1(\theta)U_1 \in \mathbb{R}^{rN \times r_c}$. Since $M_1(\theta)U_1$ has full rank rN, it can be written as

$$W_{1h} = Q(\theta)x \tag{39}$$

with $x \in \mathbb{R}^{r_c - rN}$ and where $Q(\theta) \in \mathbb{R}^{r_c \times (r_c - rN)}$ satisfies $M_1(\theta)U_1Q(\theta) = 0$. It can easily be verified that,

$$W_{1p}(\theta) = -M^T(\theta)(M(\theta)M^T(\theta))^{-1}(M_0(\theta) + M_1(\theta)U_2W_{20}) \tag{40}$$

where $M(\theta) = M_1(\theta)U_1$, is a particular solution of (38). The complete solution of (38) equals $W_{1p} = W_{1h} + W_{1p}(\theta)$ so that in (34) z_p can be written as

$$z_p = z_p(x, \theta) = U_1Q(\theta)x + U_1\tilde{W}_{1p}(\theta) + U_2W_{20} \tag{41}$$

with $x \in \mathbb{R}^{r_c - rN}$, which concludes the first part of the proof. The following property of the function $z_p(x, \theta)$ will be used in the second part of the proof

$$M_1(\theta)\frac{\partial z_p(x, \theta)}{\partial x} = M_1(\theta)U_1Q(\theta) = 0 \tag{42}$$

(ii) The second part of the proof starts with the calculation of the Fisher information matrix $Fi(x_0, \theta_0)$. Applying

$$Fi(\psi) = \mathcal{E}\left\{\left(\frac{\partial \ln f_z(z, \psi)}{\partial \psi}\right)^T\left(\frac{\partial \ln f_z(z, \psi)}{\partial \psi}\right)\right\} \tag{43}$$

to (15), with $\psi^T = [x^T\ \theta^T]$ and $z_p(x_0, \theta_0) = z_0$, gives

$$F_{xx} = \left(\frac{\partial z_p(x,\theta)}{\partial x}\right)^T C_{n_z}^+ \left(\frac{\partial z_p(x,\theta)}{\partial x}\right)\Bigg|_{x = x_0,\ \theta = \theta_0} \tag{44}$$

$$F_{x\theta} = \left(\frac{\partial z_p(x,\theta)}{\partial x}\right)^T C_{n_z}^+ \left(\frac{\partial z_p(x,\theta)}{\partial \theta}\right)\Bigg|_{x = x_0,\ \theta = \theta_0} \tag{45}$$

$$F_{\theta\theta} = \left(\frac{\partial z_p(x,\theta)}{\partial \theta}\right)^T C_{n_z}^+ \left(\frac{\partial z_p(x,\theta)}{\partial \theta}\right)\Bigg|_{x = x_0,\ \theta = \theta_0} \tag{46}$$

C_{n_z} and $C_{n_z}^+$ can be decomposed into singular values as

$$C_{n_z} = U\begin{bmatrix}\Sigma_1 & 0\\ 0 & 0\end{bmatrix}U^T = U_1\Sigma_1U_1^T \text{ and } C_{n_z}^+ = U\begin{bmatrix}\Sigma_1^{-1} & 0\\ 0 & 0\end{bmatrix}U^T = U_1\Sigma_1^{-1}U_1^T \tag{47}$$

where Σ_1 contains the non-zero singular values of C_{n_z}. Putting (44) to (46) into (19) by using (47) gives, after some calculations,

$$Fi(\theta_0) = G^T[I_{r_c} - F(F^TF)^{-1}F^T]G \tag{48}$$

with $G = \Sigma_1^{-1/2}U_1^T\partial z_p(x_0, \theta)/\partial\theta_0$ and $F = \Sigma_1^{-1/2}Q(\theta_0)$. Defining $E = M_1(\theta_0)U_1\Sigma_1^{1/2}$, it follows from (42) that $EF = 0$. Since the matrices $E \in \mathbb{R}^{rN \times r_c}$ and $F \in \mathbb{R}^{r_c \times (r_c - rN)}$ have, respectively, full rank rN and $r_c - rN$, and since $EF = 0$, it follows that the symmetric idempotent matrices $F(F^TF)^{-1}F^T$ and $E^T(EE^T)^{-1}E$ are related by

$$I_{r_c} - F(F^TF)^{-1}F^T = E^T(EE^T)^{-1}E \tag{49}$$

Hence, (48) can be simplified as

$$Fi(\theta_0) = (EG)^T(EE^T)^{-1}(EG) \tag{50}$$

Working out EG gives

$$\begin{aligned}
EG &= M_1(\theta_0)\frac{\partial U_1U_1^Tz_p(x_0,\theta)}{\partial\theta_0} && (U_1 \text{ is independent of } \theta)\\
&= M_1(\theta_0)\frac{\partial(z_p(x_0,\theta) - U_2W_{20})}{\partial\theta_0} && ((41) \text{ with } U_1^TU_1 = I_{r_c} \text{ and } U_1^TU_2 = 0)\\
&= -M_1(\theta_0)\frac{\partial(z_0 - z_p(x_0,\theta))}{\partial\theta_0} && (U_2,\ W_{20},\ z_0 \text{ are independent of } \theta)\\
&= -\frac{\partial M_1(\theta)(z_0 - z_p(x_0,\theta))}{\partial\theta_0} && (z_0 = z_p(x_0,\theta_0))\\
&= -\frac{\partial(M_0(\theta) + M_1(\theta)z_0)}{\partial\theta_0} && ((31)\text{: } M_1(\theta)z_p(x_0,\theta) = -M_0(\theta))
\end{aligned}$$

Substituting this result into (50), taking into account that $EE^T = C_e(\theta_0)$ (see (9)), gives

$$Fi(\theta_0) = \left(\frac{\partial e(\theta, z_0)}{\partial \theta_0}\right)^T C_e^{-1}(\theta_0)\left(\frac{\partial e(\theta, z_0)}{\partial \theta_0}\right) \tag{51}$$

Using $e(\theta_0, z_0) = 0$, the two other expressions in (20) follow directly.

Appendix C: Markov Estimates of the Observations for Signal Models and Transfer Function Models with Known Input

(i) For signal models, we have $M_1(\theta) = -I_{sN}$, $C_e(\theta) = C_{n_z}$ (see (9)) and $C_{n_z}^+ = C_{n_z}^{-1}$, and (11) becomes

$$\hat{z} = M_0(\hat{\theta}(z)) \tag{52}$$

(ii) For transfer function models with known input, we have

$$C_{n_z} = \begin{bmatrix} C_{n_y} & 0 \\ 0 & 0 \end{bmatrix} \Rightarrow C_{n_z} C_{n_z}^+ = \begin{bmatrix} I_{rN} & 0 \\ 0 & 0 \end{bmatrix} \tag{53}$$

Taking into account (5), (11) becomes

$$\begin{bmatrix} \hat{y} \\ 0 \end{bmatrix} = \begin{bmatrix} A^{-1}(\hat{\theta}(z))B(\hat{\theta}(z))u_0 \\ 0 \end{bmatrix} \tag{54}$$

which concludes the proof.

Appendix D: Asymptotic Distribution of the Markov Estimates without Model Errors

Expression (26) will be elaborated for the Markov estimator assuming that no model errors are present ($e(\theta_0, z_0) = 0$). Using $\varepsilon(\theta, z) = \varepsilon(\theta, z_0) + \Delta(\theta, n_z)$, with $\Delta(\theta, n_z) = \Lambda(\theta)M_1(\theta)n_z$, and $\varepsilon(\theta_0, z_0) = 0$ (see (8)), we find

$$V_N'^T(\theta_0, z) = \frac{1}{N}\left(\frac{\partial \varepsilon(\theta, z_0)}{\partial \theta_0}\right)^T \Delta(\theta_0, n_z) + v_N'^T(\theta_0, n_z) \tag{55}$$

where $v_N(\theta, n_z) = \Delta^T(\theta, n_z)\Delta(\theta, n_z)/(2N)$. Since $\mathrm{Cov}(\Delta(\theta, n_z)) = I_{rN}$ (see (8)), $\varepsilon(\theta_0, z_0) = 0$ and z_0, n_z are mutually independent, $Q_N(\theta_0)$ (26) becomes

$$\begin{aligned} Q_N(\theta_0) &= N\mathcal{E}\{ V_N'^T(\theta_0, z)V_N'(\theta_0, z)\} \\ &= V_N''(\theta_0) + N\mathcal{E}\{ v_N'^T(\theta_0, n_z)v_N'(\theta_0, n_z)\} + \\ &\quad 2\mathrm{herm}(\mathcal{E}\{ \left(\frac{\partial \varepsilon(\theta, z_0)}{\partial \theta_0}\right)^T \}\mathcal{E}\{ \Delta(\theta_0, n_z)v_N'(\theta_0, n_z)\}) \end{aligned} \tag{56}$$

Putting (56) into (26) gives (27).

Appendix E: Proof of the Asymptotic Efficiency of the Markov Estimates

The second term in the expression of $q_N(\theta_0)$ (see (27)) is a function of the third order moments of n_z and hence is zero for Gaussian errors n_z. The first term is positive semi-definite and is zero if and only if $v_N(\theta, n_z)$ is independent of θ for any n_z. From (47) and Assumption 3.1., it follows that we can write n_z as $n_z = U_1\Sigma_1^{1/2}\varepsilon_z$ with $\varepsilon_z \in N_{r_c}(0, I_{r_c})$. Hence $M_1(\theta)n_z = E(\theta)\varepsilon_z$, with $E(\theta) = M_1(\theta)U_1\Sigma_1^{1/2}$, and $C_e(\theta) = E(\theta)E^T(\theta)$, so that

$$v_N(\theta, n_z) = \varepsilon_z^T E^T(\theta)(E(\theta)E^T(\theta))^{-1}E(\theta)\varepsilon_z/(2N) \quad (57)$$

The matrix $E(\theta) \in \mathbb{R}^{rN \times r_c}$ has full rank rN ($rN \le r_c \le sN$) since $C_e(\theta)$ has full rank rN. If $r_c = rN$, then $(E(\theta)E^T(\theta))^{-1} = E^{-T}(\theta)E^{-1}(\theta)$ and $v_N(\theta, n_z) = \varepsilon_z^T\varepsilon_z/(2N)$. This concludes the proof since ε_z is independent of θ.

References

[1] Ljung L., *System Identification: Theory for the User (second edition).* Upper Saddle River: Prentice Hall, 1999

[2] Pintelon, R. and Schoukens J., *System Identification: a Frequency Domain Approach.* New York: IEEE Press, 2001.

[3] Kaplan W., *Advanced Calculus.* Reading: Adison-Wesley, 1993.

[4] Ben-Israel A. and Greville T. N. E., *Generalized Inverses: Theory and Applications.* London: Wiley, 1974.

[5] Caines P. E., *Linear Stochastic Systems.* New York: Wiley, 1988.

COX'S PROPORTIONAL HAZARDS MODEL UNDER COVARIATE MEASUREMENT ERROR

A Review and Comparison of Methods

Thomas Augustin and Regina Schwarz
Department of Statistics, University of Munich
D 80799 Munich, Germany.
thomas@stat.uni-muenchen.de

Abstract This contribution studies the Cox model under covariate measurement error. Methods proposed in the literature to adjust for measurement error are reviewed. The basic structural and functional approaches are discussed in some detail, important modifications and further developments are briefly sketched. Then the basic methods are compared in a simulation study.

Keywords: measurement error, errors-in-variables, Cox model, survival analysis.

1. Introduction

Probably the most common statistical model in biometrics is Cox's proportional hazards model for analyzing survival data. Searching in the Medline database yields about 8000 entries for the term 'proportional hazards' during the last ten years. In most biometric studies, measurement error is an important issue. Often variables of interest can not be measured without substantial error, often they are even not available in principle (like the average protein intake over the last five years). Surrogates have to be used instead. The naive estimate, which just plugs in the surrogate instead of the true covariate, may be expected to be severely biased. Therefore, several methods have been developed to remove the bias by taking the measurement error appropriately into account. This paper reviews them and compares the basic approaches underlying them. For this purpose, we proceed as follows: Section 2 collects some facts on Cox's proportional hazards model, Section 3 states precisely the basic form of the error models underlying this study. The

S. Van Huffel and P. Lemmerling (eds.), Total Least Squares and Errors-in-Variables Modeling, 179–188.

literature on covariate measurement error in the Cox model is surveyed in Section 4, where special attention is paid to those methods which have laid the foundations for further developments. In Section 5 several correction methods are compared by a simulation study.

2. Cox's proportional hazards model

According to Cox [9], for every unit $i = 1, \ldots, n$, the corresponding hazard rate $\lambda(t) = \lim_{\varepsilon \downarrow 0} \varepsilon^{-1} \cdot P(\{t \leq T_i \leq t + \varepsilon\} \mid \{T_i \geq t\})$ of the failure time T_i is related to the p-dimensional vector X_i of covariates by

$$\lambda(t|X_i) = \lambda_0(t) \cdot \exp(\beta^T X_i)\,. \tag{1}$$

The hazards of the units are proportional to each other, because the baseline hazard rate $\lambda_0(t)$ is assumed to be the same for all i. It can be left completely unspecified, making the model a semiparametric and therefore quite flexible tool.

The common random censorship model is used: rather than always observing T_i, only the pair (Y_i, Δ_i) is available where $Y_i = \min(T_i, C_i)$ and Δ_i is the indicator function of $\{T_i \leq C_i\}$. The censoring variable C_i is stochastically independent of T_i and describes the maximal time span which unit i can be in the study. Assume that no ties occur, and order the observed true failure times in increasing magnitude: $\tau_1 < \tau_2 < \ldots < \tau_j < \ldots < \tau_k$, $\tau_0 := 0$. Define for $j = 0, 1, \ldots, k$ the *risk set* $\mathcal{R}(\tau_j)$ to be the set of all units being alive immediately before τ_j.

Estimation of the parameter vector β is based on the so called *partial likelihood*, which does not involve $\lambda_0(t)$. The *partial likelihood estimate* $\hat{\beta}_{\text{PL}}$ is then obtained as the root of

$$\sum_{j=1}^{k} \left(X_j - \frac{\sum_{i \in \mathcal{R}(\tau_j)} X_i \cdot \exp\left(\beta^T X_i\right)}{\sum_{i \in \mathcal{R}(\tau_j)} \exp\left(\beta^T X_i\right)} \right) = 0\,. \tag{2}$$

For inference on the baseline hazard rate customarily the *Breslow estimate* (cf. [4, 5]) of the cumulative baseline hazard rate $\Lambda_0(t) = \int_0^t \lambda(u)du$ is used.

3. The basic error model

Unless it is explicitly mentioned the classical, homoscedastic error model in its basic form is considered throughout the paper: Take all true covariates X_i to be continuous and assume that the surrogates W_i are related to X_i by $W_i = X_i + U_i$.The measurement error U_i is required to be independent of T_j, X_j, $j = 1, \ldots, n$, as well as of $U_j, j \neq i$. Furthermore, the variables U_i are assumed to be i.i.d. normally distributed with mean zero and known (or consistently estimated) covariance matrix Σ_U.[1] The dependent variables Y_i and Δ_i are taken to be error free.

Altogether this leads to *nondifferential measurement error*: given X_i, the variables T_i and W_i are conditionally independent. This means that knowledge of X_i would make observation of W_i superfluous; W_i really is only a surrogate providing no information which is not contained in X_i.

4. Correction methods for the Cox model

Apart from some comments in Section 6 on baseline hazard rate estimation, we concentrate on measurement error corrected inference on the regression coefficients β. To structure this part of the presentation, we order the different methods according to the assumptions they require for the distribution of the true, unknown covariate X_i. We begin with the *structural approaches*, where the distribution of X_i is assumed to belong to a known class of parametric distributions. Then we turn to the *functional methods* which manage to do without any parametric assumption on the distribution law of the X_is.

4.1 Structural approaches

Generally there are two basic structural approaches: regression calibration, which will be discussed at the end of this subsection, and 'integrating the likelihood.'

The latter one uses the conditional distribution of X_i given W_i and the assumption of non-differentiality to integrate out the influence of the measurement error. Prentice [20] made it clear that applying this idea to the Cox model leads to unexpected difficulties: Under non-differential measurement error, which can be shown to be equivalent to $\lambda(t|X_i, W_i) = \lambda(t|X_i)$, one obtains in general

$$\begin{aligned}
\lambda(t|W_i) &= \lim_{\varepsilon \downarrow 0} \varepsilon^{-1} \cdot P(\{T_i \le t+\varepsilon\}|\{T_i \ge t\}, W_i) \\
&= \lim_{\varepsilon \downarrow 0} \varepsilon^{-1} \cdot \mathbf{E}\left(P(\{T_i \le t+\varepsilon\}|X_i, \{T_i \ge t\}, W_i)\big|\{T_i \ge t\}, W_i\right) \\
&= \lim_{\varepsilon \downarrow 0} \varepsilon^{-1} \cdot \mathbf{E}\left(P(\{T_i \le t+\varepsilon\}|\{T_i \ge t\}, X_i)\big|\{T_i \ge t\}, W_i\right) \\
&= \mathbf{E}(\lambda(t|X_i)|\{T_i \ge t\}, W_i),
\end{aligned}$$

[1] We allow for the border case that (some of) the rows of Σ_U are zero, so that we do not need to distinguish in notation between correctly measured and error-prone components of the covariate vector. If the j-th component $X_i[j]$ is measured without error then $U_i[j] \equiv 0$.

and for the Cox model

$$\lambda(t|W_i) = \lambda_0(t) \cdot \mathbf{E}(\exp(\beta^T X_i)|W_i, \{T_i \geq t\}) . \tag{3}$$

Note that, via the event $\{T_i \geq t\}$ appearing in the condition, the second factor, the so called induced relative risk, depends on the previous history of the process. Because of this complex dependence on the unknown baseline hazard rate, the characteristic form of (1) is lost, and partial likelihood estimation can not be directly applied any more.[2]

However, as Prentice also argued, the effect of this time dependence can be expected to be small if the failure intensity is very low, because then the condition $\{T_i \geq t\}$ is almost always satisfied. Under this so called *rare disease assumption*,

$$\begin{aligned} \lambda(t|W_i) = \mathbf{E}\left(\lambda(t|X_i)|\{T_i \geq t\}, W_i\right) &\approx \mathbf{E}\left(\lambda(t|X_i)|W_i\right) \qquad (4) \\ &= \lambda_0(t) \cdot \mathbf{E}\left(\exp(\beta^T X_i)|W_i\right) , \end{aligned}$$

and the induced relative risk can be explicitly calculated in some cases approximately. (See [19, p. 1170] for a brief discussion of the exactness of this approximation.)

If $X_i|W_i$ is normal with mean $\bar{\mu}_i$ and common covariance $\bar{\Sigma}$, then

$$\lambda(t|W_i) = \lambda_0(t) \cdot \exp(\beta^T \bar{\mu}_i + 0.5\beta^T \bar{\Sigma}\beta) =: \lambda_0^*(t) \cdot \exp(\beta^T \bar{\mu}_i) . \tag{5}$$

One important example is the situation where X_i itself is i.i.d. normally distributed, with unknown mean μ_X and non-singular covariance matrix Σ_X. Then $W_i \sim \mathcal{N}(\mu_X; \Sigma_X + \Sigma_U)$, and indeed $X_i|W_i \sim \mathcal{N}\left(\bar{\mu}_i, \bar{\Sigma}\right)$, with

$$\bar{\mu}_i = \mu_X + \Sigma_X \cdot (\Sigma_X + \Sigma_U)^{-1} \cdot (W_i - \mu_X) \tag{6}$$

and $\bar{\Sigma} = \Sigma_X - \Sigma_X \cdot (\Sigma_X + \Sigma_U)^{-1}\Sigma_X$. Now (5) reads as

$$\begin{aligned} \lambda(t|W_i) &= \lambda_0^*(t) \cdot \exp(\beta^T \mu_X + \beta^T \Sigma_X \cdot (\Sigma_X + \Sigma_U)^{-1} \cdot (W_i - \mu_X) \\ &=: \lambda_0^{**}(t) \cdot \exp(\beta^T \cdot \Sigma_X \cdot (\Sigma_X + \Sigma_U)^{-1} \cdot W_i). \end{aligned}$$

This shows that, under the assumptions stated above, the simple attenuation factor known from linear regression also applies for the Cox model; the corrected estimate $\hat{\beta}_{corr}$ is

$$\hat{\beta}_{corr} = \Sigma_X^{-1} \cdot (\Sigma_X + \Sigma_U) \cdot \hat{\beta}_{naive} . \tag{7}$$

[2]Cf., however, [19, p. 1169], who characterize a family of distributions where the adapted partial likelihood can be dealt with in a closed form.

Notice further that, given the measurement error covariance Σ_U, the nuisance parameters μ_X and Σ_X can be efficiently estimated from the observations $W_1, \ldots, W_n$.

[13] arrived at (7) as an ad-hoc proposal, motivated by the attenuation known from linear regression. In a series of simulations he observes a strong dependence of the bias on the true β and on the amount of censoring. Both issues are in accordance with the deviation given above: the lower, ceteris paribus, the true β and the higher the proportion of censored observation are, the better the rare disease assumption (4) is satisfied.

Probably the most universal tool for measurement error correction is *regression calibration* (c.f., e.g., [7, Chapter 3]). One uses the knowledge of W_i to predict X_i and replaces X_i by its expectation given W_i. In general, the estimates derived are not necessarily consistent, but the bias is considerably reduced. The main advantage of the method is its easy implementation; simply by proceeding with $\mathbf{E}(X_i|W_i)$ instead of X_i in the calculation, the estimates are obtainable by standard software.

In the presence of validation data, regression calibration can be used in a functional way, because then $\mathbf{E}(X_i \mid W_i)$ can be estimated in a non-parametric manner. [21] elaborated this idea for the Cox model. If no validation data are available, structural modeling is necessary. In the simplest case also considered above, where X_i is i.i.d. normally distributed, regression calibration coincides with the method discussed above: Substituting in Equation (2) the variable X_i by its conditional expectation $\bar{\mu}_i$ from (6) yields the estimating equation

$$\sum_{j=1}^{k} \left(\bar{\mu_j} - \frac{\sum_{i\in\mathcal{R}(\tau_j)} \bar{\mu}_i \cdot \exp\left(\beta^T \bar{\mu}_i\right)}{\sum_{i\in\mathcal{R}(\tau_j)} \exp\left(\beta^T \bar{\mu}_i\right)} \right) = 0 .$$

After simplification and multiplication with $\left(\Sigma_X \cdot (\Sigma_X + \Sigma_U)^{-1}\right)^{-1}$ one obtains

$$\sum_{j=1}^{k} \left(W_j - \frac{\sum_{i\in\mathcal{R}(\tau_j)} W_i \cdot \exp\left(\beta^T \Sigma_X \cdot (\Sigma_X + \Sigma_U)^{-1} W_i\right)}{\sum_{i\in\mathcal{R}(\tau_j)} \exp\left(\beta^T \Sigma_X \cdot (\Sigma_X + \Sigma_U)^{-1} W_i\right)} \right) = 0, \quad (8)$$

which indeed leads again to (7).

Since assumptions on the distribution of the latent variable may influence the behaviour of the estimates, more flexible models for the distribution of X_i, for instance mixtures of normals, may be very attractive. The main arguments given in [2] to develop measurement error corrected quasi-likelihood estimation carry over to the situation considered here.

In this generalized setting, both structural approaches won't be equivalent any more.

Methods related to regression calibration are studied and developed further in [19, 8, 13]. As mentioned above, [21] integrate this approach into a model, where the conditional distribution needed to adjust for measurement error is estimated from validation data. An alternative way to incorporate validation data is discussed in [23].

Under the assumption of piecewise constant hazard rates and based on numerical integration, [11] developed three likelihood based methods, which differ with respect to the modelling of the covariate distribution (nonparametric, semiparametric and parametric.)

4.2 Functional approaches

There are also several strictly functional approaches. The most popular one is [18], where Nakamura adopts his general methodology of corrected score functions (cf. [17]) to the Cox model. Nakamura's basic idea is to construct unbiased estimating equations by looking for a function in the observable quantities $\text{vec}Y := (Y_1, \ldots, Y_n)$, $\text{vec}\Delta := (\Delta_1, \ldots, \Delta_n)$ and $\mathbf{W} := (W_1, \ldots, W_n)$ such that the conditional expectation given X_i is equal to the original score function. Then, by the theorem of iterative expectation, the overall expectation of this so called *corrected score function* equals zero, from which, under mild regularity conditions, consistency and asymptotic normality of the resulting estimate can be derived. This general framework, however, can not be directly applied to Equation (2) for partial likelihood estimation in the Cox model: the fact that the denominator possesses a (complex) singularity makes the existence of a corrected score function impossible.

Nakamura [18] therefore proposes an approximate solution based on a first and second order Taylor approximation of the fraction in (2). Denoting the naive estimating function by $\Psi^-(\text{vec}Y, \text{vec}\Delta, \mathbf{W}; \beta)$ and defining

$$K_j(\mathbf{W}) = 1 - \left(\sum_{i \in \mathcal{R}(\tau_j)} \exp\left(2W_i{}^T\beta\right) \right) \cdot \left(\sum_{i \in \mathcal{R}(\tau_j)} \exp\left(W_i{}^T\beta\right) \right)^{-2},$$

one obtains the first order corrected estimating function Ψ_1 and the second order correction Ψ_2 as

$$\Psi_1(\text{vec}Y, \text{vec}\Delta; \mathbf{W}; \beta) = \Psi^-(\text{vec}Y, \text{vec}\Delta; \mathbf{W}; \beta) + \Sigma_U \cdot \beta \quad (9)$$

$$\Psi_2(\text{vec}Y, \text{vec}\Delta; \mathbf{W}; \beta) = \Psi^-(\text{vec}Y, \text{vec}\Delta; \mathbf{W}; \beta) + \sum_{j=1}^{k} (K_j(\mathbf{W}))\, \Sigma_U \beta$$

[15] study asymptotic properties of the resulting first order estimate and suggest an extension to non-normal measurement error. [3] gives a justification of (9) as an exact corrected likelihood estimate for Breslow's (cf. [4, 5]) likelihood approach to the Cox model. Based on replication data, [12] develop a nonparametric correction method which manages to do without parametric assumptions on the measurement error.

Another functional approach is studied by [6], who obtains a different unbiased score equation. [14] derives an expression for the asymptotic bias of the naive partial likelihood estimate, which can also be used for bias reduction.

5. Simulation study

By simulation we examined the behaviour of the basic estimates in the situation of one normally distributed covariate under normally distributed, homoscedastic measurement error. We compared the naive estimate, the first and second order Nakamura estimates and the two elementary structural methods, which coincide here. We varied the sample size, the distribution of the true covariate, the amount of censorship, the measurement error variance Σ_U and the underlying distribution of the survival time. Here is a brief summary of the results:

We generated Weibull distributed survival times. Varying the shape parameter had some effect on the magnitude of the bias, but did not change the phenomenological picture depicted below. As may be broadly expected, for very small measurement error the naive estimate still is superior, while for large measurement error the bias is intolerable: for $\Sigma_U = \Sigma_X$ the observed attenuation is about one half.

5.1 Structural correction

In the designs we studied, the amount of censoring did not play a very important role: this suggests that the rare disease assumption (4) underlying (7) may be interpreted liberally to some degree (c.f., however, [13], and Section 4.1.) The effect of the distribution of the true covariate X_i was surprisingly small. Misspecification (by a symmetric mixture of normals and a uniform distribution) did not substantially worsen the behaviour of the structural methods. This insensitivity, on the other hand, is also responsible for the fact that the structural methods were not able to beat Nakumara's first and second order estimates even in situations where all the assumptions on which the structural methods are based were fairly met.

5.2 The Nakamura estimates

The estimates obtained from Nakamura's correction method showed some quite remarkable features: the often observed excellent behaviour was contrasted by sometimes completely wrong results and many numerical difficulties. In a, by far not negligible, number of situations we were confronted with the problems of non-convergence or of wrong convergence, which also had been reported by some other authors. As described in [15], the estimate may not always exist because the derivative of the corrected score functions is not always negative in the neighborhood of the true β. This effect regularly happens when the measurement error gets large. Even when the measurement error variance was half of the covariates variance, Nakamura's estimates often failed to converge. We additionally want to stress that a lot of care is needed with respect to the numerical calculation of the root of the corrected estimating equations. Experimenting with different root finders we got quite often completely different estimates. This is in particular urgent for the second order estimate and if the slope of the hazard rate differs considerably from zero. The need to use numerically expensive procedures makes calculation of the estimate rather slow.

When we restricted ourselves to those situations where the solutions were apparently reliable, the bias was very small indeed. Then Nakamura's methodology proved to be very powerful for correcting the measurement error. After having removed the 'outlying values', the functional methods performed almost always better than the structural methods. As already said above, this was even true in that situation which was used to derive und justify the structural approaches.

Comparing both Nakamura estimates with each other, the second order estimate was in most situations slightly superior to the first order estimate, as long as no numerical problems appeared. Whether this gain in bias reduction is large enough to compensate the additional numerical difficulties and the higher danger to produce artefacts, has to be decided on a case-by-case basis.

6. Concluding Remarks

In this paper we reviewed methods to estimate regression parameters in the Cox model under homoscedastic measurement error. In some more detail, we discussed Nakamura's method as well as the application of the two main structural methods, which were additionally shown to be equivalent in a special case. Then the basic estimates were compared in a simulation study. The overall conclusion resulting from it is somewhat ambiguous, because the Nakamura estimates showed very extreme behaviour. On the one hand, they can lead to a lot of numerical difficulties and may produce artificial results, on the other hand, in those constellations where they behave not irregularly, they are very powerful.

We did not discuss the estimation of the baseline hazard rate under measurement error. Results on this issue can be found in [21, 16, 15, 12, 3]. A comparison of the different methods is still lacking.

Another topic of further research, also quite important for practical application, is the extension of the methods to heteroscedastic measurement error. For instance, this is of particular interest in nutritional studies, where subject matter considerations suggest comparatively high heteroscedastic measurement errors (see, e.g., [22, page 33-48].)

In the last years parametric survival models have attracted much attention, but up to now not much is known how to correct for measurement error in this context. Some results are directly available from [17] and are extended by [10]. A structural approach to measurement error correction in parametric survival models is proposed in [1].

Acknowledgments

This work was partially supported by the Deutsche Forschungsgemeinschaft (DFG) within the frame of the Sonderforschungsbereich SFB386. We thank Angela Döring, Helmut Küchenhoff and Hans Schneeweiß for valuable comments and discussions.

References

[1] T. Augustin. Correcting for measurement error in parametric duration models by quasi likelihood. *Biometrical Journal,* accepted for publication, 2001.

[2] T. Augustin. Some basic results on the extension of quasi-likelihood based measurement error correction to multivariate and flexible structural models. In W. Gaul and G. Ritter, editors, *Classification, Automation, and New Media.* Springer, 2001. To appear.

[3] T. Augustin. *An exact corrected log-likelihood for Cox's proportional hazards model,* 2001. Manuscript, University of Munich.

[4] N.E. Breslow. Contribution to Cox (1972). *Journal of the Royal Statistical Society, Series B,* 34:216–217, 1972.

[5] N.E. Breslow. Covariance analysis of censored survival data. *Biometrics,* 30:89–99, 1974.

[6] J. Buzas. Unbiased scores in proportional hazards regression with covariate measurement error. *Journal of Statistical Planning and Inference,* 67:247–257, 1998.

[7] R.J. Caroll, D. Ruppert, and L.A. Stefanski. *Measurement Error in Nonlinear Models.* Chapman and Hall, 1995.

[8] D.G. Clayton. Models for the analysis of cohort and case-control studies with inaccurately measured exposures. In J.H. Dwyer, M. Feinleib, and Lipsert et al.,

P., editors, *Statistical Models for Longitudinal Studies of Health*, pages 301–331. Oxford University Press, 1991.

[9] D.R. Cox. Regression models and life tables (with discussion). *Journal of the Royal Statistical Society, Series B*, 34:187–220, 1972.

[10] P. Gimenez, H. Bolfarine, and E.A. Colosimo. Estimation in Weibull regression model with measurement error. *Communications Statistics. Theory and Methods*, 28:495–510, 1999.

[11] P. Hu, A. Tsiatis, and M. Davidian. Estimating the parameters in the Cox model when covariate variables are measured with error. *Biometrics*, 54:1407–1419, 1998.

[12] Y. Huang and C.Y. Wang. Cox regression with accurate covariates unascertainable: A nonparametric-correction approach. *Journal of the American Statistical Association*, 45:1209–1219, 2000.

[13] M.D. Hughes. Regression dilution in the proportional hazards model. *Biometrics*, 49:1056–1066, 1993.

[14] F.H. Kong. Adjusting regression attenuation in the Cox proportional hazards model. *Journal of Statistical Planning and Inference*, 79:31–44, 1999.

[15] F.H. Kong and M. Gu. Consistent estimation in Cox proportional hazards model with covariate measurement errors. *Statistica Sinica*, 9:953–969, 1999.

[16] F.H. Kong, W. Huang, and X. Li. Estimating survival curves under proportional hazards model with covariate measurement errors. *Scandinavian Journal of Statistics*, 25:573–587, 1998.

[17] T. Nakamura. Corrected score functions for errors-in-variables models: Methodology and application to generalized linear models. *Biometrika*, 77:127–137, 1990.

[18] T. Nakamura. Proportional hazards model with covariates subject to measurement error. *Biometrics*, 48:829–838, 1992.

[19] M.S. Pepe, S.G. Self, and R. L. Prentice. Further results in covariate measurement errors in cohort studies with time to response data. *Statistics in Medicine*, 8:1167–1178, 1989.

[20] R.L. Prentice. Covariate measurement errors and parameter estimation in a failure time regression model. *Biometrika*, 69:331–342, 1982.

[21] C.Y. Wang, L. Hsu, Z.D. Feng, and R.L. Prentice. Regression calibration in failure time regression. *Biometrics*, 53:131–145, 1997.

[22] W. Willett. *Nutritional Epidemiology*. Oxford Univ. Press, 1998.

[23] H. Zhou and C.Y. Wang. Failure time regression with continuous covariates measured with error. *Journal of the Royal Statistical Society, Series B*, 62:657–665, 2000.

Part V

ERRORS-IN-VARIABLES MODELING WITH BOUNDED UNCERTAINTIES

STATE-SPACE ESTIMATION WITH UNCERTAIN MODELS

Ali H. Sayed and Ananth Subramanian
Adaptive Systems Laboratory, Electrical Engineering Department,
University of California, Los Angeles, CA 90095, USA
{sayed,msananth}@ee.ucla.edu

Abstract There are always discrepancies between design models and the actual physical systems or phenomena that they model. Regardless of their source, such perturbations can degrade the performance of otherwise optimal designs. This article discusses a design strategy for models with bounded perturbations. In comparison to other robust formulations, the resulting procedure performs data regularization as opposed to de-regularization. Applications in state-space estimation and adaptive filtering are discussed.

Keywords: regularization, least-squares, robust filter, adaptive filter, Kalman filter, parametric uncertainty.

1. Introduction

Many estimation and control problems rely on solving regularized least-squares formulations of the form (see, e.g., [1]–[3]):

$$\min_x \left[x^T Q x + (Ax-b)^T W (Ax-b)\right] \tag{1}$$

where $x^T Q x$ is a regularization term, $Q > 0$ and $W \geq 0$ are Hermitian weighting matrices, x is an unknown $n-$dimensional column vector, A is a known $N \times n$ data matrix, and b is a known $N \times 1$ measurement vector. The solution of (1) is

$$\hat{x} = \left[Q + A^T W A\right]^{-1} A^T W b \tag{2}$$

When the nominal data $\{A, b\}$ are subject to uncertainties, especially large ones, the performance of the optimal estimator (2) may degrade appreciably — see [3]. For example, if the actual data matrix were $(A + \delta A)$, for some unknown perturbation δA, then the estimator (2)

S. Van Huffel and P. Lemmerling (eds.), Total Least Squares and Errors-in-Variables Modeling, 191–202.

that is designed based on A alone and without accounting for the existence of δA, can perform poorly. In [4,5], a generalization of (1) that accounts for uncertainties in $\{A, b\}$ was introduced. This generalization leads to a robust solution that performs regularization as opposed to de-regularization; a property that is useful for on-line (iterative) schemes since it avoids the need for checking existence conditions. This chapter has two objectives. First, it provides an overview of a special case of the robust formulation of [4,5] along with an illustrative application in the context of state-space estimation. Then it extends the analysis to the case of recursive least-squares filtering and discusses an application in the context of adaptive filtering. The articles [3]–[7] provide references and connections to other related works in the literature (e.g., [11]–[13]). For reasons of space, we omit this discussion here and refer interested readers to [3]–[7] and [10] and the references therein.

Notation. For a column vector z and a positive-definite matrix W, we write $\|z\|_W^2$ to denote the weighted Euclidean norm, z^TWz. For brevity, we may sometimes write $A^TW(\cdot)$ instead of A^TWA especially when the factor A admits a long expression (see, e.g., Eq. (27)).

2. Uncertain weighted least-squares

Let $J(x, y)$ denote a cost function of the form $J(x, y) = x^TQx + R(x, y)$ with

$$R(x,y) = \Big((A+\delta A)x - (b+\delta b)\Big)^T W \Big((A+\delta A)x - (b+\delta b)\Big) \tag{3}$$

Here δA denotes an $N \times n$ perturbation to A, δb denotes an $N \times 1$ perturbation to b, and $\{\delta A, \delta b\}$ are assumed to satisfy a model of the form

$$\begin{bmatrix} \delta A & \delta b \end{bmatrix} = H\Delta \begin{bmatrix} E_a & E_b \end{bmatrix} \tag{4}$$

where Δ is an arbitrary contraction, $\|\Delta\| \le 1$, and $\{H, E_a, E_b\}$ are known quantities of appropriate dimensions (e.g., E_b is a column vector). Consider now the problem of solving

$$\hat{x} = \arg\min_x \max_{\{\delta A, \delta b\}} J(x,y) \tag{5}$$

subject to (4). We shall assume that H and $\{E_a, E_b\}$ are not identically zero since otherwise problem (5) trivializes to (1). The statement (5) can be interpreted as a constrained two-player game problem, with the designer trying to pick an estimate $\hat{x}$ that minimizes the cost while the opponent $\{\delta A, \delta b\}$ tries to maximize the cost. The following result is proven in [4,5].

Theorem 1. *The problem (4)–(5) has a unique solution $\hat{x}$ that is given by (compare with (2))*

$$\hat{x} = \left[\widehat{Q} + A^T \widehat{W} A\right]^{-1} \left[A^T \widehat{W} b + \widehat{\beta} E_a^T E_b\right] \tag{6}$$

where $\{\widehat{Q}, \widehat{W}\}$ are obtained from $\{Q, W\}$ via

$$\widehat{Q} \triangleq Q + \widehat{\beta} E_a^T E_a \tag{7}$$

$$\widehat{W} \triangleq W + WH(\widehat{\beta} I - H^T W H)^\dagger H^T W \tag{8}$$

and the scalar $\widehat{\beta}$ is determined from the optimization

$$\widehat{\beta} = \arg \min_{\beta \geq \|H^T W H\|} G(\beta) \tag{9}$$

where the function $G(\beta)$ is defined as follows:

$$G(\beta) \triangleq x^T(\beta) Q x(\beta) + \beta \|E_a x(\beta) - E_b\|^2 + [Ax(\beta) - b]^T W(\beta) [Ax(\beta) - b] \tag{10}$$

with

$$W(\beta) \triangleq W + WH(\beta I - H^T W H)^\dagger H^T W \tag{11}$$

$$Q(\beta) \triangleq Q + \beta E_a^T E_a \tag{12}$$

and

$$x(\beta) \triangleq \left[Q(\beta) + A^T W(\beta) A\right]^{-1} \left[A^T W(\beta) b + \beta E_a^T E_b\right] \tag{13}$$

We shall denote the lower bound on β in (9) by $\beta_l = \|H^T W H\|$. Compared with the solution (2) of the standard regularized least-squares problem (1), we see that the expression for $\hat{x}$ in (6) is distinct in two important ways:

a) First, the weighting matrices $\{Q, W\}$ need to be replaced by corrected versions $\{\widehat{Q}, \widehat{W}\}$. These corrections are defined in terms of a scalar parameter $\widehat{\beta}$, which is obtained as the minimizing argument of a function $G(\beta)$ over the semi-open interval $[\beta_l, \infty)$.

b) Second, the right-hand side of (6) contains an additional term that is equal to $\widehat{\beta} E_a^T E_b$. This means that, with $\widehat{\beta}$ given or fixed, the $\hat{x}$ in (6) can be interpreted as the solution to a regularized least-squares problem of the form

$$\min_x \left(\begin{bmatrix} 1 & x^T \end{bmatrix} \begin{bmatrix} \widehat{\beta}\|E_b\|^2 & -\widehat{\beta} E_b^T E_a \\ -\widehat{\beta} E_a^T E_b & \widehat{Q} \end{bmatrix} \begin{bmatrix} 1 \\ x \end{bmatrix} + (Ax - b)^T \widehat{W} (Ax - b) \right)$$

with a *cross-coupling* term between x and unity.

The complexity of the solution of Theorem 1 is therefore comparable to that of a standard regularized least-squares problem with the additional burden of determining the optimal scalar parameter $\widehat{\beta}$ by minimizing the cost function $G(\beta)$ over the interval $[\beta_l, \infty)$. It is shown in [5] that the function $G(\beta)$ has a unique global minimum (and no local minima) inside this interval, which means that the determination of $\widehat{\beta}$ can be pursued by employing standard search procedures without worrying about convergence to undesired local minima.

Before proceeding we note that in the applications described further ahead, it will be the case that the weighting matrix W is positive-definite so that we will always have $W(\beta) > 0$. Therefore, if we restrict the minimization in (9) to the open interval (β_l, ∞) – i.e., if we exclude the boundary point β_l – then the pseudo-inverse operation in (11) can be replaced by normal matrix inversion, and it will hold that

$$W^{-1}(\beta) = W^{-1} - \beta^{-1}HH^T. \tag{14}$$

We now move on to apply the above results to two applications: one is in the context of state-space estimation and the other is in the context of adaptive filtering.

3. Robust state-space estimation

As is well-known, the Kalman filter is the optimal linear least-mean-squares state estimator for linear state-space models [1]. When the underlying model is subject to parametric uncertainties, the performance of the filter may deteriorate appreciably. The mismatch between the actual model and the assumed nominal model can be due to various factors including unmodelled dynamics and approximations during the system identification process.

In [7], a robust formulation for state-space estimation that is based on Theorem 1 has been proposed. Compared with the standard Kalman filter, which is known to minimize the regularized residual norm at each iteration, the new formulation minimizes the worst-possible regularized residual norm over the class of admissible uncertainties. In addition, compared with other robust formulations, the resulting filter performs data regularization rather than de-regularization; a property that avoids the need for existence conditions. We review the main steps below.

Thus consider a state-space model of the form

$$\begin{aligned} x_{i+1} &= F_i x_i + G_i u_i, \quad i \geq 0 && (15) \\ y_i &= H_i x_i + v_i && (16) \end{aligned}$$

where $\{x_0, u_i, v_i\}$ are uncorrelated zero-mean random variables with variances

$$\mathbf{E}\begin{bmatrix} x_0 \\ u_i \\ v_i \end{bmatrix}\begin{bmatrix} x_0 \\ u_j \\ v_j \end{bmatrix}^T = \begin{bmatrix} \Pi_0 & 0 & 0 \\ 0 & Q_i\delta_{ij} & 0 \\ 0 & 0 & R_i\delta_{ij} \end{bmatrix} \tag{17}$$

that satisfy $\Pi_0 > 0$, $R_i > 0$, and $Q_i > 0$. Let further

$$\begin{aligned} \hat{x}_i &\triangleq \text{l.l.m.s. estimate of } x_i \text{ given } \{y_0, \ldots, y_{i-1}\} \\ \hat{x}_{i|i} &\triangleq \text{l.l.m.s. estimate of } x_i \text{ given } \{y_0, \ldots, y_i\} \end{aligned}$$

with corresponding error variances P_i and $P_{i|i}$, respectively. The notation l.l.m.s. stands for "linear least-mean-squares". Then $\{\hat{x}_i, \hat{x}_{i|i}\}$ can be constructed recursively via the following time- and measurement-update time-update form of the Kalman filter (see, e.g., [1]):

$$\hat{x}_{i+1} = F_i\hat{x}_{i|i}, \quad i \geq 0 \tag{18}$$

$$\hat{x}_{i+1|i+1} = \hat{x}_{i+1} + P_{i+1|i+1}H_{i+1}^T R_{i+1}^{-1} e_{i+1} \tag{19}$$

$$e_{i+1} = y_{i+1} - H_{i+1}\hat{x}_{i+1} \tag{20}$$

$$P_{i+1} = F_i P_{i|i} F_i^T + G_i Q_i G_i^T \tag{21}$$

$$P_{i+1|i+1} = P_{i+1} - P_{i+1}H_{i+1}^T R_{e,i+1}^{-1} H_{i+1} P_{i+1} \tag{22}$$

$$R_{e,i+1} = R_{i+1} + H_{i+1}P_{i+1}H_{i+1}^T \tag{23}$$

with initial conditions

$$\hat{x}_{0|0} = P_{0|0}^{-1} H_0^T R_0^{-1} y_0, \quad P_{0|0} = (\Pi_0^{-1} + H_0^T R_0^{-1} H_0)^{-1}$$

It can also be verified that these equations are equivalent to the following prediction form of the Kalman filter:

$$\hat{x}_{i+1} = F_i\hat{x}_i + K_i R_{e,i}^{-1} e_i \tag{24}$$

$$P_{i+1} = F_i P_i F_i^T + G_i Q_i G_i^T - K_i R_{e,i}^{-1} K_i^T \tag{25}$$

$$K_i = F_i P_i H_i^T, \quad R_{e,i} = R_i + H_i P_i H_i^T \tag{26}$$

with initial conditions $\hat{x}_0 = 0$ and $P_0 = \Pi_0$.

Each step (18)–(23) of the time- and measurement-update form admits a useful deterministic interpretation as the solution to a regularized least-squares problem as follows (see, e.g., [8]). Given $\{\hat{x}_{i|i}, P_{i|i} > 0, y_{i+1}\}$, consider the problem of estimating x_i again, along with u_i, by solving

$$\min_{\{x_i, u_i\}} \left((x_i - \hat{x}_{i|i})^T P_{i|i}^{-1}(\cdot) + u_i^T Q_i^{-1} u_i + (y_{i+1} - H_{i+1}x_{i+1})^T R_{i+1}^{-1}(\cdot) \right) \tag{27}$$

If we make the substitution $x_{i+1} = F_i x_i + G_i u_i$, then the cost in (27) reduces to a regularized least-squares cost of the form (1) with the identifications

$$\begin{array}{rcl} x & \longleftarrow & \text{col}\{x_i - \hat{x}_{i|i},\ u_i\}, \quad b \longleftarrow y_{i+1} - H_{i+1}F_i\hat{x}_{i|i} \\ A & \longleftarrow & H_{i+1}\left[\begin{array}{cc} F_i & G_i \end{array}\right], \quad Q \longleftarrow (P_{i|i}^{-1} \oplus Q_i^{-1}), \quad W \longleftarrow R_{i+1}^{-1} \end{array}$$

The solution of this problem can be shown to lead to (18)–(23).

Assume now that the model (15)–(16) is uncertain, say

$$x_{i+1} = (F_i + \delta F_i)x_i + (G_i + \delta G_i)u_i \tag{28}$$

$$y_i = H_i x_i + v_i \tag{29}$$

$$\left[\begin{array}{cc} \delta F_i & \delta G_i \end{array}\right] = M_i \Delta_i \left[\begin{array}{cc} E_{f,i} & E_{g,i} \end{array}\right] \tag{30}$$

for some known matrices $\{M_i, E_{f,i}, E_{g,i}\}$ and for an arbitrary contraction Δ_i. Assume further that at step i we are given an *a priori* estimate for x_i, say $\hat{x}_{i|i}$, and a positive-definite weighting matrix $P_{i|i}$. Using y_{i+1}, we may update the estimate of x_i from $\hat{x}_{i|i}$ to $\hat{x}_{i|i+1}$ by solving

$$\min_{\{x_i,u_i\}} \max_{\{\delta F_i, \delta G_i\}} \left((x_i - \hat{x}_{i|i})^T P_{i|i}^{-1}(\cdot) + u_i^T Q_i^{-1} u_i + (y_{i+1} - H_{i+1}x_{i+1})^T R_{i+1}^{-1}(\cdot) \right) \tag{31}$$

subject to (28)–(30). This problem can be seen to be the robust version of (27) in the same way that (3)–(5) is the robust version of (1). Now (31) can be written more compactly in the form (3)–(5) with the identifications:

$$\begin{array}{rcl} x & \longleftarrow & \text{col}\{x_i - \hat{x}_{i|i},\ u_i\}, \quad b \longleftarrow y_{i+1} - H_{i+1}F_i\hat{x}_{i|i} \\ \delta A & \longleftarrow & H_{i+1}M_i\Delta_i \left[\begin{array}{cc} E_{f,i} & E_{g,i} \end{array}\right] \\ \delta b & \longleftarrow & -H_{i+1}M_i\Delta_i E_{f,i}\hat{x}_{i|i}, \quad Q \longleftarrow (P_{i|i}^{-1} \oplus Q_i^{-1}) \\ W & \longleftarrow & R_{i+1}^{-1}, \ H \longleftarrow H_{i+1}M_i, \ E_a \longleftarrow \left[\begin{array}{cc} E_{f,i} & E_{g,i} \end{array}\right] \\ E_b & \longleftarrow & -E_{f,i}\hat{x}_{i|i}, \ \Delta \longleftarrow \Delta_i, \ A \longleftarrow H_{i+1}\left[\begin{array}{cc} F_i & G_i \end{array}\right] \end{array}$$

Using Theorem 1, and some considerable algebra, we arrive at the equations listed in Table 1 where we defined

$$\beta_{l,i} \triangleq \|M_i^T H_{i+1}^T R_{i+1}^{-1} H_{i+1} M_i\| \tag{32}$$

The major step in the algorithm of Table 1 is step 3, which consists of recursions that are very similar in nature to the prediction form of the Kalman filter. The main difference is that the new recursions operate on modified parameters rather than on the given nominal values. In addition, the recursion for P_i is not a standard Riccati recursion since the product $\widehat{G}_i\widehat{Q}_i\widehat{G}_i^T$ is also dependent on P_i. However, in some special cases, the recursion for P_i collapses to a Riccati recursion. In addition, alternative equivalent implementations of the robust filter of Table 1 in

Table 1. Listing of the robust state-space estimation algorithm in prediction form.

Assumed uncertain model. Eqs. (28)–(30).

Initial conditions: $\hat{x}_0 = 0$, $P_0 = \Pi_0$, and $\widehat{R}_0 = R_0$.

Step 1a. Using $\{\widehat{R}_i, H_i, P_i\}$ compute $\{R_{e,i}, P_{i|i}\}$:

$$R_{e,i} = \widehat{R}_i + H_i P_i H_i^T, \quad P_{i|i} = P_i - P_i H_i^T R_{e,i}^{-1} H_i P_i$$

Step 1b. If $H_{i+1}M_i = 0$, then set $\widehat{\beta}_i = 0$. Otherwise, determine $\widehat{\beta}_i$ by minimizing the corresponding $G(\beta)$ over the interval $(\beta_{l,i}, \infty)$.

Step 2. Compute the corrected parameters:

$$\begin{aligned}
\widehat{Q}_i^{-1} &= Q_i^{-1} + \widehat{\beta}_i E_{g,i}^T \left[I + \widehat{\beta}_i E_{f,i} P_{i|i} E_{f,i}^T\right]^{-1} E_{g,i} \\
\widehat{R}_{i+1} &= R_{i+1} - \widehat{\beta}_i^{-1} H_{i+1} M_i M_i^T H_{i+1}^T \\
\widehat{P}_{i|i} &= P_{i|i} - P_{i|i} E_{f,i}^T (\widehat{\beta}_i^{-1} I + E_{f,i} P_{i|i} E_{f,i}^T)^{-1} E_{f,i} P_{i|i} \\
\widehat{G}_i &= G_i - \widehat{\beta}_i F_i \widehat{P}_{i|i} E_{f,i}^T E_{g,i} \\
\widehat{F}_i &= (F_i - \widehat{\beta}_i \widehat{G}_i \widehat{Q}_i E_{g,i}^T E_{f,i})(I - \widehat{\beta}_i \widehat{P}_{i|i} E_{f,i}^T E_{f,i})
\end{aligned}$$

If $\widehat{\beta}_i = 0$, then simply set $\widehat{Q}_i = Q_i$, $\widehat{R}_{i+1} = R_{i+1}$, $\widehat{P}_{i|i} = P_{i|i}$, $\widehat{G}_i = G_i$, and $\widehat{F}_i = F_i$.

Step 3. Now update $\{\hat{x}_i, P_i\}$ to $\{\hat{x}_{i+1}, P_{i+1}\}$ as follows:

$$\begin{aligned}
\hat{x}_{i+1} &= \widehat{F}_i \hat{x}_i + \widehat{F}_i P_i H_i^T R_{e,i}^{-1} e_i \\
e_i &= y_i - H_i \hat{x}_i \\
P_{i+1} &= F_i P_i F_i^T - \overline{K}_i \overline{R}_{e,i}^{-1} \overline{K}_i^T + \widehat{G}_i \widehat{Q}_i \widehat{G}_i^T \\
\overline{K}_i &= F_i P_i \overline{H}_i^T, \quad \overline{R}_{e,i} = I + \overline{H}_i P_i \overline{H}_i^T
\end{aligned}$$

where $\overline{H}_i^T = \begin{bmatrix} H_i^T \widehat{R}_i^{-T/2} & \sqrt{\widehat{\beta}_i} E_{f,i}^T \end{bmatrix}$.

information form and in time- and measurement-update form are also possible (see [7]).

Observe that the algorithm of Table 1 requires, at each iteration i, the minimization of $G(\beta)$ over $(\beta_{l,i}, \infty)$. It turns out that a reasonable approximation that avoids these repeated minimizations is to choose $\widehat{\beta}_i = (1+\alpha)\beta_{l,i}$. That is, we set $\widehat{\beta}_i$ at a multiple of the lower bound — if the lower bound is zero, we set $\widehat{\beta}_i$ to zero and replace $\widehat{\beta}_i^{-1}$ by $\widehat{\beta}_i^{\dagger}$ (which is also zero). The parameter α could be made time-variant; it serves as a "tuning" parameter that can be adjusted by the designer.

4. Robust adaptive filtering

In the model (28)–(30), we did not consider uncertainties in the output equation (29). We now look at this case, an application of which arises in the context of adaptive filtering.

Thus consider noisy measurements $\{d(i)\}$ that satisfy a model of the form,

$$d(i) = u_i w^o + n(i), \tag{33}$$

where w^o is an unknown $M \times 1$ column vector that we wish to estimate, u_i is a $1 \times M$ regression vector, and $n(i)$ is measurement noise. Given a collection of $(N+1)$ such data points, $\{d(i)\}_{i=0}^{N}$, and the corresponding $(N+1)$ regressors $\{u_i\}_{i=0}^{N}$, the exponentially-weighted least-squares problem estimates w^o by solving

$$\min_{w} \left[\mu\lambda^{N+1}\|w\|^2 + \sum_{i=0}^{N} \lambda^{N-i}|d(i) - u_i w|^2 \right], \tag{34}$$

where μ is a positive scalar regularization parameter and λ is a forgetting factor satisfying $0 \ll \lambda < 1$. In this way past data are exponentially weighted less than recent data.

Let w_N denote the optimal solution of (34), and let w_{N-1} denote the solution to a least-squares problem similar to (34) with data up to time $N-1$ (and with $\mu\lambda^{N+1}$ replaced by $\mu\lambda^{N}$). The well-known recursive least-squares (RLS) algorithm allows us to update w_{N-1} to w_N as follows:

$$\begin{aligned} w_N &= w_{N-1} + g_N \epsilon(N) && (35)\\ \epsilon(N) &= d(N) - u_N w_{N-1} && (36)\\ g_N &= \lambda^{-1}\overline{P}_{N-1}u_N^T \gamma(N) && (37)\\ \gamma^{-1}(N) &= 1 + \lambda^{-1} u_N \overline{P}_{N-1} u_N^T && (38)\\ \overline{P}_N &= \lambda^{-1}\overline{P}_{N-1} - g_N \gamma^{-1}(N) g_N^T && (39) \end{aligned}$$

with initial conditions $w_{-1} = 0$ and $\overline{P}_{-1} = \mu^{-1} I$.

Now there is an intrinsic relation between exponentially-weighted RLS and Kalman filtering. More specifically, it was shown in [9] that the RLS equations (35)–(39) can be obtained directly from the Kalman filtering equations that correspond to a particular state-space model. This model is constructed as follows. Define the scaled variables

$$y(i) \triangleq d(i)/\sqrt{\lambda^i}, \qquad x_i \triangleq w^o/\sqrt{\lambda^i} \tag{40}$$

and introduce the state-space model

$$x_{i+1} = \lambda^{-1/2} x_i, \quad i \geq 0,$$

$$y(i) \quad = \quad u_i x_i + v(i), \tag{41}$$

where $\{x_0, v(i)\}$ are taken as uncorrelated zero-mean random variables with variances $\mu^{-1}\lambda^{-1}I$ and δ_{ij}, respectively. Here δ_{ij} is the Kronecker delta function that is equal to unity when $i = j$ and zero elsewhere. The Kalman filter equations in this case collapse to the following:

$$\hat{x}_{i+1} = \lambda^{-1/2}\hat{x}_i + \lambda^{-1/2}P_i u_i^T R_{e,i}^{-1}[y(i) - u_i\hat{x}_i]$$

where

$$P_{i+1} = \lambda^{-1}P_i - K_i R_{e,i}^{-1} K_i^T, \quad K_i = \lambda^{-1/2}P_i u_i^T, \quad R_{e,i} = 1 + u_i P_i u_i^T$$

with initial conditions $\hat{x}_0 = 0$ and $P_0 = \mu^{-1}\lambda^{-1}I$. It can be verified that the RLS variables $\{\overline{P}_i, w_i\}$ and the Kalman filtering variables $\{P_i, \hat{x}_{i+1}\}$ are related as follows:

$$\hat{x}_{i+1} = w_i/\sqrt{\lambda^{(i+1)}}, \qquad \overline{P}_i = \lambda P_{i+1}$$

When the regressors $\{u_i\}$ are subject to uncertainties, we can instead consider the model

$$x_{i+1} \quad = \quad \lambda^{-1/2}x_i, \quad i \geq 0, \tag{42}$$
$$y(i) \quad = \quad (u_i + \delta u_i)x_i + v(i). \tag{43}$$

This would correspond to assuming that the observations $\{d(i)\}$ in (33) actually arise from the perturbed linear regression model

$$d(i) = (u_i + \delta u_i)w^o + n(i). \tag{44}$$

The perturbations $\{\delta u_i\}$ are now row vectors and they are modeled as

$$\delta u_i \quad = \quad M_i \Delta_i E_{u,i} \tag{45}$$

where M_i is a known scaling factor, Δ_i is an unknown scalar that is bounded by unity (i.e., $|\Delta_i| \leq 1$), and $E_{u,i}$ is a known row vector. The model (44)–(45) allows us to account for uncertainties in the regression vectors u_i in different ways. Consider, for example, the case when M_i is a positive constant, say $M_i = \eta$, and choose

$$E_{u,i} = [1\ 1\ \ldots\ 1] \tag{46}$$

Then these choices for $\{M_i, E_{u,i}\}$ amount to assuming that each entry of u_i is perturbed by an amount that is bounded by η. We can also vary the degree of uncertainty in the individual entries of u_i by assigning different values to the entries of $E_{u,i}$. In addition, the fact that M_i and

Table 2. Listing of the robust adaptive algorithm.

Assumed uncertain model. Eqs. (44)–(45).

Initial conditions: $w_{-1} = 0$, $P_0 = \frac{1}{\mu\lambda} I$, and $\widehat{R}_0 = 1$.

Step 1a. Using $\{\widehat{R}_i, u_i, P_i\}$ compute $\{R_{e,i}, P_{i|i}\}$:

$$R_{e,i} = \widehat{R}_i + u_i P_i u_i^T, \quad P_{i|i} = P_i - P_i u_i^T R_{e,i}^{-1} u_i P_i$$

Step 1b. If $u_{i+1} M_{i+1} = 0$, then set $\widehat{\beta}_i = 0$. Otherwise, construct $G(\beta)$ of (10) with the identifications (48)–(53) and determine $\widehat{\beta}_i$ by minimizing $G(\beta)$ over the interval $\widehat{\beta}_i > \beta_{l,i} = M_{i+1}^2$.

Step 2. Compute the corrected parameters

$$\begin{aligned} \widehat{R}_{i+1} &= 1 - M_{i+1}^2 / \widehat{\beta}_i \\ \widehat{P}_{i|i} &= P_{i|i} - \lambda^{-1} P_{i|i} E_{u,i+1}^T (\widehat{\beta}_i^{-1} I + \lambda^{-1} E_{u,i+1} P_{i|i} E_{u,i+1}^T)^{-1} E_{u,i+1} P_{i|i} \\ \widehat{F}_i &= (I - \widehat{\beta}_i \lambda^{-1} \widehat{P}_{i|i} E_{u,i+1}^T E_{u,i+1}) \end{aligned}$$

If $\widehat{\beta}_i = 0$, then simply set $\widehat{P}_{i|i} = P_{i|i}$, $\widehat{R}_{i+1} = 1$, and $\widehat{F}_i = I$.

Step 3. Update $\{w_i, P_i\}$ as follows:

$$\begin{aligned} w_i &= \widehat{F}_i w_{i-1} + \widehat{F}_i P_i u_i^T R_{e,i}^{-1} e(i) \\ e(i) &= d(i) - u_i w_{i-1} \\ \overline{K}_i &= \lambda^{-1/2} P_i \overline{u}_i^T, \quad \overline{R}_{e,i} = 1 + \overline{u}_i P_i \overline{u}_i^T \\ P_{i+1} &= \lambda^{-1} P_i - \overline{K}_i \overline{R}_{e,i}^{-1} \overline{K}_i^T \end{aligned}$$

where

$$\overline{u}_i \triangleq \begin{bmatrix} u_i / \sqrt{\widehat{R}_i} \\ \sqrt{\widehat{\beta}_i / \lambda}\, E_{u,i+1} \end{bmatrix}$$

$E_{u,i}$ are allowed to change with time offers the designer the opportunity to modify the specification of the uncertainty with time.

As in the state-space context of Sec. 3, assume now that at step i we are given an *a priori* estimate for the state x_i of (42)–(43). We shall denote this initial estimate by $\hat{x}_{i|i}$. Assume further that we are also given a positive-definite weighting matrix $P_{i|i}$, along with the observation at time $(i+1)$, i.e., $y(i+1)$. Using this initial information, we may update the estimate of x_i from $\hat{x}_{i|i}$ to $\hat{x}_{i|i+1}$ by solving

$$\min_{\{x_i\}} \max_{\delta u_i} \left(\|x_i - \hat{x}_{i|i}\|^2_{P_{i|i}^{-1}} + \|y(i+1) - (u_{i+1} + \delta u_{i+1}) x_{i+1}\|^2 \right) \quad (47)$$

subject to (42)–(45). This problem can be written more compactly in the form (4)–(5) with the identifications:

$$
\begin{aligned}
x &\longleftarrow x_i - \hat{x}_{i|i}, \qquad b \longleftarrow y(i+1) - \lambda^{-1/2} u_{i+1} \hat{x}_{i|i} && (48)\\
A &\longleftarrow \lambda^{-1/2} u_{i+1}, \qquad \delta A \longleftarrow \lambda^{-1/2} M_{i+1} \Delta_{i+1} E_{u,i+1} && (49)\\
\delta b &\longleftarrow -\lambda^{-1/2} M_{i+1} \Delta_{i+1} E_{u,i+1} \hat{x}_{i|i}, \qquad Q \longleftarrow P_{i|i}^{-1} && (50)\\
W &\longleftarrow 1, \qquad H \longleftarrow M_{i+1} && (51)\\
E_a &\longleftarrow \lambda^{-1/2} E_{u,i+1}, \qquad E_b \longleftarrow -\lambda^{-1/2} E_{u,i+1} \hat{x}_{i|i}, && (52)\\
\Delta &\longleftarrow \Delta_{i+1} && (53)
\end{aligned}
$$

Using Theorem 1 and the correspondences (40) between the original variables and the state-space variables, we arrive at the equations listed in Table 2.

To illustrate the developed filter, we choose an implementation of order 5 with $E_{u,i} = [2.0 \quad 1.5 \quad 2.5 \quad 1.0 \quad 1.3]$, $M_i = 2.5$ for all i, and Δ_i selected uniformly within $[-1, 1]$. We also set $\lambda = 0.95$. The figure shows the average squared weight-error curves (averaged over 100 experiments) for the conventional RLS algorithm and for the robust algorithm. The improvement in performance will be more noticeable for larger uncertainties.

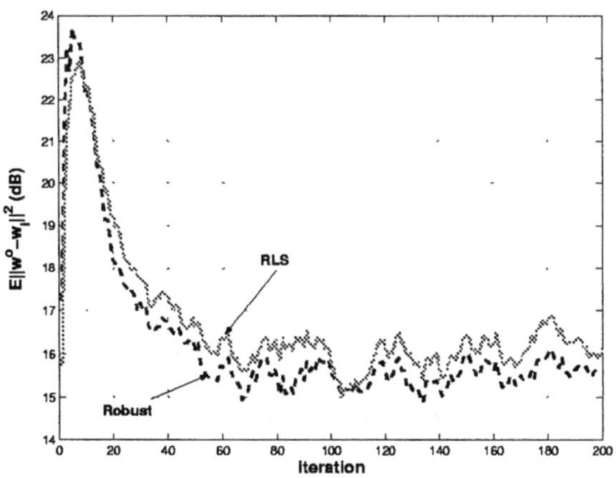

Figure 1. Averaged squared weight-error curves for RLS (solid line) and the robust filter (dashed line) for $\lambda = 0.95$.

5. Concluding remarks

The robust design procedure described in this chapter performs a local optimization step at each iteration. The optimization is not global and the parameter $\widehat{\beta}$ becomes a tuning parameter that could be adjusted by the designer. At present we are investigating designs with weighting matrices related to actual error covariance matrices, in addition to allowing for stochastic modeling of the uncertainties. These issues will be discussed elsewhere.

Acknowledgments

This material is based on work supported by the National Science Foundation under Award No. ECS-9632919.

References

[1] T. Kailath, A. H. Sayed, and B. Hassibi. *Linear Estimation.* Prentice-Hall, NJ, 2000.

[2] B. Hassibi, A. H. Sayed, and T. Kailath. *Indefinite Quadratic Estimation and Control: A Unified Approach to $\mathcal{H}_2$ and $\mathcal{H}_\infty$ Theories.* SIAM, PA, 1999.

[3] A. H. Sayed, V. H. Nascimento, and S. Chandrasekaran. Estimation and control with bounded data uncertainties. *Linear Algebra and Its Applications*, 284:259–306, Nov. 1998.

[4] A. H. Sayed and V. H. Nascimento. Design criteria for uncertain models with structured and unstructured uncertainties, in *Robustness in Identification and Control*, A. Garulli, A. Tesi, and A. Vicino, editors, 245:159–173, Springer Verlag, London, 1999.

[5] A. H. Sayed, V. H. Nascimento, and F. A. M. Cipparrone. A regularized robust design criterion for uncertain data. To appear in *SIAM. J. Matrix Anal. Appl.*

[6] A. H. Sayed and S. Chandrasekaran. Parameter estimation in the presence of multiple sources and levels of uncertainties, *IEEE Transactions on Signal Processing*, 48(3):680–692, Mar. 2000.

[7] A. H. Sayed. A framework for state space estimation with uncertain models. *IEEE Trans. Automat. Contr.*, 46(7):998–1013, July 2001.

[8] A. E. Bryson and Y.-C. Ho. *Applied Optimal Control: Optimization, Estimation, and Control.* Taylor & Francis, revised printing, 1975.

[9] A. H. Sayed and T. Kailath. A state-space approach to adaptive RLS filtering. *IEEE Signal Processing Magazine*, 11(3):18–60, July 1994.

[10] S. Chandrasekaran, G. Golub, M. Gu, and A. H. Sayed. Parameter estimation in the presence of bounded data uncertainties. *SIAM. J. Matrix Anal. Appl.*, 19(1):235–252, Jan. 1998.

[11] I. R. Petersen and A. V. Savkin. *Robust Kalman Filtering for Signals and Systems with Large Uncertainties.* Birkhauser, Boston, 1999.

[12] L. Xie, Y. C. Soh, and C. E. de Souza. Robust Kalman filtering for uncertain discrete-time systems. *IEEE Trans. Automat. Contr.*, 39(6):1310–1314, 1994.

[13] L. E. Ghaoui and H. Hebret. Robust solutions to least-squares problems with uncertain data. *SIAM J. Matrix Anal. Appl.*, 18(4):1035–1064, 1997.

MODELS FOR ROBUST ESTIMATION AND IDENTIFICATION

Shivkumar Chandrasekaran
Department of Electrical and Computer Engineering
University of California
Santa Barbara, CA 93106
shiv@ece.ucsb.edu

Keith Schubert
Department of Mathematics and Computer Science
University of Redlands
Redlands, CA 92374
schubert@uor.edu

Abstract In this paper, we will investigate estimation and identification theories with the goal of determining some new methods of adding robustness. We consider uncertain estimation problems, namely ones in which the uncertainty multiplies the quantities to be estimated. Mathematically the problem can be stated as, for system matrices and data matrices that lie in the sets $(A + \delta A)$ and $(b + \delta b)$ respectively, find the value of x that minimizes the cost $\|(A+\delta A)x - (b+\delta b)\|$. We will examine how the proposed techniques compare with currently used methods such as Least Squares (LS), Total Least Squares (TLS), and Tikhonov Regularization (TR). Several results are presented and some future directions are suggested.

Keywords: regularizaion, least squares.

1. Introduction

Every system in the real world is uncertain to one degree or another, and thus everyone who does estimation or identification must consider the assumptions made about the system and resulting problem. Consider,

S. Van Huffel and P. Lemmerling (eds.), Total Least Squares and Errors-in-Variables Modeling, 203–212.

for example the simple system described by $Ax = b$, with

$$A = \begin{bmatrix} 0.11765 & 0.12909 \\ -0.24957 & -0.26919 \end{bmatrix}, \qquad b = \begin{bmatrix} -0.074888 \\ 0.154728 \end{bmatrix}.$$

For this exact system the solution is given by

$$x = \begin{bmatrix} 0.34 \\ -0.89 \end{bmatrix}.$$

This is a nice system with reasonable condition number, but if we round A to two decimal places,

$$A = \begin{bmatrix} 0.12 & 0.13 \\ -0.25 & -0.27 \end{bmatrix},$$

we find that the new solution is

$$x = \begin{bmatrix} 1.0505 \\ -1.5457 \end{bmatrix}.$$

The best thing we can say about this is that the signs of the solution are correct. This illustrates that even innocent looking systems can exhibit bad behavior in normal situations. What can be done? Consider the general form of the regularized solution,

$$x(\psi) = \left(A^T A + \psi I\right)^{-1} A^T b, \tag{1}$$

with $\psi = 10^{-7}$. This yields a solution of

$$x\left(10^{-7}\right) = \begin{bmatrix} 0.21515 \\ -0.77273 \end{bmatrix}.$$

This is much better, but can we automate the selection of the regularization parameter? We would like to examine the one parameter family given in equation 1 to find the one closest to the true system, but that requires knowing the answer apriori. We must look for another standard. Note that frequently the exact solution is not as important as the residual so we need to take this into account. We will consider several methods for selecting the regulation parameter in this paper. We will then end with a numerical example of performance.

2. Minmax

The minmax problem was described and solved in [2, 9]. The underlying assumption is that the true system is the worst possible one in the entire set of all possible systems specified by the uncertainties in the problem. The assumptions are thus clearly pessimistic. Let's look at the results of the model. The basic problem can be stated as

$$\min_{x} \max_{\substack{||\delta A|| \leq \eta \\ ||\delta b|| \leq \eta_b}} \left(||(A + \delta A)x - (b + \delta b)|| \right).$$

After performing the maximization the resulting problem is

$$\min_{x} \left(||Ax - b|| + \eta ||x|| + \eta_b \right).$$

The problem is convex so the solution is guaranteed, but involves several cases. The basic structure of the solution can be stated as

$$x_{\min\max} \;=\; (A^T A + \alpha I)^{-1} A^T b, \qquad \alpha = \frac{\eta ||Ax - b||}{||x||}.$$

The majority of work for this problem revolves around finding the value of α. One thing we can assert about α though is that it is non-negative and in general it will be strictly positive. We note that as a result of this we can state that,

$$\frac{\sigma_n^2}{\sigma_n^2 + \alpha} ||x_{LS}|| \leq ||x_{\min\max}|| \leq \frac{\sigma_1^2}{\sigma_1^2 + \alpha} ||x_{LS}||.$$

The result follows by taking norms of the expression for the minmax solution and doing some algebra. The point of this is that the minmax returns an estimate that has less signal strength (smaller norm) than the well known LS estimator. The assumptions and results are thus clearly pessimistic.

3. Multi-column minmax

A special case of this problem was solved in [2], where one block column of the matrix A is assumed known exactly. Essentially, this problem also assumes that the real system is worse, but now we can impose column bounds, which can reflect relative certainty of different parts in the system. The problem is still pessimistic, but contains several twists on the non-partitioned problem. The general problem, with A partitioned into n block columns, can be stated as

$$\min_{x_i} \max_{\substack{||\delta A_i|| \leq \eta_i \\ ||\delta b|| \leq \eta_b}} \left(||\sum_{i=1}^{n} (A_i + \delta A_i) x_i - (b + \delta b)|| \right).$$

Using techniques similar to those in [2], we can simplify the problem to

$$\min_{x_i} \left(\|Ax - b\| + \sum_{i=1}^{n} \eta_i \|x_i\| + \eta_b \right).$$

This problem is a convex sum of euclidean norms. A large body of literature exists for solving the sum of Euclidean norms problem. The problem dates back to Fermat, who posed a special case. Various methods have been proposed which range from a sequence of linear least squares problems [15, 10, 6, 7, 14] to successive over-relaxation [13] to hyperbolic approximation procedure [8] to subgradients [11, 5]. All of these have at best linear convergence so we recommend the quadratically convergent method proposed by Michael Overton in [12]. Overton's method uses an active set and considers the projected objective function which is locally continuously differentiable. Note that Overton's method is similar to [1].

All of the methods mentioned do not take advantage of the reduction in size that can be obtained by using a secular equation as was done in the simpler non-partitioned case. We note that the basic solution when there are p partitions can be written as

$$\begin{aligned} x_i &= \frac{1}{\psi_i} A_i^T \left(I + \sum_{i=1}^{p} \frac{1}{\psi_i} A_i A_i^T \right)^{-1} b, \\ &= \frac{1}{\psi_i} A_i^T \left(I + A \Psi^{-1} A^T \right)^{-1} b \end{aligned}$$

with

$$\Psi = \begin{bmatrix} \psi_1 I & \cdots & 0 \\ \vdots & \ddots & \vdots \\ 0 & \cdots & \psi_p I \end{bmatrix}, \qquad \psi_i = \frac{\eta_p \|A\hat{x} - b\|}{\|\hat{x}_p\|} \quad \text{and} \quad x_\Psi = \begin{bmatrix} x_1 \\ \vdots \\ x_p \end{bmatrix}.$$

We note that we can also write a secular equation for each partition, which can be solved to find the values of ψ_i. The secular equations, G_i, are given by

$$G_i(\psi) = b^T F N_i F b \tag{2}$$

with

$$F = \left(I + \sum_{i=1}^{p} \frac{1}{\psi_i} A_i A_i^T \right)^{-1}, \qquad N_i = \left(A_i A_i^T - \eta_i^2 I \right).$$

While this is a smaller problem than the original, and in most cases can be solved rapidly, we do not have a proof of quadratic convergence of the root finder for the secular equations. It is thus not guaranteed to be faster, but offers a potential savings.

On a different front, we would think that the size of the multi-column partitioned min-max solution, x_Ψ should be smaller than the least squares solution, x_{LS}, since both have the same numerator and the denominator of x_Ψ is larger. This seems reasonable particularly given that this was true in the non-partitioned case, and in some sense the partitioned case reflects a problem that is more known and thus less uncertain. This is not always the case though. To demonstrate this we consider a simple problem. Consider the following A and b matrices with each column of A a separate partition,

$$A = \begin{bmatrix} 1 & 0 & 0.1 \\ 1 & -1 & 1 \\ 0 & 0 & 0.1 \\ 0 & 0 & 0 \end{bmatrix}, \qquad b = \begin{bmatrix} 1 \\ 0 \\ 0 \\ 10 \end{bmatrix}.$$

We note that the least squares solution is given by

$$x_{LS} = \begin{bmatrix} 1 & 1 & 0 \end{bmatrix}^T.$$

Now consider the case when $\eta_1 = 2, \eta_2 = 0$, and $\eta_3 = 0$. The solution, x_Ψ is given by

$$x_\Psi = \begin{bmatrix} 0 & 5 & 5 \end{bmatrix}^T.$$

It is trivial to see that $\|x_{LS}\| < \|x_\Psi\|$, and thus the idea is disproved. The question remains then as to what we can say about the size of x_Ψ and thus where it lies. The following is not tight in its bound but it does provide a good starting comparison to the non-partitioned case which always has a smaller norm than the least squares,

$$\|x_\Psi\| \leq \frac{\sigma_1^2}{\sigma_n^2} \|x_{LS}\|.$$

Additionally, in the non-partitioned problem we have the simple relation that the solution x is non-zero if and only if $\|A^T b\| > \eta \|b\|$. This is not true for the partitioned case. This is easily seen by considering the following:

$$A_1 = \begin{bmatrix} 1 \\ 1 \\ 0 \end{bmatrix}, \qquad A_2 = \begin{bmatrix} 0 \\ 1 \\ 0 \end{bmatrix}, \qquad b = \begin{bmatrix} 1 \\ 0 \\ 1 \end{bmatrix}.$$

It is readily apparent that $A_2^T b = 0$ and thus from the original problem we should have $x_2 = 0$ for all η_2. Now consider $\eta_1 = \eta_2 = \frac{1}{4}$ and we find that

$$x_1 = 1 - \frac{2}{\sqrt{11}}, \qquad x_2 = \frac{3}{\sqrt{11}} - 1.$$

Our intuition from the non-partitioned case suggests that $x_2 = 0$, but this does not hold because of the column interaction. The partitioned and non-partitioned cases are fundamentally different.

So what do we have? First, we have a quadratically convergent method for finding the solution, as provided by Overton's method. Second, we have a region which contains the solution. Third, we have the form of solution and a secular equation for each partition. Fourth, we can see that the solution is surprisingly different from the non-partitioned case, and so applying results from one case to the other is inherently dangerous.

4. Bounded errors-in-variables

This problem has a degenerate (multiple solution) case and a non-degenerate case. The non-degenerate case was solved in [3], while the degenerate case was solved in [4]. The problem assumes the underlying problem is the best possible, similar to TLS but with a bound on how far the A and b matrices can be projected. This is the first optimistic problem we consider, and it can be stated as

$$\min_{x} \min_{\substack{\|\delta A\| \leq \eta \\ \|\delta b\| \leq \eta_b}} \left(\|(A + \delta A)x - (b + \delta b)\| \right).$$

Additionally this problem is not convex. In [3], it was proven that the necessary and sufficient conditions for non-degeneracy are

1) $\eta < \sigma_n$,

2) $b^T(I - A(A^T A - \eta^2 I)^{-1}b > 0$.

The problem can be reduced to

$$\min_{x}(\|Ax - b\| - \eta\|x\|).$$

Note that in the degenerate case the additional constraint of selecting the x with minimum norm is imposed in [4] to get a unique solution. The general form of the solution for the non-degenerate case is

$$x = (A^T A - \alpha I)^{-1} A^T b \quad \text{with} \quad \alpha = \frac{\eta\|Ax - b\|}{\|x\|}.$$

The value of α is determined by a secular equation. Note that the solution in the non-degenerate case always does de-regulation. On the other hand the solution in the degenerate case is

$$x = (A^T A + \alpha I)^{-1} A^T b$$

with

$$\max(-\sigma_n^2, -\eta^2) \leq \alpha \leq \eta\sigma_1.$$

The particular value of α is given by a secular equation. Here we can see that if $\alpha > 0$ then the degenerate case will do regulation, so the degenerate case can either de-regularize (be optimistic) or regularize (be pessimistic). It is also interesting to note that the degenerate bounded errors-in-variables and the minmix models can sometimes give the same solution. In this case the solution has the best features of both methods.

5. Backward error

The final problem we will consider is the backward error model. This model contains both optimistic and pessimistic assumptions, and is non-convex. This problem is taken up in a paper to be submitted shortly for publication. The problem is given by the expression

$$\min_x \max_{\|\delta A\| \leq \eta} \frac{\|(A + \delta A)x - b\|}{\|A\|\|x\| + \|b\|}.$$

The maximization can be performed without difficulty to obtain

$$\min_x \frac{\|Ax - b\| + \eta\|x\|}{\|A\|\|x\| + \|b\|}.$$

Due to the difficulty of the problem, we pose instead an intermediate problem that demonstrates some interesting qualities of the original.

$$\min_x \frac{\|Ax - b\| + \eta\|x\|}{\|A\|\|x\|}.$$

Note that the solution to this problem is identical to

$$\min_x \frac{\|Ax - b\|}{\|A\|\|x\|}.$$

The solution is found by taking the derivative and setting equal to zero. The resulting solution form is denoted SBE for simplified backward error and is given by

$$x_{SBE} = \left(A^T A - \gamma_{SBE} I\right)^{-1} A^T b$$

with

$$\gamma_{SBE} = \frac{\|Ax_{SBE} - b\|^2}{\|x_{SBE}\|^2}.$$

The particular value of γ_{SBE} is determined by the root of a secular equation in the interval, $0 \leq \gamma_{SBE} \leq \sigma_n^2$. We can see that these problems de-regularize, and so contain optimistic assumptions. We can even tighten up the interval to show that

$$\sigma_{n+1}^2 \leq \gamma \leq \sigma_n^2,$$

where σ_{n+1} is the TLS parameter. Thus the simpler problem is more optimistic than TLS! One repercussion of the lack of the norm of b in the denominator of the cost is that it is possible for one element of the solution of the simple problem to become infinite in a particular case. The full backward error problem is thus more desirable. Generally, the smaller the regression parameter, the better the result. In most cases the full backward error produces the smallest regression parameter, and thus tends to give the best solution.

6. Numerical example

We have discussed several different problem formulations that can be used in estimation. We now want to get a feel for how these problems operate on a simple example. Consider for example a simple one dimensional "skyline" image that has been blurred. A "skyline" image is a one dimensional image that looks like a city skyline when graphed, and thus is the most basic image processing example. "Skyline" images involve sharp corners, and it is of key importance to accurately locate these corner transitions. Blurring occurs often in images. For example atmospheric conditions, dust, or imperfections in the optics can cause a blurred image. Blurring is usually modeled as a gaussian blur, which is a great smoothing filter. The gaussian blur causes greater distortion on the corners, which is exactly where we do not want it. The component of a gaussian blur with standard deviation, σ, in position, (i,j), is given by

$$G_{i,j} = e^{-\left(\frac{i-j}{\hat{\sigma}}\right)^2}.$$

If we go on the presumption that we do not know the exact blur that was applied (the standard deviation, $\hat{\sigma}$ unknown) we cannot expect to get the exact image back. While we realize that we will not be able to perfectly extract the original system, we want to see if we can get a little more information than we have now. We "know" the blur is small

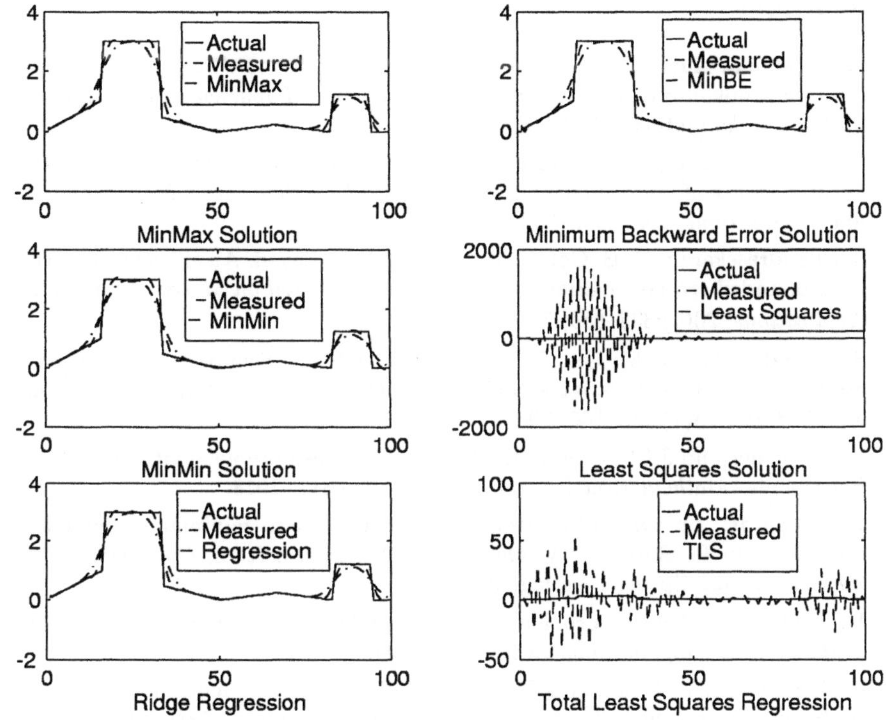

Figure 1. Skyline problem

compared to the information so we are confident that we should be able to get something.

The least squares solution fails completely, yielding a result that is about three orders of magnitude off, see Figure 1. We notice that the TLS solution is better than the LS solution, but still not acceptable. The ridge regression solution works well due to its increased robustness. The minmax performs well given its robustness. We see that the bounded errors-in-variables solution also exhibits robustness as this is one of the cases where the problem is degenerate and it can either regularize or de-regularize. In this case the solution is regularized due to the relatively large uncertainty. Most interestingly note that the backward error solution performs the best of all. It does an excellent job of finding the corners without sacrificing the edges.

Acknowledgments

The authors were partially supported by NSF grant CCR-9734290.

References

[1] P.H. Calamai and A.R. Conn. A stable algorithm for solving the multifacility location problem involving euclidean distances. *SIAM Journal on Scientific and Statistical Computing*, 1:512–526, 1980.

[2] S. Chandrasekaran, G. H. Golub, M. Gu, and A. H. Sayed. Parameter estimation in the presence of bounded data uncertainties. *SIAM Journal on Matrix Analysis and Applications*, 19(1):235–252, 1998.

[3] S. Chandrasekaran, G. H. Golub, M. Gu, and A. H. Sayed. An efficient algorithm for a bounded errors-in-variables model. *SIAM Journal on Matrix Analysis and Applications*, 20(4):839–859, 1999.

[4] S. Chandrasekaran, M. Gu, A. H. Sayed, and K. E. Schubert. The degenerate bounded errors-in-variables model. *SIAM Journal on Matrix Analysis and Applications*, 23(1):138–166, 2001.

[5] F. Cordellier and J.Ch. Giorot. On the fermat-weber problem with convex cost functionals. *Mathematical Programming*, 14:295–311, 1978.

[6] U. Eckhardt. On an optimization problem related to minimal surfaces with obstacles. In R. Bulirsch, W. Oettli, and J. Stoer, editors, *Optimization and optimal control.* Springer-Verlag, 1975.

[7] U. Eckhardt. Weber's problem and weiszfeld's algorithm in general spaces. *Mathematical Programming*, 18:186–196, 1980.

[8] J.W. Eyster, J.A. White, and W.W. Wierwille. On solving multifacility location problems using a hyperboloid approximation procedure. *AIIE Transactions*, 5:1–6, 1973.

[9] L. El Ghaoui. and H. Lebret. Robust solutions to least-squares problems with uncertain data. *SIAM Journal on Matrix Analysis and Applications*, 18(4):1035–1064, 1997.

[10] H.W. Kuhn. A note on fermat's problem. *Mathematical Programming*, 4:98–107, 1973.

[11] R.F. Love. Locating facilities in three-dimensional space by convex programming. *Naval Research Logistics Quarterly*, 16:503–516, 1969. Uses convex programming to solve sum of Euclidean norms problem.

[12] M. L. Overton. A quadratically convergent method for minimizing a sum of euclidean norms. *Mathematical Programming*, 27:34–63, 1983.

[13] S. Schechter. Minimization of a convex function by relaxation. In J. Abadie, editor, *Integer and nonlinear programming*, pages 177–190. North-Holland, 1970.

[14] H. Voss and U. Eckhardt. Linear convergence of generalized weiszfeld's method. *Computing*, 25:243–251, 1980.

[15] E. Weiszfeld. Sur le point par lequel la somme des distances de n points donns est minimum. *Tohoku Mathematics Journal*, 43:355–386, 1937.

ROBUST SOLUTIONS TO LINEAR APPROXIMATION PROBLEMS UNDER ELLIPSOIDAL UNCERTAINTY

G. Alistair Watson

Department of Mathematics, University of Dundee, Dundee DD1 4HN, Scotland

gawatson@maths.dundee.ac.uk

Abstract The problem of fitting a linear model to data, under uncertainty which can be regarded as being ellipsoidal, is considered in a very general setting. For a range of such problems, robust counterparts are established, and methods of solution are considered.

Keywords: robust fitting, ellipsoidal bounded uncertainty, linear errors-in-variables model, total l_p norm approximation.

1. Introduction

Let $A \in \mathbb{R}^{m \times n}$, $b \in \mathbb{R}^m$ contain given data. Then a fundamental problem, and the one which lies at the heart of this paper, is that of determining $x \in \mathbb{R}^n$ such that

$$Ax \approx b. \tag{1}$$

If A and b both contain significant errors, then it may be appropriate to determine x by solving the problem

$$\text{minimize } \|[E : \ c]\| \text{ subject to } (A + E)x = b + c, \tag{2}$$

where the matrix norm is one on $(m \times (n+1))$ matrices. The Frobenius norm of course gives total least squares, although various other norms and generalizations to take account of structure, for example, have also been suggested ([11], [14], [17], [20], [21]).

A consequence of this formulation, however, is that the size of errors in A or b may be exaggerated, and a modification is to incorporate bounds (assumed given) on the allowable perturbations. It is then necessary to relax the requirement that the constraints of (2) be satisfied exactly, and

S. Van Huffel and P. Lemmerling (eds.), Total Least Squares and Errors-in-Variables Modeling, 213–222.

only satisfy these *approximately*. For example we may try to determine

$$\min_{x\in\mathbb{R}^n} \max_{\|E\|\le\rho,\ \|c\|\le\rho_c} \|(A+E)x-(b+c)\|, \tag{3}$$

for appropriate matrix and vector norms, where ρ and ρ_c are given. This means that x is required to be the minimizer over all allowable perturbations in the data, and in this case the solution may be regarded as *robust*, an important requirement in many engineering applications, including system identification and signal processing. This kind of bounded error problem and variants has been considered for example in [3], [4], [5], [6], [7], [9], [10], [12], [18], [19], [24], [25]. An interpretation of (3) is that we require to minimize $\|\tilde{A}x-\tilde{b}\|$ with respect to x over the worst of all perturbations defined by

$$(\tilde{A},\tilde{b}) \in \mathcal{E} = \{(A+E, b+c) : \ \|E\| \le \rho, \ \|c\| \le \rho_c\},$$

where ρ, ρ_c and the norms are given.

An alternative is to allow perturbations from an uncertainty set typified by the restriction that the perturbed data lies in a set of the form

$$\mathcal{E} = \{(A+UP, b+Ud) : \|U\| \le 1\},$$

where P, d and the norm on U are given. With this kind of restriction on the perturbations, and variants in a similar spirit, the problem is said to have *ellipsoidal uncertainty*. Ben-Tal and Nemirovski [2] consider the use of such uncertainty sets in the context of a wide range of convex optimization problems, with emphasis on the tractability of robust counterparts.

Of interest here are robust counterparts of linear approximation problems in the context of ellipsoidal uncertainty. So far, results have mainly concentrated on dealing with particular examples; see for example [12], [15]. The purpose of this paper is to present results in a very general setting which allows previously developed results (and others) to be obtained relatively easily as special cases. Matrix norms will be needed, and these will be linked to vector norms through the idea of separability. Most commonly occurring matrix norms satisfy the following definition.

Definition 2. *A matrix norm on $m\times n$ matrices is said to be* separable *if given vectors $u \in \mathbb{R}^m$ and $v \in \mathbb{R}^n$, there are vector norms $\|.\|_A$ on $\mathbb{R}^m$ and $\|.\|_B$ on $\mathbb{R}^n$ such that*

$$\|uv^T\| = \|u\|_A\|v\|_B^*, \ \|uv^T\|^* = \|u\|_A^*\|v\|_B,$$

*where the superscript * refers to the dual norm.*

A result which holds for separable norms and will be useful subsequently is that

$$\|Mv\|_A \le \|M\|\|v\|_B : \tag{4}$$

see for example [14] or [23]. It will be convenient to define r to be

$$r = Ax - b,$$

with the dependence on x not explicitly stated.

2. Perturbations of $[A : b]$

Let $P \in \mathbb{R}^{n\times n}, d \in \mathbb{R}^n$ be given and define

$$\mathcal{E} = \{(A + UP, b + Ud) : \|U\| \le 1\},$$

where the norm on $U \in \mathbb{R}^{m\times n}$ is separable, as in the definition. Then we consider finding x to minimize $\|\tilde{A}x - \tilde{b}\|_A$ over the worst of all perturbations defined by $(\tilde{A}, \tilde{b}) \in \mathcal{E}$, that is finding

$$\min_x \max_{\|U\|\le 1} \|(A + UP)x - (b + Ud)\|_A. \tag{5}$$

Theorem 1. *The problem (5) is solved by finding*

$$\min_x \{\|Ax - b\|_A + \|Px - d\|_B\}. \tag{6}$$

Proof. Let $U \in \mathbb{R}^{m\times n}$, $\|U\| \le 1$ be arbitrary. For any $x \in \mathbb{R}^n$,

$$\|(A + UP)x - (b + Ud)\|_A \le \|r\|_A + \|Px - d\|_B,$$

using (4). Now let $U = uw^T$, where $w \in \partial\|Px - d\|_B$ (and ∂ as usual denotes the subdifferential) and where $u = r/\|r\|_A$ if $r \ne 0$ otherwise u is arbitrary but $\|u\|_A = 1$. Then $\|U\| \le 1$ and further

$$\begin{aligned} \|(A + UP)x - (b + Ud)\|_A &= \|r + UPx - Ud\|_A \\ &= \|r + uw^T(Px - d)\|_A \\ &= \|r\|_A + \|Px - d\|_B. \end{aligned}$$

The result is proved. □

If P is nonsingular, then if $y = Px - d$, it follows that (6) is equivalent to

$$\min_y \{\|A_1 y - b_1\|_A + \|y\|_B\}, \tag{7}$$

where $A_1 = AP^{-1}$, $b_1 = b - P^{-1}d$.

Perhaps the most common examples of a separable norm are the Frobenius norm or the spectral norm, when both the norms $\|.\|_A$ and

$\|.\|_B$ are l_2 norms. In this case, (7) is a special case of a problem considered in [6], [10], and efficient algorithms are available. If the separable norm is an l_p norm on the elements regarded as an extended vector, then $\|.\|_A = \|.\|_p$ and $\|.\|_B = \|.\|_q$, where $1/p+1/q=1$. If the separable norm is an operator norm, then equality holds in (4), which fixes the norms in that case. In particular when both norms are the same, then those same norms occur in (6) or (7). For these cases involving l_p norms ($p \neq 2$, $1 < p < \infty$), algorithms based on Newton's method can be used [24]. For combinations of l_1 and l_∞ norms, the problem can be posed as a linear programming problem.

3. Perturbations of the rows of $[A : b]$

We can consider a slightly different underlying problem here, where rather than considering $\|\tilde{A}x - \tilde{b}\|$ we deal with $h(\tilde{A}x - \tilde{b})$, where h is a given function satisfying

$$|w_i| \leq |y_i|, \ i = 1, \dots, m \ \Rightarrow \ h(w) \leq h(y).$$

Note this means that $h(y) = h(|y|)$. This is less general than before in that if h is a norm then that norm must be monotonic (though this includes most norms of interest including all l_p norms). However, it also includes important functions which are not norms, for example the Huber M-estimator.

Let $A^T = [a_1, a_2, \dots, a_m]$ and define

$$\mathcal{E}_i = \{(a_i + P_i^T u, b_i + d_i^T u) : \ \|u\|_A \leq 1\}, \ i = 1, \dots, m,$$

where $\|.\|_A$ is an arbitrary norm on $\mathbb{R}^k$, and each $P_i \in \mathbb{R}^{k \times n}$ and $d_i \in \mathbb{R}^k$ is given. Consider finding:

$$\min_x \max_{\|u_i\|_A \leq 1, i=1,\dots,m} h(y), \tag{8}$$

where

$$y_i = (a_i + P_i^T u_i)^T x - (b_i + d_i^T u_i), \ i = 1, \dots, m.$$

Theorem 2. *The problem (8) is solved by finding*

$$\min_x h(z), \tag{9}$$

where

$$z_i = |a_i^T x - b_i| + \|P_i x - d_i\|_A^*, \ i = 1, \dots, m.$$

Proof. Let x be arbitrary and let $\|u_i\|_A \leq 1, \ i = 1, \dots, m$. Then

$$|y_i| \leq |a_i^T x - b_i| + \|P_i x - d_i\|_A^*, \ i = 1, \dots, m.$$

This gives an upper bound for the max term in (8). Next we show that this is attained. Let $i, 1 \leq i \leq m$, and let $\theta_i = \text{sign}(a_i^T x - b_i)$, $u_i = \theta_i w_i$, where $w_i \in \partial \|P_i x - d_i\|_A^*$. Then $\|u_i\|_A = 1$, and

$$y_i = \theta_i |a_i^T x - b_i| + \theta_i \|P_i x - d_i\|_A^*.$$

Therefore

$$|y_i| = |a_i^T x - b_i| + \|P_i x - d_i\|_A^*, \ i = 1, \ldots, m,$$

and the result follows. □

Note that if the ith row of $[A : b]$ is exact, then we can simply take $P_i = 0$, $d_i = 0$ so that $y_i = r_i$.

Here there is no matrix norm involved, so the notion of separability is irrelevant. However, there are two "measures", those defined by h and by $\|.\|_A$, which may or may not be the same. The case when $\|.\|_A$ is the l_2 norm is of particular interest, and we will confine attention first to that case. Suppose first that h is also the l_2 norm. Then the problem can be restated as

$$\text{minimize } s \text{ subject to } \ \|t\|_2 \leq s,$$

$$|a_i^T x - b_i| + \|P_i x - d_i\|_2 \leq t_i, \ i = 1, \ldots, m.$$

This is an example of a second order cone program (the minimization of a linear function over the intersection of an affine set and the product of quadratic cones), which can be efficiently solved by primal-dual interior point methods [13]. When h is the l_1 or the l_∞ norm, then the problems can be similarly reformulated. For example the l_1 case [15] becomes

$$\text{minimize} \sum_{i=1}^{m} t_i \text{ subject to}$$

$$|a_i^T x - b_i| + \|P_i x - d_i\|_2 \leq t_i, \ i = 1, \ldots, m.$$

An appropriate reformulation of the case when h is the Huber function is also considered in [15]. For h an l_p norm for another value of p, again an obvious reformulation exists. However, the resulting problem is no longer a convex second order cone program, although generally a smooth (convex) problem.

Recalling that the l_1 and l_∞ norms are dual, then if h and $\|.\|_A$ are l_1 or l_∞ norms then the problems can be reformulated as linear programming problems. In particular, if h is the l_1 norm and $\|.\|_A$ is the l_∞ norm, then the problem is just a linear l_1 problem.

4. Perturbations of the columns of $[A : b]$

Suppose that instead of dealing separately with rows, it is appropriate to deal separately with the *columns* but in an analogous way. We return to the original (arbitrary) norm problem here. Let

$$\mathcal{E}_i = \{Ae_i + P_i u : \ \|u\|_A \leq 1\}, \ i = 1, \ldots, n,$$

$$\mathcal{E}_{n+1} = \{b + P_{n+1} u : \ \|u\|_A \leq 1\},$$

where $P_i \in \mathbb{R}^{m \times m}$ are nonsingular symmetric matrices, and e_i is the ith coordinate vector. Then we would like to find

$$\min_x \max_{\|u_i\|_A \leq 1, i=1,\ldots,n+1} \|y\|_A, \tag{10}$$

where

$$y = \sum_{i=1}^{n} x_i (Ae_i + P_i u_i) - (b + P_{n+1} u_{n+1}).$$

An alternative which is more tractable is to define

$$\rho_i = \|P_i\|, \ i = 1, \ldots, n+1,$$

where

$$\|P\| = \max \frac{\|Pz\|_A}{\|z\|_A},$$

and redefine $\mathcal{E}_i$ to be the larger set

$$\mathcal{E}_i = \{Ae_i + P_i u : \ \|P_i u\|_A \leq \rho_i\}, \ i = 1, \ldots, n,$$

$$\mathcal{E}_{n+1} = \{b + P_{n+1} u : \ \|P_{n+1} u\|_A \leq \rho_{n+1}\}.$$

Now consider finding:

$$\min_x \max_{\|P_i u_i\|_A \leq \rho_i, i=1,\ldots,n+1} \|y\|_A. \tag{11}$$

Because the set over which maximization is required is larger in the second case, this may be overly conservative. Indeed its solution corresponds to the minimization of an upper bound for the maximum in (10), not however attained. By contrast, (11) is a tractable problem, as we now show. The replacement of an intractable robust counterpart by its tractable approximation is a natural way to proceed, and is necessary in other problems, see for example [1].

Theorem 3. *The problem (11) is solved by finding*

$$\min_x \{\|Ax - b\|_A + \sum_{i=1}^{n} |x_i| \rho_i\}. \tag{12}$$

Proof. Let x be arbitrary and let $\|P_i u_i\|_A \leq \rho_i$, $i = 1, \ldots, n+1$. Then

$$\begin{aligned} \|y\|_A &= \|Ax - b + \sum_{i=1}^{n} x_i P_i u_i - P_{n+1} u_{n+1}\|_A \\ &\leq \|Ax - b\|_A + \sum_{i=1}^{n} |x_i| \rho_i + \rho_{n+1} \end{aligned}$$

This gives an upper bound for the max term in (11). We show that this is attained. Let

$$P_i u_i = \theta_i \rho_i u, i = 1, \ldots, n,$$

$$P_{n+1} u_{n+1} = -\rho_{n+1} u,$$

where $\theta_i = \text{sign}(x_i), i = 1, \ldots, n$ and where $u = r/\|r\|_A$ if $r \neq 0$ otherwise u is arbitrary but $\|u\|_A = 1$. Then $\|P_i u_i\|_A \leq \rho_i, i = 1, \ldots, n+1$, and also

$$\|Ax - b + \sum_{i=1}^{n} x_i P_i u_i - P_{n+1} u_{n+1}\|_A = \|Ax - b\|_A + \sum_{i=1}^{n} |x_i| \rho_i + \rho_{n+1}.$$

Because the presence of the constant ρ_{n+1} has no effect on the minimizing x, the result follows. □

If any of the columns of $[A : b]$ are exact, then again the corresponding P_i is zero, so $\rho_i = 0$. When $\|.\|_A$ is the l_2 norm, then the objective function (12) has a familiar form, and arises in many other contexts, for example support vector regression [22] and basis pursuit [8]. The problem can be reformulated as a smooth constrained optimization problem. Such methods apply to other cases, for example when $\|.\|_A$ is an l_p norm, $1 < p < \infty$. The extreme cases when $p = 1$ or $p = \infty$ can be posed as linear programming problems.

5. Structured perturbations

If there is structure to the perturbations, which must be preserved (for example sparsity, or Toeplitz form), then the perturbation to A may take the form

$$\sum_{j=1}^{k} u_j A_j,$$

where $A_j \in \mathbb{R}^{m \times n}, j = 1, \ldots, k$, have the required structure. Similarly the perturbation to b might take the form

$$\sum_{j=1}^{k} u_j b_j,$$

where $b_j \in \mathbb{R}^m, j = 1, \ldots, k$. Then in the spirit of the previous problems we require to work with the set

$$\mathcal{E} = \{(A + \sum_{j=1}^{k} u_j A_j, b + \sum_{j=1}^{k} u_j b_j) : \|u\|_B \leq 1\},$$

where the norm on u is an arbitrary norm on $\mathbb{R}^k$. The problem can then be stated as that of finding

$$\min_x \max_{\|u\|_B \leq 1} h(y), \tag{13}$$

where h is as in Section 2 and where

$$y = (A + \sum_{j=1}^{k} u_j A_j)x - (b + \sum_{j=1}^{k} u_j b_j).$$

Therefore

$$\begin{aligned} y_i &= e_i^T((A + \sum_{j=1}^{k} u_j A_j)x - (b + \sum_{j=1}^{k} u_j b_j)) \\ &= r_i + \sum_{j=1}^{k} u_j e_i^T(A_j x - b_j) = r_i + u^T(P_i x - d_i), \end{aligned}$$

where each P_i is a $k \times n$ matrix with jth row $e_j^T P_i = e_i^T A_j$, $j = 1, \ldots, k$, and each d_i is defined by $e_j^T d_i = e_i^T b_j$, $j = 1, \ldots, k$, for $i = 1, \ldots, m$. Therefore the structured problem is similar to that of Section 2, with this choice of P_i's and d_i's. In other words we have (13) where

$$y_i = r_i + u^T(P_i x - d_i), \ i = 1, \ldots, m,$$

which differs from (8) in that there is only one vector u.

Replacing $h(y)$ by $\|y\|_A$, where the norm is an arbitrary norm on $\mathbb{R}^m$, the problem (13) then involves finding

$$\min_x \max_{\|u\|_B \leq 1} \|Cu + r\|_A, \tag{14}$$

where the ith row of C is given by

$$e_i^T C = (P_i x - d_i)^T, \ i = 1, \ldots, m.$$

This is equivalent to finding

$$\min_x \max_{u \in Ext(W)} \|Cu + r\|_A,$$

where $Ext(W)$ denotes the extreme points of the unit ball $\|u\|_B \leq 1$ [16]. When $\|.\|_B$ is a polyhedral norm, then the number of extreme points is finite. This means that if $\|.\|_A$ is also a polyhedral norm, the problem can be formulated as a linear programming problem. For the l_1 norm we have

$$Ext(W) = \{\pm e_1, \ldots, \pm e_k\},$$

where e_i is the ith coordinate vector in $\mathbb{R}^k$, and for the l_∞ norm we have

$$Ext(W) = \{u \in \mathbb{R}^k : u_i = \pm 1, i = 1, \ldots, k\}.$$

For further details of the l_1 and l_∞ cases see [12]. When both norms are l_2 norms, then it is shown in [10] how the problem can be reformulated as a semi-definite programming problem, for which good algorithms are available.

References

[1] A. Ben-Tal and A. Nemirovski. Robust convex optimization. *Mathematics of Operations Research*, 23: 769–803, 1998.

[2] A. Ben-Tal and A. Nemirovski. Robust solutions to uncertain linear programs via convex programming. *Operations Research Letters*, 25: 1-13, 1999.

[3] S. Chandrasekaran, M. Gu, A. H. Sayed and K. E. Schubert. The degenerate bounded errors-in-variables model. *SIAM J. Matrix Anal. Appl.* (to appear)

[4] S. Chandrasekaran, G. H. Golub, M. Gu and A. H. Sayed. Efficient algorithms for least squares type problems with bounded uncertainties. In S. Van Huffel, editor, *Recent Advances in Total Least Squares Techniques and Errors-in-Variables Modeling*, pages 171–180, SIAM, Philadelphia, PA, 1997.

[5] S. Chandrasekaran, G. H. Golub, M. Gu and A. H. Sayed. Parameter estimation in the presence of bounded modeling errors. *IEEE Signal Proc. Let.*, 4: 195–197, 1997.

[6] S. Chandrasekaran, G. H. Golub, M. Gu and A. H. Sayed. Parameter estimation in the presence of bounded data uncertainties. *SIAM J. Matrix Anal. Appl.* 19: 235–252, 1998.

[7] S. Chandrasekaran, G. H. Golub, M. Gu and A. H. Sayed. An efficient algorithm for a bounded errors-in-variables model. *SIAM J. Matrix Anal. Appl.* 20: 839–859, 1999.

[8] S. S. Chen, D. L. Donoho and M. A. Saunders. Atomic decomposition by basis pursuit. *SIAMJ Sci. Comp.* 20: 33–61, 1998.

[9] L. El Ghaoui and H. Lebret. Robust solutions to least squares problems with uncertain data. In S. Van Huffel, editor, *Recent Advances in Total Least Squares Techniques and Errors-in-Variables Modeling*, pages 161–170, SIAM, Philadelphia, PA, 1997.

[10] L. El Ghaoui and H. Lebret. Robust solutions to least squares problems with uncertain data. *SIAM J. Matrix Anal. Appl.* 18: 1035–1064, 1997.

[11] G. H. Golub and C. F. Van Loan. An analysis of the total least squares problem. *SIAM J. Num. Anal.* 17: 883–893, 1980.

[12] H. A. Hindi and S. P. Boyd. Robust solutions to l_1, l_2 and l_∞ linear approximation problems using convex optimization. Technical Report, Stanford University, 1997.

[13] M. S. Lobo, L. Vandenberghe, S. Boyd and H. Lebret. Applications of second-order cone programming. *Lin. Alg. Appl.* 284: 193–228, 1998.

[14] M. R. Osborne and G. A. Watson. An analysis of the total approximation problem in separable norms, and an algorithm for the total l_1 problem. *SIAM J. Sci Stat. Comp.* 6: 410–424, 1985.

[15] M. C. Pinar. Linear data fitting problems under ellipsoidal uncertainty. Preprint, 2000.

[16] R. T. Rockafellar. *Convex Analysis.* Princeton University Press, Princeton, 1970.

[17] J. B. Rosen, H. Park and J. Glick. Total least norm formulation and solution for structured problems. *SIAM J. Matrix Anal. Appl.* 17: 110–126, 1996.

[18] A, H. Sayed and S. Chandrasekaran. Estimation in the presence of multiple sources of uncertainties with applications. in *Proceedings of the Asilomar Conference*, Pacific Grove, CA, pages 1811–1815, 1998.

[19] A. H. Sayed, V. H. Nascimento and S. Chandrasekaran. Estimation and control in the presence of bounded data uncertainties. *Linear Alg. Appl.* 284: 259–306, 1998.

[20] S. Van Huffel and J. Vandewalle. *The Total Least Squares Problem: Computational Aspects and Analysis.* in Frontiers in Applied Mathematics 9, SIAM. Philadelphia, PA, 1991.

[21] S. Van Huffel (ed.). *Recent Advances in Total Least Squares Techniques and Errors-in-Variables Modeling,* SIAM, Philadelphia, PA, 1997.

[22] V. Vapnik, S. E. Golowich and A. Smola. Support vector method for function approximation, regression estimation and signal processing. in M. C. Mozer, M. I. Jordan and T. Petsche, editors, *Advances in Neural Information Processing Systems 9,* MIT Press, Cambridge, MA, 1997.

[23] G. A. Watson. Choice of norms for data fitting and function approximation. *Acta Numerica 1998*; 337–377, 1998.

[24] G. A. Watson. Solving data fitting problems in l_p norms with bounded uncertainties in the data. in D. F. Griffiths and G. A. Watson, editors, *Numerical Analysis 1999, Proceedings of the Dundee Conference*, Chapman and Hall/CRC Research Notes in Mathematics No 420, pages 249–265, 2000.

[25] G. A. Watson. Data fitting problems with bounded uncertainties in the data. *SIAM J. Matrix Anal. Appl.* 22: 1274–1293, 2001.

Part VI

CURVE FITTING

QR FACTORIZATION OF THE JACOBIAN IN SOME STRUCTURED NONLINEAR LEAST SQUARES PROBLEMS

Åke Björck

Department of Mathematics, Linköping University, S-581 83, Linköping, Sweden

akbjo@math.liu.se

Abstract For solving nonlinear least squares problems a Gauss–Newton trust region method is often employed. In the case of orthogonal distance regression it has been believed that solving the resulting linear problem at each trust region step by computing the QR factorization of the full Jacobian matrix would be very inefficient. By taking full advantage of the structure of the sparse blocks in the QR factorization of the Jacobian, we derive here an algorithm with the same overall complexity, which uses a QR factorization of the full Jacobian matrix. The same observation applies also to sparse structured total least squares problems, where similarly structured Jacobian matrices occur.

Keywords: orthogonal distance regression, nonlinear errors-in-variables, structured total least squares, nonlinear least squares.

1. Structured nonlinear least squares problems

As the main example we consider *orthogonal distance regression* (ODR), which is the problem of fitting a curve given by a parametric function $y = f(x, \beta)$, $\beta \in \mathbb{R}^p$, to observations $x_i \in \mathbb{R}^{n_x}$, $y_i \in \mathbb{R}^{n_y}$, $i = 1, \ldots, m$, subject to errors $\bar{\delta}_i$ and $\bar{\epsilon}_i$, respectively. The orthogonal distance from (x_i, y_i) to the curve equals

$$d_i = \min(\|\epsilon_i\|_2^2 + \|\delta_i\|_2^2) \quad \text{subject to} \quad y_i + \epsilon_i = f(x_i + \delta_i, \beta). \qquad (1)$$

If the errors $\bar{\epsilon}_i$ and $\bar{\delta}_i$ are independent random variables with zero mean and constant covariance matrix $\sigma^2 I$, then the parameters β should be chosen to minimize the sum of squares $\sum_{i=1}^{m} d_i^2$. In a more general setting weighted norms can be substituted in (1). Independent of statistical considerations, the orthogonal distance measure has natural applications in fitting geometrical elements; see [6, 14].

S. Van Huffel and P. Lemmerling (eds.), Total Least Squares and Errors-in-Variables Modeling, 225–234.

By eliminating ϵ_i in (1) the ODR problem can be formulated as a nonlinear least squares problem in the parameters $\delta_1, \ldots, \delta_n, \beta$:

$$\min_{\delta,\beta} \left\| \begin{bmatrix} r_1(\delta) \\ r_2(\delta,\beta) \end{bmatrix} \right\|_2^2, \tag{2}$$

where

$$r_1(\delta)_i = \delta_i, \qquad r_2(\delta,\beta)_i = f(x_i + \delta_i, \beta) - y_i, \quad i = 1, \ldots, m. \tag{3}$$

are residual vectors. The problem (2)–(3) has $(mn_x + p)$ unknowns δ_i, $i = 1, \ldots, m$, and β, compared to only p in ordinary least squares fitting. In applications usually $mn_x \gg p$, so accounting for the errors in δ_i in x_i considerably increases the size of the problem and makes the use of standard software for nonlinear least squares very inefficient. Algorithms for ODR, which have a computational complexity of the same order as for an ordinary least squares fit have been developed by Boggs, Byrd, and Schnabel [2, 3], and Schwetlick and Tiller [12, 13].

A similarly structured nonlinear least squares problem occurs in the *structured total least squares* (STLS) problems. Given the data matrix A and the vector b the total least squares problem is (see [8]) to find E and x such that

$$\min \|[E \quad r]\|_F, \quad \text{such that} \quad r = b - (A+E)x. \tag{4}$$

If A is sparse we may want to impose the same sparsity structure on E. Following [11] an affine structure is imposed on E by defining a matrix X such that $X\delta = Ex$, where $\delta \in \mathbb{R}^q$ is a vector containing the nonzero elements of E and the elements of the matrix $X \in \mathbb{R}^{m\times q}$ consists of the elements of the vector x with a suitable repetition. This STLS problem can now be written as an unconstrained nonlinear least squares problem

$$\min_{\delta,x} \left\| \begin{bmatrix} \delta \\ r(\delta,x) \end{bmatrix} \right\|, \tag{5}$$

where

$$r(\delta, x) = Ax - b + X\delta.$$

In case E is general sparse the special structure of the Jacobian will be similar to that in the ODR problem (see [11]).

The main purpose of this paper is to show how the QR factorization of the full Jacobian matrix can be used in computing the trust region step without increasing the computational complexity. This is achieved by taking full advantage of the special structure of the Jacobian matrix and storing part of the QR factors implicitly.

An outline of the paper is as follows. In Section 2 we first review the solution of the linear subproblems in a trust region Gauss–Newton type algorithm, in the case when the Jacobian matrix is sparse. Then we consider the special structure of the Jacobian matrix in ODR and outline (for the case $n_y = 1$) how the trust region step is computed in [2]. Using a simple result on the structure of the QR factors of a Hessenberg matrix consisting of a diagonal plus a full row, a new algorithm is derived in Section 3.

2. Gauss–Newton's method and regularization of sparse problems

The Gauss–Newton method with a trust region stabilization is the method of choice for solving nonlinear least squares problems. It requires solving a *sequence* of regularized linear least squares problems

$$\min_p \left\| \begin{bmatrix} J \\ \mu D \end{bmatrix} p - \begin{bmatrix} f \\ 0 \end{bmatrix} \right\|_2, \tag{6}$$

where J is the Jacobian and D a diagonal matrix. When J is an $m \times n$ dense matrix a two-stage method is normally used (see Moré [10]):

1. Compute a QR factorization of J with column pivoting

$$Q_1^T J P = \begin{bmatrix} R \\ 0 \end{bmatrix}, \qquad Q_1^T f = \begin{bmatrix} g \\ h \end{bmatrix}. \tag{7}$$

where Q_1 is orthogonal and P a permutation matrix.

2. For a sequence of values of μ compute the QR factorization

$$Q^T \begin{bmatrix} R \\ \mu D \end{bmatrix} = \begin{bmatrix} R_\mu \\ 0 \end{bmatrix}, \qquad Q^T \begin{bmatrix} g \\ 0 \end{bmatrix} = \begin{bmatrix} s_\mu \\ t_\mu \end{bmatrix}. \tag{8}$$

The solution p to (6) is obtained from the triangular system $R_\mu p = s_\mu$.

Stage 1 is done only once at a cost of about $(mn^2 - n^3/3)$ flops. Stage 2, which requires about $n^3/3$ flops, is performed for each μ.

When J is dense a more efficient algorithm has been given by Eldén [4]. In this J is reduced to bidiagonal form in Stage 1 at a cost of $\min\{mn^2 + n^3, 2(mn^2 - n^3/3)\}$ flops (see [1, Sec. 2.6.2]). This is at most twice the work of the QR factorization. The cost of the second stage, which is repeated for each value of μ, can then be reduced to $11n$ flops.

When the Jacobian J is large and sparse none of the above two schemes is efficient. Reducing J to bidiagonal form requires two-sided orthogonal transformations, which inevitably will tend to severely destroy sparsity. In the first scheme, it is likely that R computed in Stage 1 will contain *more* nonzero elements than J. Hence for a sparse Jacobian Stage 2 usually is *more costly* than Stage 1 and it is more efficient to simply compute the QR factorization

$$Q_\mu^T \begin{bmatrix} J \\ \mu D \end{bmatrix} = \begin{bmatrix} R_\mu \\ 0 \end{bmatrix}, \qquad Q_\mu^T \begin{bmatrix} f \\ 0 \end{bmatrix} = \begin{bmatrix} s_\mu \\ t_\mu \end{bmatrix}, \tag{9}$$

from scratch for each value of μ.

Another point to notice in the sparse case is that (9) the (block) column ordering of the matrix can greatly influence the sparsity of R and the work needed in the QR factorization. Further, although R does not depend on the (block) row ordering this ordering may affect intermediate fill-in and hence the work required for the factorization. For example, assume that J is banded with row bandwidth $w \ll n$ and that a row sequential Givens reduction is used. Then with the ordering shown in (9) the QR factorization will take $O(wn^2)$ flops. This can be reduced to an optimal $O(w^2n)$ flops provided the rows are first suitably reordered; see Eldén [5] and Björck [1, Sec. 6.6.3].

2.1 The trust region step in ODR.

We now consider how to take advantage of the structure in the ODR problem (2)–(3). (The STLS problem in Section 1 can be treated similarly.) The Jacobian matrix has the 2×2 block form

$$\mathcal{J} = \begin{bmatrix} D & 0 \\ V & J \end{bmatrix} \in \mathbb{R}^{(mn+n)\times(mn+p)}. \tag{10}$$

Here $D > 0$ is a diagonal matrix of order mn_x with elements related to the variance in δ. The zero block reflects the fact that r_1 does not depend on β. Further, assuming $n_y = 1$, $V = \mathrm{diag}(v_1^T, \dots, v_m^T) \in \mathbb{R}^{m\times mn_x}$, and $J \in \mathbb{R}^{m\times p}$, where

$$v_i^T = \frac{\partial f(x_i + \delta_i, \beta)}{\partial x}, \qquad J_{ij} = \frac{\partial f(x_i + \delta_i, \beta)}{\partial \beta_j}.$$

Note that $\mathcal{J}$ is sparse and highly structured. In case $n_y > 1$ the Jacobian (10) will have a similar structure with $V = \mathrm{diag}(V_1^T, V_2^T, \dots, V_m^T)$, where $V_i^T \in \mathbb{R}^{n_y \times n_x}$.

In a trust region Gauss–Newton step we compute corrections $\Delta\delta_k$ and $\Delta\beta_k$ to the current approximations, as the solution of the linear least squares problem

$$\min_{\Delta\delta, \Delta\beta} \left\| \mathcal{J}_\mu \begin{bmatrix} \Delta\delta \\ \Delta\beta \end{bmatrix} - r \right\|_2, \tag{11}$$

where

$$\mathcal{J}_\mu = \begin{bmatrix} D & 0 \\ V & J \\ \mu T & 0 \\ 0 & \mu S \end{bmatrix}, \qquad r = \begin{bmatrix} r_1 \\ r_2 \\ 0 \\ 0 \end{bmatrix}, \tag{12}$$

for several different values of the parameter μ, where S and T are non-singular diagonal regularization matrices.

In (11) the variables naturally split into two groups. Consider now a general problem of the form

$$\min_{\delta,\beta} \left\| [B_1 \ B_2] \begin{bmatrix} \delta \\ \beta \end{bmatrix} - b \right\|_2^2. \tag{13}$$

By a geometrical argument (cf. [9]) it is easy to show that the solution can be obtained in two steps as follows: First compute β as solution to the problem

$$\min_\beta \| P_{\mathcal{N}(B_1^T)} (B_2\beta - b) \|_2^2, \tag{14}$$

where $P_{\mathcal{N}(B_1^T)}$ is the orthogonal projector onto $\mathcal{N}(B_1^T)$. Here one can, for example, use $P_{\mathcal{N}(B_1^T)} = I - P_{\mathcal{R}(B_1)} = I - B_1(B_1^T B_1)^{-1} B_1^T$, which follows from the normal equations. When β is known we set $r_2 = b - B_2\beta$ and compute δ by solving

$$\min_\delta \| B_1\delta - r_2 \|_2. \tag{15}$$

A QR factorization can be used to solve (14). and the normal equations to solve (15).

Boggs, Byrd, and Schnabel [2] gave a method to carry out the computations in a trust region step, so that the cost of a step is of the same order as for a standard least squares fit. It is claimed in [2] that computing a QR factorization of $\mathcal{J}_\mu$ would require $O((mn_x + p)^2)$ operations for each μ. For this reason the $\Delta\delta$ variables are instead eliminated using the normal equations, as outlined above, combined with the Sherman–Morrison–Woodbury formula. The reduced least squares problem for $\Delta\beta$ can then be written $\min \| \tilde{J}\Delta\delta - \tilde{r}_2 \|_2$, where

$$\tilde{J} = \begin{bmatrix} MJ \\ \mu S \end{bmatrix}, \qquad \tilde{r}_2 = \left[-M(r_1 - VE^{-2}Dr_2) \ 0 \right],$$

and $E^2 = D^2 + \mu^2 T^2$. An explicit expression for the diagonal matrix M is given in [2] in terms of E and V.

Schwetlick and Tiller [13] instead use a trust region path defined by a nonstandard scaling matrix to avoid the cost of computing the QR factorization.

3. A structured QR factorization

In this section we derive an alternative algorithm for computing the trust region step in ODR and sparse STLS, which only employs orthogonal transformations applied to $\mathcal{J}_\mu$. For ODR this new algorithm has the same leading cost as the algorithm in [2]. We achieve this by noting that because of the special structure only part of the nonzero elements in the factors Q and R need to be stored. Also, as explained in Section 2.1, we do not use the two stage procedure (7)–(8).

To factor a Jacobian matrix of the form (12), in the first step Givens rotations are used to merge the two diagonal blocks D and μT.

$$G \left[\begin{array}{cc|c} D & 0 & r_1 \\ V & J & r_2 \\ 0 & \mu S & 0 \\ \mu T & 0 & 0 \end{array}\right] = \left[\begin{array}{cc|c} D_\mu & 0 & \tilde{r}_1 \\ V & J & r_2 \\ 0 & \mu S & 0 \\ 0 & 0 & r_4 \end{array}\right]. \tag{16}$$

where $D_\mu = (D^2 + \mu^2 T^2)^{1/2}$ is diagonal. The rotations are also applied to the right hand side vector. This step does not affect the second block column and the last block row in $\mathcal{J}_\mu$.

We next consider the orthogonal triangularization of the first block column in (16). This is the key step, and it will not affect the last two block rows in (16). A straightforward reduction using Givens transformations will *not* be efficient. We now show how to take full advantage of the structure in the blocks D_μ and V.

For ease of notation we assume here (as in [2]) that $n_y = 1$ and put $n_x = n$. We make the observation that after a row permutation the matrix has the block angular structure (see Björck [1, Sec. 6.3]),

$$\Pi \begin{bmatrix} D_\mu & 0 \\ V & J \end{bmatrix} = \left[\begin{array}{cccc|c} W_1 & & & & J_1 \\ & W_2 & & & J_2 \\ & & \ddots & & \vdots \\ & & & W_m & J_m \end{array}\right], \tag{17}$$

where

$$W_i = \begin{bmatrix} D_i \\ v_i^T \end{bmatrix}, \qquad J_i = \begin{bmatrix} 0 \\ e_i^T J \end{bmatrix}. \tag{18}$$

Here $W_i \in \mathbb{R}^{(n+1)\times n}$ is a rank one modification of a diagonal matrix and lower Hessenberg. J_i consists of n zero rows followed by the ith row of J. Thus this step splits into m subproblems each of which involves a

QR factorization of a matrix of the form ($n = 4$):

$$Q_i^T W_i = Q_i^T \begin{bmatrix} x & & & \\ & x & & \\ & & x & \\ & & & x \\ x & x & x & x \end{bmatrix} = \begin{bmatrix} x & x & x & x \\ & x & x & x \\ & & x & x \\ & & & x \\ 0 & 0 & 0 & 0 \end{bmatrix} = \begin{bmatrix} U_i \\ 0 \end{bmatrix}.$$

Note that although the triangularization is achieved by a product of only n Givens rotations U_i will be a full upper triangular matrix. However, using the following lemma we can avoid the explicit computation of the off-diagonal elements of U_i.

Lemma 1. *Let the matrix $W \in \mathbb{R}^{(n+1)\times n}$ have the lower Hessenberg form*

$$W = \begin{bmatrix} D \\ v^T \end{bmatrix}, \quad D = \operatorname{diag}(d_1, \ldots, d_n),$$

where $v \in \mathbb{R}^n$. Then the QR factorization is

$$Q^T W = \begin{bmatrix} U \\ 0 \end{bmatrix},$$

where $Q_i^T = G_n \cdots G_2 G_1$, $G_j = \left[\begin{smallmatrix} c_j & s_j \\ -s_j & c_j \end{smallmatrix}\right]$, and G_j is a Givens rotation involving the rows $(j, n+1)$.

The off-diagonal elements of U are obtained from the last column of Q^T, $\operatorname{triu}(U) = \operatorname{triu}(qv^T)$, $Q^T e_{n+1} = \left[\begin{smallmatrix} q \\ \gamma \end{smallmatrix}\right]$, where $\operatorname{triu}(A)$ denotes the strictly upper triangular part of A.

Proof. To reconstruct the off-diagonal part of the upper triangular matrix U we write

$$Q^T W = Q^T \begin{bmatrix} D \\ 0 \end{bmatrix} + Q^T e_{n+1} v^T = Q^T \tilde{D} + \begin{bmatrix} q \\ \gamma \end{bmatrix} v^T = \begin{bmatrix} U \\ 0 \end{bmatrix}. \tag{19}$$

Here, $Q^T \tilde{D}$ by construction is lower triangular and hence the strictly upper triangular part of U must equal the upper triangular part of the rank one matrix qv^T. The last column in Q_i^T is obtained by performing the QR factorization

$$Q^T \begin{bmatrix} W & e_{n+1} \end{bmatrix} = \begin{bmatrix} U & q \\ 0 & \gamma \end{bmatrix},$$

where $q = [s_1,\ s_2 c_1,\ \ldots\ , s_n c_{n-1} \cdots c_1]^T$ and $\gamma = c_n c_{n-1} \cdots c_1$. □

We remark that after a diagonal scaling Lemma 1 follows from Theorem A2 in [7].

The orthogonal triangularization of the block diagonal part of the matrix (17) is achieved by premultiplying with $\Pi^T \text{diag}\left(Q_1^T, \ldots, Q_m^T\right)$. From Lemma 1 it follows that in the QR factorization of W_i we need only compute and store the *diagonal elements* of U_i and the *last column* of Q_i, provided that we save v_i^T. Further, because of the structure of J_i (see (18)) the product $Q_i^T J_i$ can be computed from the last column of Q_i. This will also suffice for carrying out all later steps. Applying Q_i^T to the block J_i we get

$$Q_i^T J_i = Q_i^T \begin{bmatrix} 0 \\ e_i^T J \end{bmatrix} = Q_i^T e_{n+1} e_i^T J = \begin{bmatrix} q_i \\ \gamma_i \end{bmatrix} e_i^T J. \tag{20}$$

It follows that the matrix in (16) is transformed according to

$$\Pi^T Q^T \Pi \begin{bmatrix} D_\mu & 0 \\ V & J \end{bmatrix} = \begin{bmatrix} U & K \\ 0 & \tilde{J} \end{bmatrix}, \tag{21}$$

where

$$U = \text{diag}\left(U_1, \ldots, U_m\right), \qquad \tilde{J} = \text{diag}\left(\gamma_1, \ldots, \gamma_m\right) J, \tag{22}$$

$$K = \begin{bmatrix} K_1 \\ \vdots \\ K_m \end{bmatrix}, \qquad K_i = q_i(e_i^T J). \tag{23}$$

Note that K_i are rank one matrices and K a full matrix of size $mn \times p$. However, K is never explicitly required.

Using the above observations we only need to perform the orthogonal triangularization on the submatrix

$$\begin{bmatrix} D_\mu & 0 & \tilde{r}_1 \\ V & e & r_2 \end{bmatrix},$$

where $e = [1, 1, \ldots, 1]^T$ is a vector of all ones. The extra second column here is an auxiliary vector of length $mn + m$ used to accumulate the information needed about the Givens rotations. It will also be used to recompute the off-diagonal elements in U_i. The right hand side vector is simultaneously transformed. After reversing the row permutation the last two columns have been transformed according to

$$\Pi^T Q^T \Pi \begin{bmatrix} 0 & \tilde{r}_1 \\ e & r_2 \end{bmatrix} = \begin{bmatrix} q & t \\ \gamma & \tilde{r}_2 \end{bmatrix}, \tag{24}$$

where $\gamma^T = [\gamma_1, \ldots, \gamma_m]$, $q^T = [q_1^T, \ldots, q_m^T]$.

This step requires the construction of mn Givens rotations and their application to two vectors of length $mn + n$. Hence the total cost is $O(mn)$ operations. Storage is only needed for one extra vector of length $mn + n$.

Having obtained $\tilde{J}$ and $\tilde{r}_2$ we next solve the subproblem from (21)

$$\min_{\Delta\beta} \left\| \begin{bmatrix} \tilde{J} \\ \mu S \end{bmatrix} \Delta\beta - \begin{bmatrix} \tilde{r}_2 \\ 0 \end{bmatrix} \right\|_2 ,$$

for $\Delta\beta$, having added the diagonal regularization matrix μS in (16). This step has the same complexity as for an ordinary least squares fit and will dominate the computational complexity.

The final step is to obtain $\Delta\delta$. From (21) and (24) it follows that

$$U\Delta\delta = t - K\Delta\beta,$$

which can be solved by back-substitution. This task again splits into m small subproblems, and using (23) we have for $i = 1, \ldots, m$,

$$U_i\Delta\delta_i = w_i, \quad w_i = t_i - K_i\Delta\beta = t_i - (e_i^T J\Delta\beta)q_i, \tag{25}$$

Here the reduced right hand side vector can be computed from q_i, t_i, and $J\Delta\beta$ and thus, as remarked before, K is not explicitly required. By Lemma 1 we have $U_i = \text{diag}\,(U_i) + \text{triu}(q_i v_i^T)$, and thus we can solve m upper triangular systems

$$\text{diag}\,(U_i)\,\Delta\delta_i = w_i - \text{triu}(q_i v_i^T)\Delta\delta_i, \quad i = 1, \ldots, m, \tag{26}$$

where the diag (U_i) has been saved. Also this step can be done in $O(mn)$ operations by successively accumulating the inner product $v_i^T\Delta\delta_i$ during the backsolve.

We remark that the outlined algorithm is also very storage efficient; only one extra vector of length $mn + n$ is needed.

The algorithm can be generalized in a straightforward way to the case when $n_y > 1$. Then the matrix W_i is a diagonal matrix modified by n_y rows,

$$W_i = \begin{bmatrix} D_i \\ V_i^T \end{bmatrix}, \quad V_i^T \in \mathbb{R}^{n_y \times n_x}.$$

Lemma 1 can easily be generalized to matrices of this form. The main difference is that we now have to save the n_y last columns of each orthogonal transformation Q_i^T. The off-diagonal elements of U are then obtained from the upper triangular part of a rank p matrix. However, note that this makes sense only if $n_y \ll n_x$.

4. Conclusions

We have shown how to efficiently compute QR factorization of the full Jacobian matrix in some structured nonlinear least squares problems such as orthogonal distance regression and sparse structured TLS problems. This was achieved using a result on the structure of the QR factors of a Hessenberg matrix consisting of a diagonal plus a full row.

Acknowledgments

The author wants to thank Michael Saunders for helpful comments and for pointing out Theorem A2 in [7].

References

[1] Å. Björck. *Numerical Methods for Least Squares Problems.* SIAM, Philadelphia, PA, 1996.

[2] P. T. Boggs, R. H. Byrd, and R. Schnabel. A stable and efficient algorithm for nonlinear orthogonal regression. *SIAM J. Sci. Statist. Comput.*, 8:1052–1078, 1987.

[3] P. T. Boggs, J. R. Donaldson, R. H. Byrd, and R. B. Schnabel. Algorithm 676. ODRPACK: Software for weighted orthogonal distance regression. *ACM Trans. Math. Software*, 15:348–364, 1989.

[4] L. Eldén. Algorithms for the regularization of ill-conditioned least squares problems. *BIT*, 17:134–145, 1977.

[5] L. Eldén. An algorithm for the regularization of ill-conditioned banded least squares problems. *SIAM J. Sci. Statist. Comput.*, 5:237–254, 1984.

[6] W. Gander, G. H. Golub, and R. Strebel. Least-squares fitting of circles and ellipses. *BIT*, 34:558–578, 1994.

[7] P. E. Gill, W. Murray, and M. A. Saunders. Methods for computing and modifying the *ldv* factors of a matrix. *Math. Comp.*, 29:1051–1077, 1975.

[8] G. H. Golub and C. F. Van Loan. An analysis of the total least squares problem. *SIAM J. Numer. Anal.*, 17:883–893, 1980.

[9] G. H. Golub and V. Pereyra. The differentiation of pseudoinverses and nonlinear least squares problems whose variables separate. *SIAM J. Numer. Anal.*, 10:413–432, 1973.

[10] J. Moré. The Levenberg–Marquardt algorithm: Implementation and theory. In G. Watson, editor, *Numerical Analysis. Proc. Biennial Conf. Dundee 1977*, number 30 in Lecture Notes in Mathematics, pages 105–116. Springer-Verlag, 1978.

[11] J. B. Rosen, H. Park, and J. Glick. Total least norm formulation and solution for structured problems. *SIAM J. Matrix. Anal. Appl.*, 17:110–126, 1996.

[12] H. Schwetlick and V. Tiller. Numerical methods for estimating parameters in nonlinear models with errors in the variables. *Technometrics*, 27:17–24, 1985.

[13] H. Schwetlick and V. Tiller. Nonstandard scaling matrices for trust region Gauss–Newton methods. *SIAM J. Sci Statist. Comput.*, 10:654–670, 1989.

[14] J. Varah. Least squares data fitting with implicit functions. *BIT*, 36:842–854, 1996.

NEURAL MINOR COMPONENT ANALYSIS AND TLS

Giansalvo Cirrincione
exin@dag.it
University of Picardie CREA, 33 rue Saint Leu, 80039 Amiens, France

Maurizio Cirrincione
nimzo@cerisep.pa.cnr.it
CERISEP-CNR, viale delle Scienze snc, 90128 Palermo, Italy

Abstract The TLS problem can be solved by using neural networks. However, this method can be better described in the framework of iterative numerical methods for TLS. This is justified by the separation between the neural and the numerical experts. The inexperience of the neural experts in numerical analysis also explains the confusion in the neural literature about the Minor Component Analysis (MCA) and the TLS. This paper shows the exact difference between MCA and TLS techniques and the algorithmic consequences. The MCA neural networks are presented, analysed and compared, above all for large sets of linear equations. The superiority of the MCA EXIN neural network is emphasized. In the end the TLS neural networks are described. The TLS EXIN neural network (derived from MCA EXIN) is the best technique and has very important properties, like the existence of universal initial conditions, the possibility to be accelerated and to solve the nongeneric TLS problem automatically.

Keywords: minor component analysis, neural computing, total least squares.

1. Introduction

Given the system $Ax = b$ where $A \in \mathbb{R}^{mxn}$ and $b \in \mathbb{R}^m$, the TLS solution mimimizes the following cost function:

$$E_{TLS}(x) = \frac{(Ax-b)^T(Ax-b)}{1+x^Tx} = \frac{\left\|[A;b]\left[x^T;-1\right]^T\right\|_2^2}{\left\|\left[x^T;-1\right]^T\right\|_2^2} \tag{1}$$

S. Van Huffel and P. Lemmerling (eds.), Total Least Squares and Errors-in-Variables Modeling, 235–250.

which is the Rayleigh Quotient of $[A;b]^T[A;b]$ constrained to the TLS hyperplane, defined by $x_{n+1} = -1$. Hence, the TLS solution is parallel to the right singular vector ($\in \mathbb{R}^{n+1}$) corresponding to the minimum singular value of $[A;b]$, that is its Minor Component (MC). In the neural literature we can distinguish two approaches:

- At first the MC of the autocorrelation matrix of the input data R (*equivalent* to $\frac{[A;b]^T[A;b]}{m}$) is computed; then the MC has to be normalized in order to have the last component equal to -1.
- The estimation at each iteration is constrained to lie on the TLS hyperplane.

According to the approach, we can define [1] two families of neural networks: the MCA neural networks which use the first approach and the TLS neural networks which exploit the second one. In this paper we will analyze these two families both theoretically and experimentally.

2. The MCA neural networks

The only nonlinear network for MCA is the Hopfield network by Mathew and Reddy [12]. The authors develop a constrained energy function, using a penalty function, to minimize the RQ. The neurons have sigmoidal activation functions. Unfortunately, the structure of the network is problem-dependent (the number of neurons is equal to the dimension of the eigenvectors); furthermore, it is necessary to estimate the trace of the covariance matrix for selecting appropriate penalty factors. All other existing neural networks are made up of *one* simple *linear* neuron.

2.1 MCA linear neurons

Consider a linear neuron

$$y(t) = \sum_{i=1}^{n+1} \Psi i(t)\, \xi_i(t) = \Psi^T(t)\, \xi(t) \tag{2}$$

where $\Psi \in \mathbb{R}^{n+1}$ is the weight vector. The training set (TS) is composed of the rows of $[A;b]$. Assume $\xi(t)$ is a bounded continuous valued stationary ergodic data vector with finite second order moments. The existing learning laws for the MCA of the autocorrelation matrix $R = E\left[\xi(t)\,\xi^T(t)\right]$ of the input vector $\xi(t)$ are listed below.

2.1.1 The Oja's learning laws. Changing the Oja's learning law for PCA into a constrained anti-Hebbian rule, by reversing the sign, the following rule (**OJA**) [15] is given

$$\Psi(t+1) = \Psi(t) - \alpha(t)\, y(t) \left[\xi(t) - y(t)\, \Psi(t)\right] \tag{3}$$

being $\alpha(t)$ the learning rate. Its explicitly normalized version (**OJAn**) [15] is

$$\Psi(t+1) = \Psi(t) - \alpha(t)\, y(t) \left[\xi(t) - \frac{y(t)\, \Psi(t)}{\Psi^T(t)\, \Psi(t)}\right] \tag{4}$$

Under certain assumptions, using the techniques of the stochastic approximation theory, the corresponding averaging differential equations are, respectively,

$$\frac{dw(t)}{dt} = -Rw(t) + \left[\Psi^T(t)\, Rw(t)\right] \Psi(t) \tag{5}$$

$$\frac{dw(t)}{dt} = -Rw(t) + \underbrace{\frac{\Psi^T(t)\, Rw(t)}{\Psi^T(t)\, \Psi(t)}}_{\textit{Rayleigh Quotient}} \Psi(t) \tag{6}$$

Another Oja's learning rule [13] (**OJA+**) is the following :

$$\begin{aligned}\Psi(t+1) &= \Psi(t) - \alpha(t) \left[y(t)\, \xi(t) - \left(y^2(t) + \right.\right. \\ &\quad \left.\left. +1 - \|\Psi(t)\|_2^2 \right) \Psi(t) \right]\end{aligned} \tag{7}$$

The corresponding averaging ODE is given by :

$$\begin{aligned}\frac{dw(t)}{dt} &= -Rw(t) + \left[\Psi^T(t)\, Rw(t)\right] \Psi(t) \\ &\quad +\Psi(t) - \|\Psi(t)\|_2^2\, \Psi(t)\end{aligned} \tag{8}$$

2.1.2 The Luo-Unbehauen-Cichocki's learning law. Luo, Unbehauen and Cichocki [11] propose the following rule (**LUO**) :

$$\begin{aligned}\Psi(t+1) &= \Psi(t) - \alpha(t) \left[\Psi^T(t)\, \Psi(t)\, y(t)\, \xi(t) + \right. \\ &\quad \left. -y^2(t)\, \Psi(t)\right]\end{aligned} \tag{9}$$

with the corresponding averaging ODE

$$\frac{dw(t)}{dt} = -\Psi^T(t)\, \Psi(t)\, Rw(t) + \left[\Psi^T(t)\, Rw(t)\right] \Psi(t) \tag{10}$$

2.1.3 The Feng-Bao-Jiao's learning law. In [7] the following rule (**FENG** or **FENG1**[1]) is introduced :

$$\Psi(t+1) = \Psi(t) - \alpha(t) \left[\Psi^T(t)\, \Psi(t)\, y(t)\, \xi(t) - \Psi(t)\right] \tag{11}$$

[1] Another rule is also given in the paper, but it works worse.

The corresponding averaging ODE is given by :

$$\frac{dw(t)}{dt} = -\Psi^T(t)\,\Psi(t)\,Rw(t) + \Psi(t) \tag{12}$$

2.1.4 The MCA EXIN learning law.

In [1] the following rule (**MCA EXIN**) is proposed:

$$\Psi(t+1) = \Psi(t) - \frac{\alpha(t)\,y(t)}{\|\Psi(t)\|_2^2}\left[\xi(t) - \frac{y(t)\,\Psi(t)}{\|\Psi(t)\|_2^2}\right] \tag{13}$$

The corresponding averaging ODE is given by :

$$\frac{dw(t)}{dt} = -\frac{1}{\|\Psi(t)\|_2^2}\left[R - \frac{\Psi^T(t)\,Rw(t)}{\|\Psi(t)\|_2^2}\right]\Psi(t) \tag{14}$$

and represents the exact gradient flow of the Rayleigh Quotient of R.

2.2 Rayleigh quotient gradient flows

All MCA learning laws are RQ gradient flows. In [1] it is proved that the ODE's of LUO, OJAn and MCA EXIN are equivalent, because they only differ from the Riemannian metric (the other ODE's are approximately equivalent). This fact implies a similar stability analysis except for the critical points. Indeed, the Rayleigh Quotient critical points are *degenerate* because in these points the RQ Hessian matrix is not invertible. Hence, the phase portrait of the gradient flow has only degenerate straight lines (*critical directions*) in the direction of the RQ eigenvectors, i.e. the critical points are not isolated. It implies that the local stability properties (local phase portrait) of the gradient flow around the critical points change with the Riemannian metric.

2.3 Error cost analysis

All MCA learning laws have approximately or exactly (MCA EXIN) the RQ of R as error cost. In [4] the following assertions are proved (see Figure 1 for the 3d case).

- If using a gradient flow, it is better to work with low weights in order to avoid the flatness of the RQ cost landscape.
- The error cost has n critical directions. Among these, all the directions associated with the non-minor components are saddle directions and correspond to bigger and bigger values of the cost

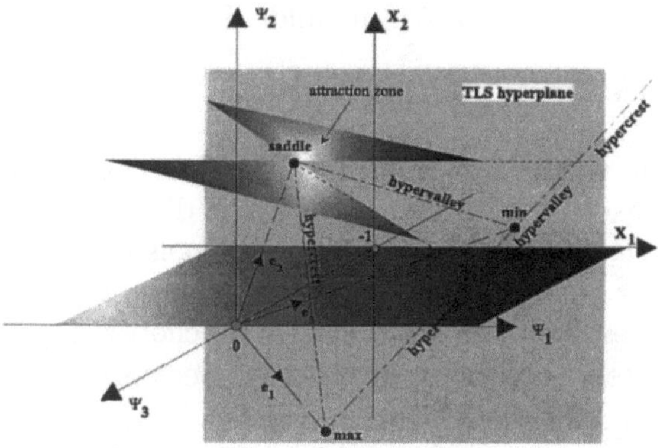

Figure 1. Cost stability analysis for MCA EXIN and TLS EXIN in the space $\mathbb{R}^3$.

function as the corresponding eigenvalues increase. They are minima in any direction of the space spanned by bigger eigenvectors (in the sense of eigenvectors associated with bigger eigenvalues) but are maxima in any direction of the space spanned by smaller eigenvectors: while getting deeper (i.e. toward the minimum) in the cost landscape, the number of dimensions of escape from a saddle point becomes smaller. The critical direction associated to the minimum eigenvalue is the only global minimum.

- The hyperplane through the origin and perpendicular to the minimum direction is made up of points which are maxima along the minimum direction (hypercrest along the minimum direction). The hyperplane through the origin and perpendicular to the maximum direction is made up of points which are minima along the maximum direction (hypervalley along the maximum direction).The hyperplane through the origin and perpendicular to a saddle direction is made up of two kinds of points: minima along the saddle direction (hypervalley along the saddle direction) and maxima along the saddle direction (hypercrest along the saddle direction).

2.4 Dynamics

Let R be positive semidefinite with minimum eigenvalue λ_n of multiplicity one and corresponding unit eigenvector z_n. If $\Psi^T(0)\, z_n \neq 0$, then, *in the limits of validity of the ODE approximation* [4],

1) in the first part of the weight evolution, $\|\Psi(t)\|_2^2 \approx \|\Psi(0)\|_2^2$;

2) for OJA, MCA EXIN and LUO, asymptotically $\Psi(t)$ is parallel to z_n;

3) for OJA+ (if $\lambda_n < 1$) and OJAn, it holds $\lim_{t\to\infty} \Psi(t) = \pm z_n$

1) for FENG it holds $\lim_{t\to\infty} \Psi(t) = \pm\frac{1}{\sqrt{\lambda_n}} z_n$.

If the initial weight vector has modulus less than unity, the MCA EXIN weight vector reaches the minor component direction faster than OJAn which is faster than LUO. The contrary happens if the modulus of the initial weight vector is more than unity. It may seem that OJAn and LUO can give better results than MCA EXIN by choosing large initial conditions. Unfortunately this choice is not good because of the flatness of the RQ landscape [4]. The literature only takes in account the ODE approximation. In [4] the dynamical analysis is directly derived from the stochastic discrete laws. The behaviour of the weight vector, except for OJA+ and FENG, can be described in the following way:

1) an initial transient;

2) fluctuations around the locus of constant modulus, but with an increasing bias; the fluctuations are function of the learning rate as it will be shown later;

3) arrival in the desired direction;

4) the fluctuations, by increasing the modulus, push the weight to another critical point and so on until ∞.

LUO, OJAn and MCA EXIN do not converge. MCA EXIN has the slowest divergence; LUO has the fastest divergence. However, for LUO the divergence happens in a *finite* time t_∞ (*sudden divergence,* [14]), which depends on the spreading of the eigenvalue spectrum of R; if the eigenvalues of R are clustered, the sudden divergence appears late. Furthermore, t_∞ is proportional to the inverse of λ_n (high λ_n means noisy data). For initial conditions of norm greater than one, OJA also suffers from the sudden divergence. Unfortunately, it is not always possible to choose the initial conditions, e.g. in slowly varying systems for real time applications. Furthermore, a matrix R with high λ_n (noise in the data) worsens the sudden divergence problem. OJA+ is not divergent and it does not suffer from the sudden divergence, but, unfortunately, it requires the assumption that λ_n has to be less than unity. If it cannot be assumed in advance (e.g. for noisy data), OJA+ may suffer from the sudden divergence. The FENG learning law, even if it converges to the solution, has some unacceptable problems, namely very high oscillations around the solution, which prevent the choice of a reliable stop criterion, and the difficulty to control the rate of convergence (see [4]).

The choice of the learning rate may generate instability. This is called *(dynamic) instability divergence* in [4] where a complete analysis for all neurons is given.

From the point of view of the variance/bias dilemma the MCA EXIN learning law is a high variance/low bias algorithm. Indeed, choosing low initial conditions implies a high variance in the first temporal part of the weight time evolution, but not at the expense of the final accuracy, because in the divergence phase the fluctuations decrease to zero. OJAn and LUO haven't this feature: they need low initial conditions (which implies high variance) and have low bias, but with high oscillations around the solution for a longer time than MCA EXIN. On the contrary, OJA, OJA+ and FENG are algorithms with low variance and high bias. However, FENG has larger fluctuations and is unreliable for near singular matrices. Obviously, for all the algorithms, the presence of outliers worsens the dynamic stability. However, there also exists a robust variant (NMCA EXIN [3]), which yields accurate results.

Resuming the analysis in [4], the following conclusions can be stated.

LUO - It has a slow convergence to the MC direction and then diverges (order $O\left(\alpha^2\right)$). It also suffers from the sudden divergence, which can be anticipated by noisy data. It cannot be reliably stopped and is very sensitive to outliers. It is a high variance/low bias algorithm, but it oscillates too much around the solution. It works badly for medium dimensional data.

OJAn - It has a slow convergence (faster than LUO) to the MC direction and then diverges (order $O\left(\alpha^2\right)$, but slower than LUO). The rate of divergence depends exponentially on the level of noise in the data (λ). It cannot be reliably stopped and is very sensitive to outliers. It is a high variance/low bias algorithm, but it oscillates too much around the solution. It works badly for medium dimensional data

OJA - It has a very slow convergence to the MC direction and its square weight modulus change is of order $O\left(\alpha\right)$. The weights decrease to 0 (in case of initial conditions of modulus greater than 1, there is sudden divergence, which is anticipated in case of noisy data). It cannot be reliably stopped and is very sensitive to outliers. It is a low variance/high bias algorithm and the computational cost per iteration is the best. It works very badly for medium dimensional data

OJA+ - It has a very slow convergence to the MC direction and its square weight modulus change is of order $O\left(\alpha\right)$. The weights converge only for $\lambda_n < 1$ (in case of initial conditions of modulus greater than 1 and $\lambda_m > 1$, there is sudden divergence, which is anticipated in case of noisy data). It cannot be reliably stopped and is very sensitive to outliers. It is a low variance/high bias algorithm. It works very badly for medium dimensional data.

FENG - It is the worst learning law because of the too large oscillations (the dynamic stability doesn't improve when the MC direction is reached because it only depends on p and the learning rate). It has a slow convergence (same as LUO) to the MC direction and is very sensitive to the outliers. It cannot be reliably stopped and can diverge for near singular matrices. It works very badly for medium dimensional data.

MCA EXIN - It is by far the best MCA learning law. It has the best convergence to the MC direction, the slower divergence, doesn't have problems neither of sudden nor of instability divergence. It works very well in high dimensional spaces and has been used in real applications [4]. It can be reliably stopped. It is a high variance/low bias algorithm with a very good dynamics because of its inertia for large weights. It has the same computational cost as the other MCA laws, except OJA.

2.5 Numerical considerations

2.5.1 Computational cost. The MCA learning laws are iterative algorithms and have a different computational cost per iteration, which is here evaluated in floating point operations and shown in tab.1. OJA has the lowest cost. All costs depend on the dimensionality of the data. For high dimensional data, all learning laws have the same cost except OJA which has a saving per iteration of 33% in flops.

Table 1. Cost per iteration of the MCA learning laws

	OJA+	LUO	EXIN	OJAn	FENG	OJA
flops / iteration	$8n+3$	$8n+1$	$8n+1$	$8n$	$8n-1$	$6n$

2.5.2 Quantization errors. The limited precision (quantization) errors can degradate the solution of the gradient based algorithms w.r.t. the performance achievable in infinite precision. These errors accumulate in time without bound, leading, in the long term (ten of millions of iterations) to an eventual overflow. This kind of divergence is called *numerical divergence*in [4]. The degradation of the solution is proportional to the conditioning of the input. Decreasing the learning rate in the infinite precision algorithm leads to improved performance. Nevertheless, this decrease increases the deviation from infinite precision performance. Increasing the learning rate can also magnify numerical errors, so there is a trade off in terms of numerical effects on the size of the learning rate. A technique known as *leakage* is proposed in [4] to prevent the overflow at the expense of some increase in cost of implementation and at the expense of a further small degradation in performance.

2.5.3 Acceleration techniques. The MCA learning laws are sequential algorithms. Several methods, used in neural theory, can be applied (momentum, bold driver, delta-bar-delta, quickprop, see [4]). However, the batch techniques yield a better acceleration. If the data are collected in *blocks* and then fed to the neuron, which changes its weights only after the whole block presentation, all classical *batch* techniques can be used. However, it is not practical to use the Newton and quasi-Newton descent techniques because the RQ Hessian matrix at the minimum is singular and then the inverse does not exist. On the other hand the conjugate gradient approach can deal with the singular H_r [10, pp.256-259] and can be used for block learning [1].

2.6 Simulations

The following simulations use, as data, a zero mean Gaussian random vector $x(t)$ generated by a matrix R whose spectrum is chosen in advance. The initial weight vector components are chosen randomly in $[0, 1]$. λ_n is always equal to 1. The other eigenvalues are given by the law $\lambda_i = n - i$; then the condition number $\kappa_2(R) = \frac{\lambda_1}{\lambda_n}$ increases with n, but R always remains a well conditioned matrix. Tab.2 shows, for four MCA laws, the best results over several experiments, in terms of flops, for each value of n. Except MCA EXIN, all other laws diverge for low values of n. This problem can be explained by the choice of the initial conditions: increasing the number of components, the initial weight modulus increases and quickly becomes greater than 1. About OJA, there is sudden divergence. It has been noticed [4] that, for increasing n, the sudden divergence is anticipated. There can also be instability divergence. About FENG, there is instability divergence. Indeed [4] the fluctuations only depend on the weight modulus and the data. The modulus is large and remains so because of the too large oscillations which prevent the weights from approaching the MC direction. As a consequence, this generates more and more instability until the finite time divergence. For increasing n, the divergence is anticipated. About LUO, two explanations are possible: sudden and instability divergence.

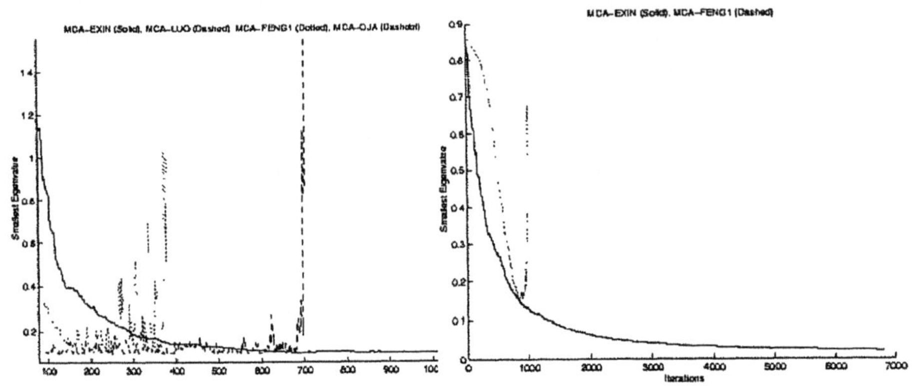

Figure 2. Computation of the smallest eigenvalue of a 11x11 (left) and a 100x100 (right) autocorrelation matrix.

However, in these experiments the divergence is of the instability kind, because it is accompanied by very large oscillations and certainly anticipates the sudden divergence [4]. Figure 2 shows two experiments for, respectively, $n = 11$ (left) and 100 (right). The figure on the left confirms that FENG undergoes the instability divergence (notice the peaks before the divergence). Also OJA diverges (here there is instability divergence). The figure on the right shows the good results obtained by MCA EXIN. Other simulations can be found in [4].

Table 2. Total cost of the MCA learning laws for autocorrelation matrices of increasing size (* stands for inaccurate result, div for divergence)

n	κ_2	EXIN	LUO	FENG	OJA
5	40	65966	59131	60345	42942
7	60	183033	640074	742346	div
8	70	205968	706075	830000	div
9	80	228681	965544	*	div
10	90	252002	1061098	div	div
15	140	366255	div	div	div
50	490	3725868	div	div	div
100	990	6258001	div	div	div

2.6.1 Application to the parameter estimation of induction motors. An interesting application of MCA EXIN can be found in [6]. It deals with the on-line estimation of the parameters of an induction electrical motor, which is very important for high-performance ac drives with vector control or direct torque control where inaccurate rotor time constant estimation causes a failure of the control strategy or induction motors directly connected to the utility grid where estimation of stator and rotor resistances can give information about the temperature of the windings, which is in general beneficial for the improvement of the service. One result is visualized in Figure 3, that shows the comparison with the Ordinary Least Squares technique which is the current technique in this area. Here, the accuracy parameter ρ is defined as $\rho = \frac{\sum_{i=1}^{n}\left(w_i(t)-w_i^*\right)^2}{n}$ ($w_i(t)$ is the ith component of the neuron weight vector and w_i^* is the ith component of the desired parameter vector w^*).

3. The TLS neural networks

If the weight vector is constrained to lie on the TLS hyperplane (i.e. losing a degree of freedom), it can be argued that all the convergence problems, which are the consequence of the RQ degeneracy property, are no more valid, because, on the TLS hyperplane, the only critical points are the intersections with the critical straight lines, and therefore they are *isolated* (it implies *the global asymptotic stability* of the solution on the hyperplane plus the eventual solution ∞). This reasoning is illustrated in Figure 1 for the three dimensional Ψ-space. Furthermore, no normalization operation would be used. The only two existing TLS neural networks are TLS EXIN [1], [2] and TLS GAO [8], [9].

3.1 The TLS EXIN learning law

From eq.(1), it holds

$$E_{TLS}(x) = \sum_{i=1}^{m} E^{(i)}(x)$$

where

$$E^{(i)}(x) = \frac{\left(a_i^T x - b_i\right)^2}{1 + x^T x} = \frac{\sum_{j=1}^{n}\left(a_{ij}x_j - b_i\right)^2}{1 + x^T x} = \frac{\delta^2}{1 + x^T x} \tag{15}$$

being i the index of the ith-row of $[A; b]$. Hence,

$$\frac{dE^{(i)}}{dx} = \frac{\delta a_i}{1 + x^T x} - \frac{\delta^2 x}{\left(1 + x^T x\right)^2} \tag{16}$$

and the corresponding steepest descent discrete time formula is given by:

$$x(t+1) = x(t) - \alpha(t)\gamma(t)a_i + \left[\alpha(t)\gamma^2(t)\right]x(t) \tag{17}$$

where

$$\gamma(t) = \frac{\delta(t)}{1 + x^T(t)x(t)} \tag{18}$$

This is the TLS EXIN learning law and $E^{(i)}$ is its cost function. It can also be derived by the MCA EXIN law (13), by replacing $\Psi(t)$ with $\left[x^T(t); -1\right]^T$, $\xi(t)$ with $\left[a_i^T; b_i\right]^T$ and taking only the first n components. TLS EXIN is a linear unit with n inputs (vector a_i), n weights (vector x), one output (scalar $y_i = x^T a_i$) and one training error (scalar $\delta(t)$). In [1] it is proved that the Hessian matrix of the TLS cost function is positive definite (which implies the nonsingularity) at the cost function minimum. Then, unlike the MCA neurons, in block and batch mode it can be accelerated by the Newton and quasi-Newton (e.g. BFGS) techniques; in [1] it is shown that BFGS works much better than the conjugate gradient (CG) techniques.

TLS EXIN works in a lower dimensional (n) weight space than the MCA neurons. If, from one side it is an advantage both from a stability (it has $n+1$ critical points: one minimum, $n-1$ saddles and one maximum) and a numerical point of view, from the other side it implies a *domain of convergence*, i.e. it does not converge for every possible choice of the initial conditions. This domain has been analytically studied in [1]. Figure 3 shows, for a two-dimensional TLS problem analysed in [1] the weight vector evolution versus several values of the initial conditions. All the trajectories beginning from the plus points go to the straight line through the minimum and the saddle at the right side of the saddle and then converge to the minimum. All the trajectories beginning from the cross points go to the same straight line at the left side of the saddle and then diverge in the direction of the straight line. In [1] it is proved the fundamental statement that the TLS origin always belongs to the TLS domain of convergence. Hence, using null initial conditions is the best choice if no a priori information is given. It is also proved that for this choice TLS EXIN automatically solves the nongeneric TLS problem. As a consequence, null initial conditions always guarantee the correct TLS solution, independently of the rank of A !

3.2 The TLS GAO learning law

The $(n+1)$th-dimensional TLS GAO learning discrete law is:

$$\Psi(t+1) = \Psi(t) - \alpha(t)\, y(t)\left[\xi(t) + \Psi(t)\,\xi_{n+1}(t)\right] \tag{19}$$

In [9, th.2] it is shown that, taking as initial condition $\Psi_{n+1}(0) = -1$, then $\Psi(t)$ always lies in the TLS hyperplane and therefore only the first n components of eq.(19) is meaningful (the same can be proved for TLS EXIN, see [1]). Considering the derivation of TLS GAO in [9, p.722] as a gradient descent algorithm using a linearization (it implies the assumptions of small gains and, above all, weight norms much smaller

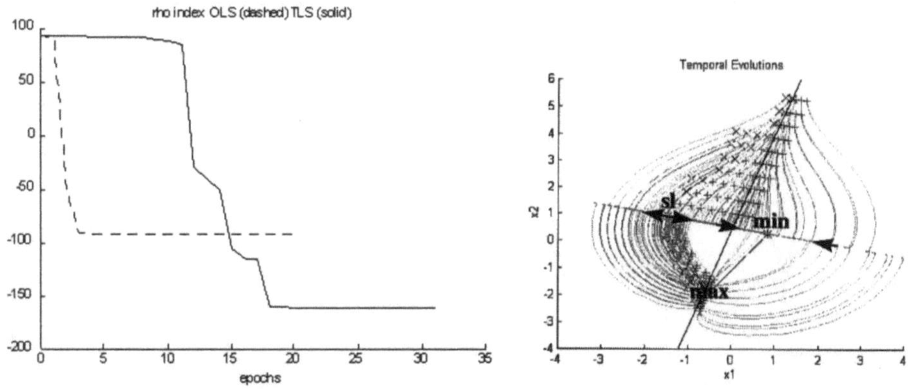

Figure 3. Left: The temporal evolution of the sequential TLS EXIN neuron for a generic two-dimensional TLS problem; Right: Comparison between the accuracy obtained by MCA EXIN (TLS) and the OLS method (w.r.t. epochs, because both use block techniques).

than one) of the local gradient vector of the instantaneous estimation of the TLS cost function, it follows [1] that TLS GAO can be derived as linearization of TLS EXIN for weight norms much smaller than one. Comparing the two learning laws, it can be seen that the operation cost per iteration is smaller for TLS GAO; the TLS GAO learning law cannot be accelerated (unlike TLS EXIN) by methods like CG and BFGS because it does not represent the gradient flow of an error function; the dependence of TLS GAO on strongly constraining assumptions limits its dynamic behaviour. Indeed, the violation of the small gain assumption, typical in the transient, implies that TLS EXIN has a better transient and a faster and more accurate response than TLS GAO. The violation of the small weight assumption limits the initial conditions field of validity of TLS GAO, in the sense of accuracy and dynamical behaviour.

3.3 Simulations

In [2] examples are given in the case of the TLS parameter estimation of an IIR filter. The initial conditions are null[2] and an accuracy parameter ρ is defined as $\frac{\sum_{i=1}^{n}(w_i(t)-w_i^*)}{n}$ ($w_i(t)$ are the n components of the neuron weight vector and w_i^* are the n components of the desired parameters). The TLS EXIN neuron is compared with the TLS GAO neuron and the Recursive Least-Squares (RLS) algorithm. The noise in the data is uniformly distributed over $\left[-\frac{\kappa}{2}, \frac{\kappa}{2}\right]$. Table 3 ($\sigma^2$ = variance)

[2]This choice agrees with the TLS GAO assumptions.

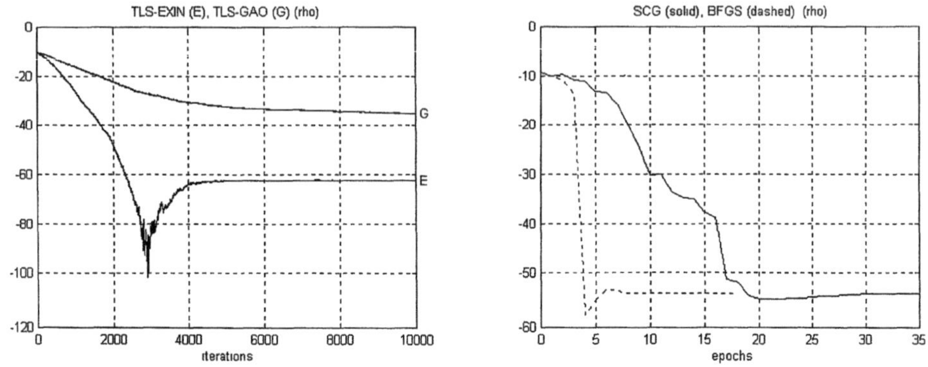

Figure 4. Left: For $\kappa = 0.4$; averaged on 60 independent trials for ρ (in dB); Right: TLS EXIN acceleration methods for $\kappa = 1$; an epoch represents a block of 1200 training vectors

resumes the results for $\kappa = 0.4$. Figure 4 plots the index parameter ρ vs. time. An improvement of more than 50 dB is obtained w.r.t. TLS GAO. Despite a certain computational burden, TLS EXIN outperforms the other two methods. Figure 4 shows, for $\kappa = 1$, the very good performance of the TLS EXIN acceleration methods[3] (RLS and TLS GAO yield, respectively, -8 dB and -26 dB). In particular, BFGS has a much superior accuracy. The epochs are actually the above mentioned *blocks,* in order to use these methods on line. Interestingly, as the noise variance increases, TLS GAO diverges. This phenomenon can be explained considering that, because of its learning law linearization, during the transient and in presence of the high noise, the weights are so far to the solution to find themselves outside the convergence domain. TLS EXIN is still convergent, but better results are obtained by using the acceleration techniques.

Table 3. Best solutions for the adaptive IIR filtering; m.e.=mean error

	s_1	s_2	c_0	c_1	c_2
Desired	1.1993	−0.5156	0.0705	0.1410	0.0705
RLS m.e.	0.6862	0.5516	0.0002	0.0483	0.1155
RLS σ^2	0.0072	0.0064	0.0024	0.0025	0.0028
GAO m.e.	0.0768	0.0617	0.0087	0.0071	0.0155
GAO σ^2	0.0260	0.0192	0.0219	0.0241	0.0241
EXIN m.e.	0.0503	0.0353	0.0007	0.0029	0.0061
EXIN σ^2	0.0032	0.0024	0.0002	0.0002	0.0002

[3]SCG stand for Scaled CG, see [1].

4. Conclusions

What is the best neural network for TLS ? In [1] TLS EXIN was born as an improvement of MCA EXIN, which is the best MCA neuron. With regard to MCA EXIN, TLS EXIN does not suffer from the problem of critical lines, works in a lower dimensional weight space, can be accelerated better and solves both generic and nongeneric TLS without needing any information about the rank of A. There also exists a very powerful variant, GeTLS EXIN [1], which solves also the OLS and DLS problems, is faster and has better dynamics. On the contrary, MCA EXIN always converges to the MC direction, while TLS EXIN needs null initial conditions if no a priori information is given. Hence, working on the TLS hyperplane means introducing a domain of convergence. However, in [1] it is shown that the TLS hyperplane origin is placed in a very favorable position, between the maximum and the minimum of the error cost, without saddles in between. Interestingly, GeTLS can be used as an MCA neuron. The resulting neuron is called MCA EXIN+ and works much better than MCA EXIN [1].

Why to use a neural network for TLS ? These neural learning laws are simply iterative algorithms, and so have all the advantages and drawbacks of these techniques w.r.t. the direct methods. However, they often yield better results than the TLS iterative methods (see [5] for a detailed comparison with these techniques) and can be easily integrated with other similar neural modules (see [1] for a computer vision fully neural module).

References

[1] G. Cirrincione. *A Neural Approach to the Structure from Motion Problem.* PhD thesis, LIS INPG Grenoble, December 1998.

[2] G. Cirrincione and M. Cirrincione. Linear system identification using the tls exin neuron. In *NEURAP 98*, Marseille (France), March 1998.

[3] G. Cirrincione and M. Cirrincione. Robust total least squares by the nonlinear mca exin neuron. In *Proc. IEEE Intern. Joint Symposia on Intelligence and Systems*, pages 295–300, Rockville, Maryland (USA), May 1998.

[4] G. Cirrincione, M. Cirrincione, J. Herault, and S. Van Huffel. The mca exin neuron for the minor component analysis. *IEEE Trans. on Neural Networks*, 2001. (to appear).

[5] G. Cirrincione, M. Cirrincione, and S. Van Huffel. Generalization and scheduling of tls problems. *IEEE Transactions on Signal Processing*, 2001. Submitted.

[6] G. Cirrincione, M. Cirrincione, and G. Vitale. Identificazione di parametri di azionamenti elettrici con tecniche combinate neurali e tls. *10 Seminario Interattivo su Azionamenti Elettrici, Innovazioni Tecnologiche e Problematiche Emergenti (ANAE)*, March 1999. Bressanone (BZ), Italy.

[7] D.Z. Feng, Z. Bao, and L.C. Jiao. Total least mean squares algorithm. *IEEE Trans. on Signal Processing*, 46(8):2122–2130, August 1998.

[8] K. Gao, M.O. Ahmad, and M.N. Swamy. Learning algorithm for total least-squares adaptive signal processing. *Electronic Letters*, 28(4):430–432, February 1992.

[9] K. Gao, M.O. Ahmad, and M.N. Swamy. A constrained anti-hebbian learning algorithm for total least-squares estimation with applications to adaptive fir and iir filtering. *IEEE Trans. on Circuits and Systems-II: Analog and Digital Processing*, 41(11):718–729, November 1994.

[10] M. Hestenes. *Conjugate Direction Methods in Optimization.* Springer-Verlag New York, 1980.

[11] F. Luo, R. Unbehauen, and A. Cichocki. A minor component analysis algorithm. *Neural Networks*, 10(2):291–297, 1997.

[12] G. Mathew and V. Reddy. Development and analysis of a neural network approach to pisarenko's harmonic retrieval method. *IEEE Trans. on Signal Processing*, (42):663–667, 1994a.

[13] E. Oja. Principal components, minor components and linear neural networks. *Neural Networks*, 5:927–935, 1992.

[14] A. Taleb and G. Cirrincione. Against the convergence of the minor component analysis neurons. *IEEE Transactions on Neural Networks*, 10(1):207–210, January 1999.

[15] L. Xu, E. Oja, and C. Suen. Modified hebbian learning for curve and surface fitting. *Neural Networks*, 5:441–457, 1992.

ON THE STRUCTURAL LINE SEGMENT MODEL

Ori Davidov, Alexander Goldenshluger and Roman Reidman
Department of Statistics, University of Haifa, Haifa 31905, Israel
{davidov, goldensh, reidman}@stat.haifa.ac.il

Abstract We introduce and study the structural line segment model, which is motivated by applications in the field of computer vision. The method of moments estimator (MME) is investigated under three different sets of assumptions. We derive the asymptotic distribution of the MME and study its performance numerically.

Keywords: structural relationship, measurement error, method of moments, computer vision, line segment fitting.

1. Introduction and formulation

In this paper we consider the problem of reconstructing a line segment from noisy data. This is a basic problem in computer vision with numerous practical applications. Line segment fitting is frequently used in the transition from numerical to symbolic description of images. For example, many procedures for edge detection and image segmentation require accurate fitting of segments to form the boundary between distinct regions of the image. Fitting line segments and other patterns, such as circles and ellipses, has been extensively studied in the computer vision literature (e.g., [4], [8], [5], [9]). Although a variety of different algorithms and methods have been developed, their statistical properties have not been thoroughly addressed. Moreover, the underlying modeling assumptions have not been clearly formulated. Consequently the performance of seemingly natural and widely used algorithms may be poor in practice. Many algorithms for fitting lines, and line segments are based on variants of the method of Total Least Squares (TLS). A recent example is [1]. The method of TLS has received considerable attention in numerical analysis, computer science, statistics and other fields (see [7], [6]).

S. Van Huffel and P. Lemmerling (eds.), Total Least Squares and Errors-in-Variables Modeling, 251–260.

Formally, we consider the model

$$x_i = \lambda_i \xi_x + (1-\lambda_i)\eta_x + \varepsilon_i, \quad (1)$$
$$y_i = \lambda_i \xi_y + (1-\lambda_i)\eta_y + v_i, \quad (2)$$

where $(\varepsilon_1, v_1), \ldots, (\varepsilon_n, v_n)$ is a sequence of independent normally distributed random vectors with mean zero and diagonal variance matrix $\text{diag}(\sigma_\varepsilon^2, \sigma_v^2)$. The unobservable quantities $\lambda_1, \ldots, \lambda_n$ take values in the interval $[0,1]$. The goal is to estimate $\eta = (\eta_x, \eta_y)$ and $\xi = (\xi_x, \xi_y)$, which are the endpoints of the interval using the observations $(x_1, y_1), \ldots, (x_n, y_n)$. There are two general approaches to model $\lambda_1, \ldots, \lambda_n$; (a) they may be treated as random variables; or (b) as unknown constants. Assuming that λ_i are random variables with distribution G, known or unknown, leads to the so called structural model. Alternatively considered as unknown constants the quantities $\lambda_1, \ldots, \lambda_n$ are nuisance parameters. Note that the dimension of the nuisance vector grows with the sample size. This model is known as the functional model. The model (1)–(2) is closely related to regression problems with error in variables, e.g., [2]. In [3] the model (1)–(2) has been studied assuming that λ_i are random variables with known mean and variance, μ_λ and σ_λ^2, respectively. In addition it was assumed that the error variances σ_ε^2 and σ_v^2 are known in advance. The quantity $\mathcal{S} = \{\theta, \sigma_\varepsilon^2, \sigma_v^2, G\}$ where $\theta = [\eta, \xi]$ is called the structure of the model. It was shown in [3] that the structure, $\mathcal{S}$, is identifiable if and only if 0 and 1 belong to the support of G. A method of moments estimator (MME) has been proposed and its large sample properties were explored.

In this paper we relax assumptions of [3] and consider three sets of more practical conditions. In particular, we derive and explore estimators in the following cases: (*a*) Ratio of Error Variances Known (Case A): here we assume that μ_λ, σ_λ^2 and $\rho = \sigma_\varepsilon^2/\sigma_v^2$ are known; (*b*) One of the Error Variances is Known (Case B): here we assume that μ_λ, σ_λ^2 and one of the error variances, say, σ_ε^2 is known; (*c*) The Length of the Line Segment is Known (Case C): here we assume that $\mu_\lambda, \sigma_\varepsilon^2, \sigma_v^2$ and the length of the line segment $\delta = \sqrt{(\eta_x - \xi_x)^2 + (\eta_y - \xi_y)^2}$ are known.

2. Results

In this section we investigate the method of moments estimator (MME) for the structural parameter $\theta = [\eta, \xi]$ indexing the model (1-2) for the three cases described in the previous section.

2.1 Derivation of the MME

Equating the sample and theoretical moments, up to the second order, results in the following system of equations

$$\bar{x} = (\xi_x - \eta_x)\mu_\lambda + \eta_x \tag{3}$$
$$\bar{y} = (\xi_y - \eta_y)\mu_\lambda + \eta_y \tag{4}$$
$$s_{xx} = (\xi_x - \eta_x)^2\sigma_\lambda^2 + \sigma_\epsilon^2 \tag{5}$$
$$s_{yy} = (\xi_y - \eta_y)^2\sigma_\lambda^2 + \sigma_\upsilon^2 \tag{6}$$
$$s_{xy} = (\xi_x - \eta_x)(\xi_y - \eta_y)\sigma_\lambda^2 \tag{7}$$

Let $\delta = (\delta_x, \delta_y) = (\xi_x - \eta_x, \xi_y - \eta_y)$ denote the signed length of the line segment along the x and y axis respectively. The unsigned or positive lengths are $\delta_x^+ = |\delta_x|$ and $\delta_y^+ = |\delta_y|$. Without loss of generality we assume that $\xi_x \geq \eta_x$. Therefore we label the interval endpoint with the smaller x coordinate as η . If $\eta_x = \xi_x$, then the point with smaller y coordinate is denoted η. Note that $\mathrm{Cov}[x, y] \neq 0$ if and only if $\xi_x \neq \eta_x$ and $\xi_y \neq \eta_y$. Hence non zero covariance precludes intervals with horizontal or vertical orientations and intervals degenerating to a point. The estimators and their properties when $\mathrm{Cov}[x, y] \neq 0$ are investigated first. The case $\mathrm{Cov}[x, y] = 0$ is discussed later.

The solution to (3)–(7) is the MME and is denoted $\hat{\theta}$. In [3] the estimator of θ has been derived assuming that σ_ε^2 and σ_υ^2 are known. The solution is

$$\begin{aligned} \hat{\eta}_x &= \bar{x} - \mu_\lambda\hat{\delta}_x^+, \\ \hat{\eta}_y &= \bar{y} - \mu_\lambda\hat{\delta}_y^+\mathrm{sgn}(s_{xy}), \\ \hat{\xi}_x &= \bar{x} + (1 - \mu_\lambda)\hat{\delta}_x^+, \\ \hat{\xi}_y &= \bar{y} + (1 - \mu_\lambda)\hat{\delta}_y^+\mathrm{sgn}(s_{xy}), \end{aligned} \tag{8}$$

where

$$\begin{aligned} \hat{\delta}_x^+ &= \sigma_\lambda^{-1}\sqrt{(s_{xx} - \sigma_\epsilon^2)_+}, \\ \hat{\delta}_y^+ &= \sigma_\lambda^{-1}\sqrt{(s_{yy} - \sigma_\upsilon^2)_+}, \end{aligned} \tag{9}$$

$\mathrm{sgn}(x)$ is the sign of x and $(x)_+ = \max(0; x)$. Note that the actual MME does not involve the $(\cdot)_+$ notation, and is well defined only if $s_{xx} \geq \sigma_\epsilon^2$ and $s_{yy} \geq \sigma_\upsilon^2$. It can be easily shown that, if $\mathrm{Cov}[x, y] \neq 0$ then the probability that the inequalities $s_{xx} \geq \sigma_\epsilon^2$ and $s_{yy} \geq \sigma_\upsilon^2$ hold tends to one as n goes to infinity. With the above natural modification, the estimator is well–defined for any set of observations.

2.1.1 Ratio of error variances is known (A).

In this case we have an additional equation

$$\sigma_\epsilon^2 = \rho\sigma_\upsilon^2 \tag{10}$$

and two additional unknowns, σ_ϵ^2 and σ_v^2. One can show that the MME has the form (8)–(9), with σ_ϵ^2 and σ_v^2 replaced by their estimates. We briefly describe their derivation. Using (10) and (5)–(7) we derive

$$\rho\sigma_v^4 - (\rho s_{yy} + s_{xx})\sigma_v^2 + (s_{yy}s_{xx} - s_{xy}^2) = 0.$$

This is a quadratic equation in σ_v^2, with two real roots. It can be shown that only the smaller root satisfies both $\hat{\sigma}_v^2 \geq 0$ and $\hat{\sigma}_v^2 \leq s_{yy}$. The first condition is a positivity condition for the estimator of a variance. The second condition follows immediately from (6). Consequently we have,

$$\hat{\sigma}_v^2 = \frac{1}{2\rho}[(\rho s_{yy} + s_{xx}) - \sqrt{(\rho s_{yy} - s_{xx})^2 + 4\rho s_{xy}^2}]$$

and $\hat{\sigma}_\epsilon^2 = \rho\hat{\sigma}_v^2$, which are substituted in (9) to obtain the MME. Consequently the MME is always well defined.

2.1.2 One of the error variances is known (B) . In this case it is assumed that σ_ϵ^2 is known whereas σ_v^2 is unknown. Hence we have a system of five equation in five unknowns. It follows from (5) that the estimators of η_x and ξ_x are exactly as in (8). Using (6) and (7) we find that

$$\hat{\sigma}_v^2 = s_{yy} - \frac{s_{xy}^2}{s_{xx} - \sigma_\epsilon^2}.$$

as an estimator for σ_v^2. The MME in this case is also of the form (8), where we substitute

$$\hat{\delta}_y = \frac{s_{xy}}{\hat{\delta}_x \sigma_\lambda^2}$$

instead of $\delta_y^+ \text{sgn}(s_{xy})$. Note that the MME is well-defined here if $s_{xx} > \sigma_\epsilon^2$.

2.1.3 The length of the line segment is known (C). In this case we assume that $\delta = \sqrt{\delta_x^2 + \delta_y^2}$ is known. Hence, there is an additional equation and an additional unknown parameter σ_λ^2 which we estimate. From (5) and (6) deduce that

$$\hat{\sigma}_\lambda^2 = \frac{s_{xx} + s_{yy} - \sigma_\epsilon^2 - \sigma_v^2}{\delta^2}. \tag{11}$$

The MME is given by (8), where we substitute $\hat{\sigma}_\lambda^2$ in (11) for σ_λ^2. The MME is defined only if $s_{xx} > \sigma_\epsilon^2$ and $s_{yy} > \sigma_v^2$.

2.2 Large sample distribution

In the following it is shown that the MME is consistent and its asymptotic distribution is derived assuming $\mathrm{Cov}[x,y] \neq 0$. For $\mathrm{Cov}[x,y] = 0$ see Section 2.3.

Theorem 1. *Assuming Case* $I = A, B, C$ *and* $\mathrm{Cov}[x,y] \neq 0$ *the MME,* $\hat{\theta}$*, is consistent and asymptotically multivariate normal with mean* θ *and covariance matrix* $V_\theta = M_I \Sigma_\theta M_I^T$ *where the matrix* Σ_θ *is given in the Appendix,*

$$M_A = \begin{bmatrix} 1+\frac{\mu_\lambda^2}{\sigma_\lambda^2} & 0 & -\frac{\delta_x\mu_\lambda}{\psi} & \frac{\rho\delta_x}{\psi} & -\frac{2\rho\delta_y\mu_\lambda}{\psi} \\ 0 & 1+\frac{\mu_\lambda^2}{\sigma_\lambda^2} & \frac{\delta_y\mu_\lambda}{\psi} & -\frac{\rho\delta_y\mu_\lambda}{\psi} & -\frac{2\delta_x\mu_\lambda}{\psi} \\ 1-\frac{\mu_\lambda(1-\mu_\lambda)}{\sigma_\lambda^2} & 0 & \frac{\delta_x(1-\mu_\lambda)}{\psi} & \frac{\rho\delta_x(1-\mu_\lambda)}{\psi} & \frac{2\rho\delta_y(1-\mu_\lambda)}{\psi} \\ 0 & 1-\frac{\mu_\lambda(1-\mu_\lambda)}{\sigma_\lambda^2} & \frac{\delta_y(1-\mu_\lambda)}{\psi} & \frac{\rho\delta_y(1-\mu_\lambda)}{\psi} & \frac{2\delta_x(1-\mu_\lambda)}{\psi} \end{bmatrix},$$

$\psi = 2\sigma_\lambda^2(\rho\delta_y^2 + \delta_x^2)$, *and*

$$M_B = \begin{bmatrix} 1+\frac{\mu_\lambda^2}{\sigma_\lambda^2} & 0 & -\frac{\mu_\lambda}{2\delta_x\sigma_\lambda^2} & 0 & 0 \\ 0 & 1+\frac{\mu_\lambda^2}{\sigma_\lambda^2} & \frac{\mu_\lambda\delta_y}{2\delta_x^2\sigma_\lambda^2} & 0 & -\frac{\mu_\lambda}{\delta_x\sigma_\lambda^2} \\ 1-\frac{(1-\mu_\lambda)\mu_\lambda}{\sigma_\lambda^2} & 0 & \frac{1-\mu_\lambda}{2\delta_x\sigma_\lambda^2} & 0 & 0 \\ 0 & 1-\frac{(1-\mu_\lambda)\mu_\lambda}{\sigma_\lambda^2} & -\frac{(1-\mu_\lambda)\delta_y}{2\delta_x^2\sigma_\lambda^2} & 0 & \frac{1-\mu_\lambda}{\delta_x\sigma_\lambda^2} \end{bmatrix},$$

$$M_C = \begin{bmatrix} 1+\frac{\delta_y^2\mu_\lambda^2}{\delta^2\sigma_\lambda^2} & -\frac{\delta_x\delta_y\mu_\lambda^2}{\delta^2\sigma_\lambda^2} & -\frac{\delta_y^2\mu_\lambda}{2\delta_x\delta^2\sigma_\lambda^2} & \frac{\delta_x\mu_\lambda}{2\delta^2\sigma_\lambda^2} & 0 \\ -\frac{\delta_x\delta_y\mu_\lambda^2}{\delta^2\sigma_\lambda^2} & 1+\frac{\delta_x^2\mu_\lambda^2}{\delta^2\sigma_\lambda^2} & \frac{\delta_y\mu_\lambda}{2\delta^2\sigma_\lambda^2} & -\frac{\delta_x^2\mu_\lambda}{2\delta_y\delta^2\sigma_\lambda^2} & 0 \\ 1-\frac{\delta_y^2(1-\mu_\lambda)\mu_\lambda}{\delta^2\sigma_\lambda^2} & \frac{\delta_x\delta_y(1-\mu_\lambda)\mu_\lambda}{\delta^2\sigma_\lambda^2} & \frac{\delta_y^2(1-\mu_\lambda)}{2\delta_x\delta^2\sigma_\lambda^2} & -\frac{\delta_x(1-\mu_\lambda)}{2\delta^2\sigma_\lambda^2} & 0 \\ \frac{\delta_x\delta_y(1-\mu_\lambda)\mu_\lambda}{\delta^2\sigma_\lambda^2} & 1-\frac{\delta_x^2(1-\mu_\lambda)\mu_\lambda}{\delta^2\sigma_\lambda^2} & -\frac{\delta_y(1-\mu_\lambda)}{2\delta^2\sigma_\lambda^2} & \frac{\delta_x^2(1-\mu_\lambda)}{2\delta_y\delta^2\sigma_\lambda^2} & 0 \end{bmatrix}.$$

Sketch of proof. We sketch the proof only for Case A. The proofs for the Case B and C are similar and omitted for brevity. Note that the model is location invariant and the MME is location equivariant. Hence its variance matrix depends on η and ξ only through δ. Therefore we compute Σ_θ, the variance matrix of $S_i = (x_i, y_i, x_i^2, y_i^2, x_i y_i)$, assuming $\eta = 0$. Define $\bar{S}_n = 1/n \sum_i S_i$ and $u_n = I_{\{s_{xy} \geq 0\}}$. Applying the central limit theorem to $\bar{S}_n$ yields,

$$\sqrt{n}(\bar{S}_n - \nu_\theta) \xrightarrow{d} N(0, \Sigma_\theta)$$

Let $g \equiv g(\bar{S}_n)$ be the estimator of σ_v^2 written as a function of $\overline{S}_n$. Define

$$H_1(\overline{S}_n) = \begin{bmatrix} h(\bar{x}_1, \bar{x}_2; -\mu_\lambda/\sigma_\lambda, \rho g) \\ h(\bar{y}_1, \bar{y}_2; -\mu_\lambda/\sigma_\lambda, g) \\ h(\bar{x}_1, \bar{x}_2; (1-\mu_\lambda)/\sigma_\lambda, \rho g) \\ h(\bar{y}_1, \bar{y}_2; (1-\mu_\lambda)/\sigma_\lambda, g) \end{bmatrix}$$

and

$$H_0(\overline{S}_n) = \begin{bmatrix} h(\bar{x}_1, \bar{x}_2; -\mu_\lambda/\sigma_\lambda, \rho g) \\ h(\bar{y}_1, \bar{y}_2; \mu_\lambda/\sigma_\lambda, g) \\ h(\bar{x}_1, \bar{x}_2; (1-\mu_\lambda)/\sigma_\lambda, \rho g) \\ h(\bar{y}_1, \bar{y}_2; -(1-\mu_\lambda)/\sigma_\lambda, g) \end{bmatrix},$$

where $h(x, y; \alpha, \beta) = x + \alpha\sqrt{y - x^2 - \beta}$. The MME may be written as

$$\hat{\theta} = u_n H_1(\overline{S}_n) + (1 - u_n) H_0(\overline{S}_n). \tag{12}$$

Also $u_n \xrightarrow{p} u$ as $n \to \infty$ where u equals 1 if $\text{Cov}[x, y] > 0$ zero if $\text{Cov}[x, y] < 0$. Since $\overline{S}_n$ converges in probability to its mean $\nu_\theta = E[\overline{S}_n]$ and $h(\overline{S}_n)$ is continuous we see that $H_u(\overline{S}_n) \xrightarrow{p} H_u(\nu_\theta)$ for $u = 0, 1$. Therefore from (12) if $\text{Cov}[x, y] > 0$ then $\hat{\theta} \xrightarrow{p} H_1(\nu_\theta) = \theta$. Alternatively if $\text{Cov}[x, y] < 0$ then $\hat{\theta} \xrightarrow{p} H_0(\nu_\theta) = \theta$. Hence $\hat{\theta}$ is consistent. Let

$$M_u = \left.\frac{dH_u(\overline{S}_n)}{d\overline{S}_n}\right|_{\overline{S}_n = \nu_\theta}, \qquad u = 0, 1.$$

It is easily verified that $M_0 = M_1 = M_A$ where M_A is given in the statement of the theorem. Applying standard Taylor series expansion, *i.e.*, the delta method, and Slutzky's theorem to (12) results in

$$\sqrt{n}(\hat{\theta} - \theta) \xrightarrow{d} N(0, V_\theta)$$

where $V_\theta = M_A \Sigma_\theta M_A^T$.

2.3 Modified MME

As we saw earlier, the form of the MME, $\hat{\theta}$, depends on the value of $\text{Cov}[x, y]$. When $\text{Cov}[x, y] \neq 0$ the general form of the MME is given by (8). If, for example, the segment is vertical, *i.e.*, $\delta_x = 0$, or horizontal, *i.e.*, $\delta_y = 0$ then $\text{Cov}[x, y] = 0$ and only a subset of equations (3)–(7) need be considered. Note that in these cases θ belongs to a lower dimensional subspace. Estimators under these conditions for all cases are easily derived. It can be shown that they follow, asymptotically, a multivariate normal distribution with easily computed covariance matrix.

However in general the value of $\mathrm{Cov}[x,y]$ is unknown in advance. Suppose $\delta_x = 0, \delta_y \neq 0$ and the general form of $\hat{\theta}$, wrongly assuming $\mathrm{Cov}[x,y] \neq 0$, given by (8), is used to estimate θ. To limit the length of the discussion we focus on Case A only. It is easily established that

$$\hat{\theta} - \theta = \begin{bmatrix} \bar{\epsilon} - \mu_\lambda \hat{\delta}_x^+ \\ \bar{v} + \bar{\lambda}\delta_y - \mu_\lambda \hat{\delta}_y^+ \mathrm{sgn}(s_{xy}) \\ \bar{\epsilon} + (1-\mu_\lambda)\hat{\delta}_x^+ \\ \bar{v} - (1-\bar{\lambda})\delta_y + (1-\mu_\lambda)\hat{\delta}_y^+ \mathrm{sgn}(s_{xy}) \end{bmatrix}$$

where $\bar{\epsilon} = n^{-1}\sum_{i=1}^{n} \epsilon_i$ and $\bar{v} = n^{-1}\sum_{i=1}^{n} v_i$, which equals to

$$\begin{bmatrix} 0 \\ \mu_\lambda \delta_y (1-v) \\ 0 \\ (1-\mu_\lambda)\delta_y (v-1) \end{bmatrix} + o_p(1), \tag{13}$$

and $\mathrm{sgn}(s_{xy}) \xrightarrow{p} v$ as $n \to \infty$, v equals ± 1 with probability $1/2$. Moreover, if $v = 1$ it follows from (13) that $\hat{\theta} = \theta + o_p(1)$, otherwise $\hat{\theta} = \theta^* + o_p(1)$ where θ^* is obtained from θ by switching its second and fourth elements, *i.e.*, the points are simply relabeled. In this case the general MME is not consistent. Fortunately the remedy to this problem is simple. Define

$$z_{xy} = I_{\{\hat{\delta}_x^+ > 0, \hat{\delta}_y^+ > 0\}} \mathrm{sgn}(s_{xy})$$

and let $\hat{\theta}^{mod}$ be the MME for Case A with z_{xy} replacing $\mathrm{sgn}(s_{xy})$.

Theorem 2. *The modified MME, $\hat{\theta}^{mod}$, is asymptotically equivalent to the MME $\hat{\theta}$ for all values of* $\mathrm{Cov}[x,y]$, *i.e., $\hat{\theta}^{mod} = \hat{\theta} + o_p(1)$ as $n \to \infty$.*

The proof is straightforward in light of the above remarks and the behavior of z_{xy}. Clearly $z_{xy} \xrightarrow{p} z$ as $n \to \infty$ where

$$z = \begin{cases} 1 & \text{if} \quad \mathrm{Cov}[x,y] > 0 \\ z_0 & \text{if} \quad \delta_x = 0 \quad \text{and/or} \quad \delta_y = 0 \\ -1 & \text{if} \quad \mathrm{Cov}[x,y] < 0 \end{cases}$$

where $P[|z_0| \leq 1] = 1$, which guarantees proper convergence. Hence $\hat{\theta}^{mod}$ naturally accommodates the four possible values for (δ_x, δ_y). Consequently the modified MME $\hat{\theta}^{mod}$ provides consistent asymptotically normal estimation regardless of the value of δ and $\mathrm{Cov}[x,y]$. Similar results hold in Case C.

3. Numerical experiments

The small sample and asymptotic performance of the estimators was evaluated numerically. Our simulation results indicate that the convergence of the MME to its asymptotic distribution is rapid in most cases. The rate of convergence depends primarily on the distribution G of λ. Naturally the rate of convergence is faster for distribution functions G with large mass near the endpoints. For fixed G the length and orientation of the segment also effect the rate of convergence.

We compared the asymptotic efficiency of the MME's in Cases A and B relative to the maximum likelihood estimator (MLE). We consider four models for G: (a) Bin$(0, 1/2)$, (b) Uniform$(0, 1)$, (c) Beta$(3, 3)$, (d) Beta$(3, 1)$. The elements of the Fisher information matrix I_θ were computed numerically using Monte–Carlo integration.

The asymptotic relative efficiency (ARE) of the MME compared with the MLE is defined as

$$\mathrm{ARE} = \left[\frac{\det(I_\theta^{-1})}{\det(V_\theta^K)}\right]^{1/4},$$

where $K = A, B$. Note that the asymptotic distribution of the MME $\hat{\theta}$ in Case C is singular because the line segment is uniquely determined by any three components of θ combined with the knowledge of the length. Figure 1 displays the effect of the length and the orientation of the segment on the ARE for Cases A and B respectively. Their effect of the length is studied assuming an orientation of 45^0, and the effect of the orientation is studied assuming unit length. Qualitatively the effect of the length of the segment is similar in both cases. Overall the ARE in reasonable cases is quite high. In Case A the orientation has little effect on the ARE which is determined primarily by the distribution of λ, whereas the reverse happens in Case B.

4. Summary

We provide an easily computable estimator for the structural line segment model. The asymptotic distribution is computed under different assumptions. The MME is shown to have high relative efficiency compared to the MLE in most cases.

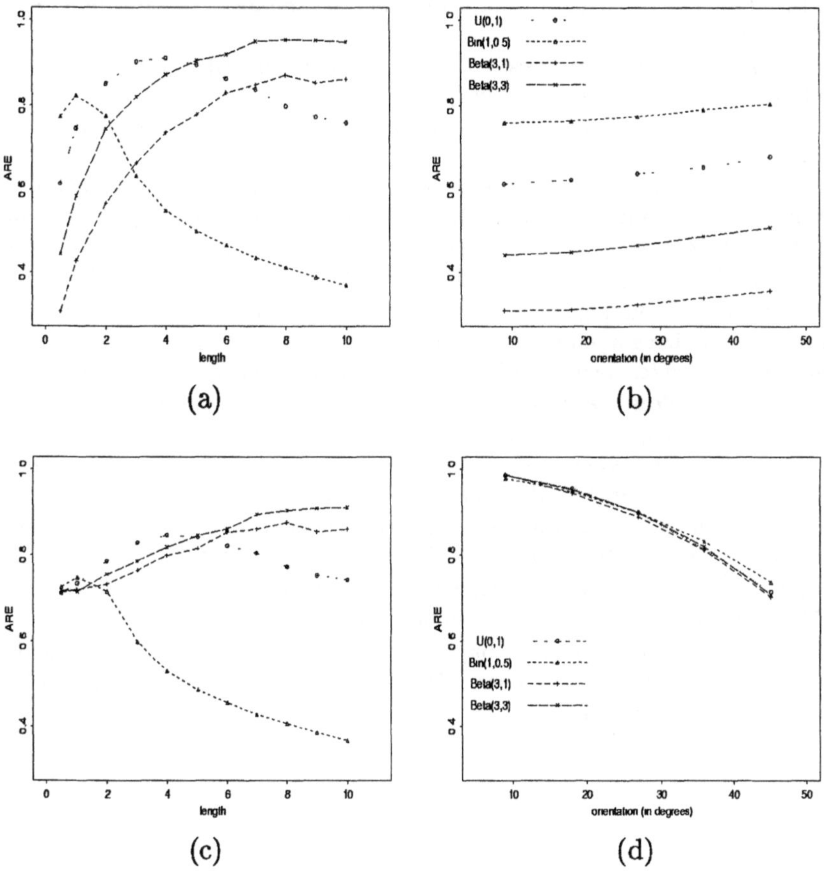

Figure 1. Asymptotic relative efficiency of the MME relative to the MLE for various lengths and orientations for Cases A (figures a, b) and B (figures c, d).

Appendix

The matrix Σ_θ is symmetric and its upper–triagonal elements are:

$$
\begin{aligned}
\sigma_{11} &= \mu_{[1,1]}\delta_x^2 + \sigma_\epsilon^2, \sigma_{12} = \mu_{[1,1]}\delta_x\delta_y, \sigma_{13} = \mu_{[1,2]}\delta_x^3 + 2\mu_{[1]}\delta_x\sigma_\epsilon^2,\\
\sigma_{14} &= \mu_{[1,2]}\delta_x\delta_y^2, \sigma_{15} = \mu_{[2,1]}\delta_x^2\delta_y + \mu_{[1]}\delta_y\sigma_\epsilon^2, \sigma_{22} = \mu_{[1,1]}\delta_y^2 + \sigma_v^2,\\
\sigma_{23} &= \mu_{[1,2]}\delta_x^2\delta_y, \sigma_{24} = \mu_{[1,2]}\delta_y^3 + 2\mu_{[1]}\delta_y\sigma_v^2, \sigma_{25} = \mu_{[2,1]}\delta_y^2\delta_x + \mu_{[1]}\delta_x\sigma_v^2,\\
\sigma_{33} &= \mu_{[2,2]}\delta_x^4 + 4\mu_{[2]}\delta_x^2\sigma_\epsilon^2 + 2\sigma_\epsilon^4, \sigma_{34} = \mu_{[2,2]}\delta_x^2\delta_y^2, \sigma_{35} = \mu_{[2,2]}\delta_x^3\delta_y + \mu_{[2]}\delta_x\delta_y\sigma_\epsilon^2,\\
\sigma_{44} &= \mu_{[2,2]}\delta_y^4 + 4\mu_{[2]}\delta_y^2\sigma_v^2 + 2\sigma_v^4, \sigma_{45} = \mu_{[2,2]}\delta_y^3\delta_x + 5\mu_{[2]}\delta_x\delta_y\sigma_v^2,\\
\sigma_{55} &= \mu_{[2,2]}\delta_x^2\delta_y^2 + \mu_{[2]}(\delta_x^2\sigma_v^2 + \delta_y^2\sigma_\epsilon^2) + \sigma_\epsilon^2\sigma_v^2
\end{aligned}
$$

where $\mu_{[i,j]} = E[\lambda^{i+j}] - E[\lambda^i]E[\lambda^j]$, $\mu_{[i]} = E[\lambda^i]$.

References

[1] K. Bruntjen and H. Späth. Incomplete total least squares. *Numer. Math.*, 81:521–538, 1999.

[2] C. Cheng and J.W. Van Ness. *Statistical Regression with Measurement Error.* Oxford University Press, New York, 1999.

[3] O. Davidov and A. Goldenshluger. Fitting a line segment to noisy data. Submitted.

[4] C.B. Madsen and H.I. Cristensen. Modeling and testing the stability of edge segments: length and orientation. In G. Borgefors, editor, *Theory & Applications of Image Processing II - Selected papers from the 9'th Scandinavian Conference on Image Analysis*, pages 1–15. World Scientific Press, Singapore, 1995.

[5] C.V. Stewart, K. Bubna, and A. Perera. Estimating model parameters and boundaries by minimizing a joint, robust objective function. In *Proceedings of the IEEE Conf. on Computer Vision and Pattern Recognition*, pages 387–393, 1999.

[6] S. Van Huffel. *Recent Advances in Total Least Squares Techniques and Errors-in-Variables Modeling.* SIAM, Philadelphia, 1997.

[7] S. Van Huffel and J. Vandewalle. *The Total Least Squares Problem: Computational Aspects and Analysis.* SIAM, Philadelphia, 1991.

[8] G. Vosselman and R. M. Haralick. Performance analysis of line and circle fitting in digital images. In H.I. Cristensen, W. Förstner, and C.B Madsen, editors, *Proceedings of the Workshop on Performance Characteristics of Vision Algorithms*, 1996. Available at http://www.vision.auc.dk/ hic/perf-proc.html.

[9] M. Werman and D. Keren. A bayesian method for fitting parametric and non-parametric models to noisy data. *IEEE Trans. Pattern Analysis and Machine Intelligence*, 23:528–534, 2001.

MODEL FITTING FOR MULTIPLE VARIABLES BY MINIMISING THE GEOMETRIC MEAN DEVIATION

Chris Tofallis
University of Hertfordshire Business School
Dept. of Statistics, Economics, Accounting and Management Systems
Mangrove Rd, Hertford, SG13 8QF, United Kingdom
c.tofallis@herts.ac.uk

Abstract We consider the problem of fitting a linear model for a number of variables but without treating any one of these variables as special, in contrast to regression where one variable is singled out as being a dependent variable. Each of the variables is allowed to have error or natural variability but we do not assume any prior knowledge about the distribution or variance of this variability. The fitting criterion we use is based on the geometric mean of the absolute deviations in each direction. This combines variables using a product rather than a sum and so allows the method to naturally produce units-invariant models; this property is vital for law-like relationships in the natural or social sciences.

Keywords: geometric mean functional relationship, least area criterion, least volume criterion, measurement error, reduced major axis.

1. Properties of the geometric mean functional relation

In two dimensions the method we are interested in involves fitting a line by minimising the total area of the triangles defined by the data points and the line (see Figure 1). This is equivalent to minimising the sum of the geometric means of the squared deviations in each dimension.

This simple geometric approach provides a line with a number of very useful or interesting properties. These are gathered together here from papers scattered in the literatures of different fields.

1) The resulting line is symmetric with respect to the two variables i.e. whichever variable is plotted on the vertical axis the resulting

S. Van Huffel and P. Lemmerling (eds.), Total Least Squares and Errors-in-Variables Modeling, 261–267.

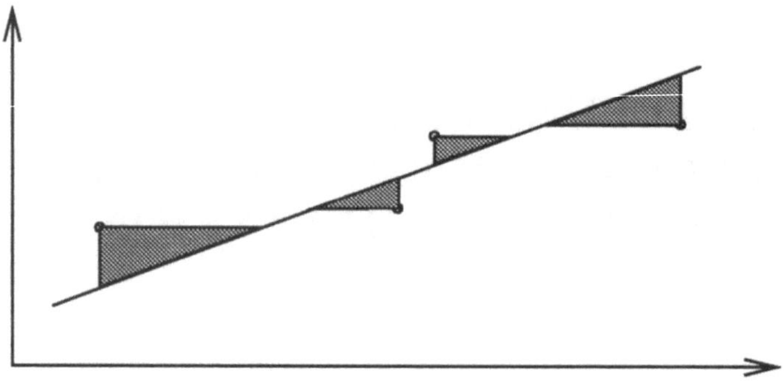

Figure 1. Fitting a line by minimising the total area of the triangles defined by the data points and the line

models will be mutually equivalent i.e. consistent with each other. This is because switching the axes does not affect the areas.

2) The fitted line is units invariant (or scale invariant). This means that if we change the units of measurement from say cm to mm we obtain an equivalent model. Surprisingly this is not true of some line-fitting procedures, notably orthogonal regression (also known as total least squares).

3) The slope of the line is the geometric mean of the two least squares regressions. In other words if the regression of y on x gives slope b_1 and the regression of x against y gives slope b_2 (meaning the rate of change of y with x) then our least areas procedure will give a slope of $(b_1 b_2)^{1/2}$. A proof of this appears in [11] and in [10]. For this reason this type of line is sometimes called the geometric mean functional relationship.

4) Minimising the areas implies the objective function involves the product of the deviations in each dimension for each point. Hence as already stated, we are also minimising the sum of the geometric means of the squared deviations in the x and y direction.

5) The expression for the magnitude of the slope of the line is perhaps the simplest of any available fitting method, it is merely the standard deviation of the y-values divided by the standard deviation of the x-values. (Proved in [11]). The sign of the slope is given by the correlation.

6) Suppose the relative frequency function, or probability density function $f(x, y)$ is symmetrical in the two variables after allowing for re-scaling i.e. $f(x, y) = f(y/c, cx)$. Then [3] has proved that

our line is *unique* in pairing off values of X and Y such that the proportion of x-scores less than X equals the proportion of y-scores less than Y. In other words it pairs off scores with equal percentiles (or quartiles or more generally any quantile or cumulative measure of relative ranking). The bivariate normal distribution is one case that satisfies the symmetry condition stated.

7) By definition the least squares regression line of y on x has the lowest mean squared error of estimation of y. Likewise the regression of x on y has the lowest mean squared error of estimation of x. No single line can simultaneously achieve both these minima so we must expect an increase in these errors. If we require that the proportional increase in each of these errors be the same (i.e. for x and y) then [3] has proved that our line is the *only* line which achieves this.

8) For the set of all possible line-fitting procedures that depend on standard deviations and correlations (i.e. second moments), ours is unique in behaving correctly i.e. invariant to switching of axes, and change of scale. This was proved by Nobel laureate Paul Samuelson in 1942 [9].

9) When the data are divided by their standard deviations then we obtain a 45 degree line which bisects the two least squares line (regression of y on x and vice versa). This line minimises the sum of squares of perpendicular distances from the data points to the line [7]. This arises from the fact that the triangles (Figure 1) become isosceles in this case.

For the case of two variables this method has appeared in the research literature of many disciplines throughout the twentieth century, but under different names. In astronomy it is known as Stromberg's impartial line. In biology it is the line of organic correlation. In economics it is the method of minimised areas or diagonal regression. In statistics it is sometimes referred to as the "standard or reduced major axis". This derives from the fact that if the data are standardised by dividing by their standard deviation, then the fitted line corresponds to the major (i.e. principal) axis of the ellipse of constant probability for the bivariate normal distribution.

Kruskal [5] proved that of all line-fitting procedures which depend only on first and second moments this is the only one which behaves correctly under translation and change of scale. Moreover, in a comparison of 34 fitting methods Riggs et. al. [8] found that in overall performance, judged both by root mean square error and percent bias, this method was almost always the best when the ratio of error variances is unity. In cases where this ratio is unknown, Ricker [6] concluded that this is the best available estimate of the functional relationship where the variability in both variates is due to measurement error or due to natural or inherent variability (e.g. as arises in biology). For these reasons it seems worthwhile to investigate this method when it has been extended to the case of multiple variables.

2. Extension to multiple dimensions

We now intend to fit a linear function of the form $\sum a_j x_j = c$ to data in p dimensions ($j = 1, \ldots, p$). Of course this is not uniquely specified because we can divide through by any non-zero constant. Thus we are free to impose a constraint on the coefficients. For later convenience the constraint we choose will involve the product of the coefficients: $\prod a_j$.

One obvious way of generalising the least area procedure to higher dimensions is to minimise the volumes (or hypervolumes). Each data point will have associated with it a "volume deviation" which is simply the product of its deviations from the fitted plane in each dimension. We must take care to make all these non-negative by taking the absolute values. For the i'th data point this volume deviation V_i is proportional to:

$$\left| \frac{\left(\sum_{j=1}^{p} a_j x_{ij} - c\right)^p}{\prod_{j=1}^{p} a_j} \right|$$

Hence one possible objective might be to minimise $\sum V_i$. This is an unconstrained fractional programming problem. An alternative objective that has already been investigated is the minimisation of $\sum V_i^{2/p}$ [2]. This is the route to generalisation based on property 4) above i.e. the geometric mean of the squared deviations. Their method of solution involved treating the problem as a nonlinear weighted least squares problem, and they list a short piece of S-plus code for achieving this.

The numerator of V_i looks like an L_p norm, but with the degree p varying as the number of variables. The L_1 norm would then correspond to taking the p'th root of the above volume deviation before summing and minimising i.e. $\min \sum V_i^{1/p}$, so we are minimising the geometric mean of the absolute deviations in each dimension. Let us see how this particular case can be formulated. The denominator remains the same after summing over the data points. One way of dealing with fractional programmes is to set the denominator in the objective function to a particular value, say unity, and then optimise the numerator. Thus we can make good use of our condition on the coefficients: Set $\prod a_j = 1$

and minimise the numerator subject to this condition. In fact we have to solve the problem a second time with $\prod a_j = -1$ to ensure that all combinations of signs are considered. To deal with the absolute values of the volume deviations we introduce pairs of non-negative variables u, v. The problem then becomes:

Find values of the coefficients a_j and c so as to

$$\begin{array}{ll} \text{minimise} & \sum u_i + v_i \\ \text{subject to} & \sum a_j x_{ij} + u_i - v_i = c, \quad i = 1, \dots, n \\ \text{and} & \prod a_j = \pm 1. \end{array}$$

where n is the number of data points. Notice that the objective function is linear and that we only have one non-linear constraint.

For fitted functions that are linear in the coefficients the optimisation method for their estimation can be shown to be a convex fractional programme so that any optimum is both unique and global. Whilst there are solution algorithms for fractional programmes, the unique global property implies that general convex solution methods can be used.

3. Numerical test

We shall now apply the least geometric mean deviation approach to try and uncover the coefficients from data, which have been generated from a known underlying model with some randomness thrown in. In order to make this a difficult test we shall choose data that suffers from multicollinearity. It is taken from Belsley's comprehensive monograph [1] on this problem. The generating model is

$$y = 1.2 - 0.4x_1 + 0.6x_2 + 0.9x_3 + \varepsilon$$

with ε normally distributed with zero mean and variance 0.01.

Two very similar data sets (A, B) are tabulated in Belsley based on this model. For set A ordinary least squares (OLS) gives:

$$y = 1.26 + 0.97x_1 + 9.0x_2 - 38.4x_3$$

The fit as measured by R^2 is very good at 0.992 but the underlying model is far from being uncovered. In particular, the coefficient of x_2 is 15 times too high and two of the coefficients have the wrong sign. Getting the signs wrong is very serious if one is trying to understand how variables are related to each other.

Turning to the least geometric mean deviation we find:

$$y = 1.28 - 0.44x_1 + 0.33x_2 + 1.74x_3$$

We now have all the correct signs and the magnitudes are also reasonable.

Repeating this for data set B:

$$\begin{array}{ll} \text{OLS:} & y = 1.275 + 0.25x_1 + 4.5x_2 - 17.6x_3 \\ \text{Geometric mean:} & y = 1.29 - 0.44x_1 + 0.42x_2 + 1.98x_3 \end{array}$$

Once again the geometric mean produces a superior model. Moreover it is also worth noting that the two OLS models are very different from each other whereas the geometric mean models seem to be more stable to small variations in the data. This is noteworthy because of how similar the two data sets were: the y-values were identical for sets A and B, and the x-values never varied by more than one in the third digit. Thus our method seems to be much more stable than OLS. Of course a comprehensive set of Monte Carlo simulations is required to fully explore this aspect. Furthermore, although it is not our intention to use this method for forecasting a particular dependent variable we have found that the mean absolute deviation for y is 10% lower than that from using OLS for both data sets.

4. Conclusion

Regression is not only one of the most used quantitative methods, it is also one of the most abused. In the simplest two variable case it is well suited to providing an expected value for the y-variable for a given x-value. Unfortunately it is also widely, and incorrectly, applied to finding law-like relationships, or estimating parameters within them.

This paper presents the idea of measuring deviations from the fitted function by means of the product of the deviations in each dimension. More generally we have a whole family of fitting norms based upon this volume deviation in much the same way we have L_p norms. A key advantage of this criterion is that the fitted models are scale-invariant i.e. they are unaffected by the units of measurement. This is an essential property if one is seeking law-like relationships in the natural sciences or social sciences.

For the case of two variables this method has appeared in the research literature of many disciplines throughout the twentieth century, but under different names. This paper has extended the technique to multiple variables. The method can be used when there is random variation (due to natural individuality or measurement errors) in the variables, but we do not assume any information is known about this variation e.g. we make no assumptions regarding the distributions of these errors or the variance of the errors (in which case maximum likelihood estimation could be applied).

A numerical test showed it to be superior to least squares in two regards: firstly, in that it was able to more accurately estimate the coefficients of the data-generating model, and secondly it showed itself to be more stable to small variations in the data. This preliminary evidence commends the method for further investigation.

References

[1] D. A. Belsley. *Conditioning Diagnostics.* Wiley, New York, 1991.

[2] N. R. Draper and Y. Yang. Generalization of the geometric mean functional relationship. *Computational Statistics and Data Analysis*, (23):355–372, 1997.

[3] P. D. Greenall. The concept of equivalent scores in similar tests. *British J. of Psychology: Statistical Section 2*, pages 30–40, 1949.

[4] E. Grosse. A catalogue of algorithms for approximation. In *Algorithms for Approximation II*, editor J. C. Mason and M. Cox, 1989, Also available at **www.hensa.ac.uk/netlib/a/catalog.html**

[5] W. H. Kruskal. On the uniqueness of the line of organic correlation. *Biometrics*, (9):47–58, 1953.

[6] W. E. Ricker. Linear regressions in fishery research. *J. Fisheries Research Board of Canada*, pages 409–434, 1973.

[7] W. E. Ricker. Computation and uses of central trend lines. *Canadian J. of Zoology*, (62):1897–1905, 1984.

[8] D. S. Riggs, J. A. Guarnieri, and S. Adelman. Fitting straight lines when both variables are subject to error. *Life Sciences*, (22):1305–1360, 1978.

[9] P. A. Samuelson. A note on alternative regressions. *Econometrica*, 10(1):80–83, 1942.

[10] G. Teissier. La relation d'allometrie: sa signification statistique et biologique. *Biometrics*, 4(1):14–48, 1948.

[11] E. B. Woolley. The method of minimized areas as a basis for correlation analysis. *Econometrica*, 9(1):38–62, 1941.

Part VII

ERRORS-IN-VARIABLES ESTIMATION IN SYSTEM IDENTIFICATION

PERSPECTIVES ON ERRORS-IN-VARIABLES ESTIMATION FOR DYNAMIC SYSTEMS

Torsten Söderström
Department of Systems and Control, Information Technology, Uppsala University, P O Box 27, SE-75103 Uppsala, Sweden
ts@SysCon.uu.se

Umberto Soverini
Department of Electronics, Information Science and Systems, University of Bologna, Viale Risorgimento 2, IT-40136 Bologna, Italy
usoverini@deis.unibo.it

Kaushik Mahata
Department of Systems and Control, Information Technology, Uppsala University, P O Box 27, SE-75103 Uppsala, Sweden
km@SysCon.uu.se

Abstract The paper gives an overview of various methods for identifying dynamic errors-in-variables systems. Several approaches are classified by how the original information in time-series data of the noisy input and output measurements is condensed before further processing. For some methods, such as instrumental variable estimators, the information is condensed into a nonsymmetric covariance matrix as a first step before further processing. In a second class of methods, where a symmetric covariance matrix is used instead, the Frisch scheme and other bias-compensation approaches appear. When dealing with the estimation problem in the frequency domain, a milder data reduction typically takes place by first computing spectral estimators of the noisy input-output data. Finally, it is also possible to apply maximum likelihood and prediction error approaches using the original time-domain data in a direct fashion. This alternative will often require quite high computational complexity but yield good statistical efficiency.

Keywords: dynamic error-in-variables, system identification, instrumental variables, Frisch scheme, prediction error.

S. Van Huffel and P. Lemmerling (eds.), Total Least Squares and Errors-in-Variables Modeling, 271–280.

1. Introduction

Identification of linear dynamic systems from noise-corrupted output measurements is a fundamental research problem which has been investigated in the past decades. Many different solutions have been presented, see, for example, [14, 22]. On the other hand, the estimation of the parameters of linear dynamic systems when also the input is affected by noise, is recognized as a more difficult problem which only recently has received increasing attention. Representations where errors or measurement noise are present on both the inputs and outputs are usually called "errors-in-variables" models and play an important role when the identification purpose is to determine the inner laws that describe the process, rather than the prediction of its future behaviour.

The class of scientific disciplines which make use of such a kind of representations is very broad, as proved by the several applications collected in [27], such as time series modelling, array signal processing for direction-of-arrival estimation, blind channel equalization, multivariate calibration in analytical chemistry, image processing, astronomical data reduction, etc.

In case of static systems, errors-in-variables representations are closely related to other well-known topics such as *latent variables* models and *factor* models [10], [16]. The extension to the dynamic case has been considered in the literature by several authors with different approaches; in fact the problem can be examined from different point of views. For example, we can distinguish between time-domain [14], [22], and frequency-domain methods [15] approaches, exploiting deterministic signals (e.g. periodic inputs [9]) or stochastic processes, methods that make use of second-order statistics only or based on higher order cumulant statistics [26]. When only second-order statistics are exploited it is a well-known result that, in general, the identification of errors-in-variables models cannot admit a single solution [1], [2].

Also the stochastic input-output noise can be represented in different ways by assuming different model structures. For example, in [5] the analysis of the Frisch scheme, originally related to the static case, has been extended to linear dynamic SISO systems under the assumption of white and uncorrelated additive noise with unknown variances.

Some comparison between different approaches are given in [17] and [23]. In this paper we will extend the overviews and give further perspectives and comparisons. More details can be found in [20].

2. Basic setup

As a typical model example, consider the following system with noise-corrupted input and output measurements. The noise-free input is denoted by $u_o(t)$ and the undisturbed output by $y_o(t)$. They are linked through the linear difference equation

$$A(q^{-1})\, y_o(t) = B(q^{-1})\, u_o(t), \tag{1}$$

where $A(q^{-1})$ and $B(q^{-1})$ are polynomials of the type

$$\begin{aligned} A(q^{-1}) &= 1 + a_1\, q^{-1} + \cdots + a_n\, q^{-n} \\ B(q^{-1}) &= b_1\, q^{-1} + \cdots + b_n\, q^{-n} \end{aligned} \tag{2}$$

and q^{-1} is the backward shift operator, i.e. $q^{-1}\, x(t) = x(t-1)$. It is not restrictive to assume that the polynomials $A(q^{-1})$, $B(q^{-1})$ have equal degree n, which represents the order of the system.

In an errors-in-variables environment we assume that the observations are corrupted by additive measurement noises $\tilde{u}(t)$ and $\tilde{y}(t)$, at the input and output respectively. Therefore, the available signals are of the form

$$\begin{aligned} u(t) &= u_o(t) + \tilde{u}(t) \\ y(t) &= y_o(t) + \tilde{y}(t) \end{aligned} \quad . \tag{3}$$

In this work the following assumptions are introduced.

A1. The dynamic system (1) is asymptotically stable, i.e. $A(z)$ has all zeros outside the unit circle; all the system modes are observable and controllable, i.e. $A(z)$ and $B(z)$ have no common factors. Moreover, the order n of the system is *a priori* known. [Most of the methods to be discussed can be extended to multivariable systems.]

A2. The true input $u_o(t)$ is a zero-mean stationary ergodic random signal. [When using a prediction error approach, we will also need that the spectral density of $u_o(t)$ is a rational function. This means that $u_o(t)$ is an ARMA process.]

A3. The sequences $\tilde{u}(t)$ and $\tilde{y}(t)$ are zero-mean white noise with unknown variances λ_u and λ_y, respectively. These processes are mutually uncorrelated and uncorrelated with the noise-free signals $u_o(t)$ and $y_o(t)$. [The assumptions can sometimes be a bit relaxed. The measurement noise may be somewhat correlated. Some estimation methods can handle that $\tilde{u}(t)$ is a moving average process, $\tilde{y}(t)$ is an ARMA process, or that $\tilde{u}(t)$ and $\tilde{y}(t)$ are correlated. The problem will be significantly simplified if the noise variances λ_u and λ_y are known, or even only the ratio λ_y/λ_u.]

The problem under investigation can be stated as follows.

Problem 1. Let the available measurements $u(t)$ and $y(t)$, for $t = 1, \ldots, N$, be generated in accordance with the previous assumptions. Determine the system characteristics, i.e. the transfer function

$$G(q^{-1}) = \frac{B(q^{-1})}{A(q^{-1})} \tag{4}$$

and the variances λ_u and λ_y of the noise sequences.

Estimation methods can be classified and organised in different ways. Here, we have found it useful to group methods together, based on an initial data compression step. After a first 'pre-processing' of the data, some reduced information is set up and used for the final computation of the parameter estimates. The different groups of methods that will be discussed in the sections to follow, differ in the way the data compression is carried out. The following cases will be treated:

a) Using a nonsymmetric covariance matrix

b) Using a symmetric covariance matrix

c) Using the input-output spectrum

d) Using the time-series data

Very roughly, the cases are listed in order of increasing computational complexity, and also in order of increasing statistical accuracy. For more details, see [20].

3. Estimation methods based on a covariance matrix

In this section we will use the parameter vector

$$\theta = (a_1 \dots a_n \; b_1 \dots b_n)^T \tag{5}$$

and the extended parameter vector

$$\overline{\theta} = \begin{bmatrix} 1 & \theta^T \end{bmatrix}^T . \tag{6}$$

Similarly we introduce the regressor vectors

$$\begin{aligned} \varphi(t) &= (-y(t-1) \cdots - y(t-n) \; u(t-1) \dots u(t-n))^T & (7)\\ \overline{\varphi}(t) &= (-y(t) \cdots - y(t-n) \; u(t-1) \dots u(t-n))^T . & (8) \end{aligned}$$

Further, we will use the conventions: θ_o denotes the true parameter vector, and $\hat{\theta}$ denotes its estimate; $\varphi_o(t)$ denotes the noisefree part of the regressor vector; $\tilde{\varphi}(t)$ denotes the noise-contribution to the regressor vector.

Introduce a vector $z(t)$ of dimension $2n$ or higher, that is uncorrelated with

$$\tilde{\overline{\varphi}}^T(t) = (-\tilde{y}(t) \ldots . - \tilde{y}(t-n) \; \tilde{u}(t-1) \ldots \tilde{u}(t-n)). \tag{9}$$

Then one can show that

$$\mathbf{E}z(t)\overline{\varphi}^T(t)\overline{\theta}_o \triangleq R_{z\overline{\varphi}}\overline{\theta}_o = 0 \; . \tag{10}$$

The matrix in (10) can easily be estimated from the data as $\widehat{R_{z\overline{\varphi}}} = \frac{1}{N}\sum_{t=1}^{N} z(t)\overline{\varphi}^T(t)$. Due to the relation (10) one can derive several estimators from the approximate relation

$$\widehat{R_{z\overline{\varphi}}}\hat{\overline{\theta}} \approx 0 \; . \tag{11}$$

One possible approach leading to an instrumental variable (IV) estimator (cf [17], [22]) is to partition the matrix in (11) as

$$\begin{bmatrix} \hat{r} & \widehat{R_{z\varphi}} \end{bmatrix} \begin{bmatrix} 1 \\ \hat{\theta} \end{bmatrix} = \hat{r} + \widehat{R_{z\varphi}}\hat{\theta} \approx 0 \; . \tag{12}$$

A weighted least squares estimator may be taken, which gives the extended IV estimator, see [21], [22],

$$\hat{\theta}_{IV} = -(\widehat{R_{z\varphi}}^T W \widehat{R_{z\varphi}})^{-1}(\widehat{R_{z\varphi}}^T W \hat{r}) \tag{13}$$

where W is a positive definite weighting matrix. An alternative approach for exploiting (11) is to note that the relation means that the matrix $\widehat{R_{z\overline{\varphi}}}$ is close to singular. Let $\hat{w}$ denote its right singular vector, associated with its smallest singular value. Then, except for a normalization, $\hat{w}$ should coincide with $\hat{\overline{\theta}}$. This is precisely the total least squares solution, [28], to equation (11). Using MATLAB notations the estimate reads

$$\hat{\theta}_{TLS} = \hat{w}_{2:2n+1}/\hat{w}_1 \; . \tag{14}$$

For the case when $\dim z(t) = 2n$ the estimators $\hat{\theta}_{IV}$ and $\hat{\theta}_{TLS}$ coincide. For arbitrary dim $z(t)$ it can be shown, see [18], that they have the same asymptotic covariance matrix in the following sense. It holds in both cases that for large N

$$\sqrt{N}(\hat{\theta} - \theta_o) \overset{N\to\infty}{\longrightarrow} \mathcal{N}(0, P) \tag{15}$$

$$P = \sigma(R_{z\varphi}^T W R_{z\varphi})^{-1} R_{z\varphi}^T W \mathrm{cov}\left(C(q^{-1})z(t)\right) W R_{z\varphi}(R_{z\varphi}^T W R_{z\varphi})^{-1} \tag{16}$$

and $W = I$ for the estimator (14). The filter $C(q^{-1})$ in (16) is obtained through the spectral factorization

$$C(z)C(z^{-1})\rho \equiv A(z)A(z^{-1})\lambda_y + B(z)B(z^{-1})\lambda_u. \tag{17}$$

Some comments: One of the principal advantages of the IV method is its applicability under fairly general noise conditions. It is rather inexpensive from a computational point of view. However, with this procedure a poor accuracy of the parameter estimates is often obtained. The main idea of the IV technique has been subsequently developed and generalized in several ways, e.g. by combining it with a weighted subspace fitting approach [24]. This combined instrumental variable weighted subspace fitting, or IV-WSF, can be viewed as an extension to the approach discussed above. The matrix $R_{z\overline{\varphi}}$ has to be full rank for the IV based estimates to exist. This is a persistence-of-excitation like condition to be satisfied by the noise-free input signal. IV based subspace algorithms employing the state space models are proposed in [7]. These algorithms can be applied to MIMO systems operating in open or closed loop, where one has to account for the process noise also.

4. Bias-compensating schemes

Next return to (11) but substitute $z(t)$ by $\varphi(t)$. Set further

$$r = -\mathbf{E}\varphi(t)y(t) . \tag{18}$$

Now, (10) will no longer hold, but instead

$$R_{\varphi\overline{\varphi}}\overline{\theta}_o = r + R_{\varphi\varphi}\theta_o = r + [R_{\varphi_o\varphi_o} + R_{\tilde{\varphi}\tilde{\varphi}}]\theta_o = R_{\tilde{\varphi}\tilde{\varphi}}\theta_o . \tag{19}$$

The matrix

$$R_{\tilde{\varphi}\tilde{\varphi}} = \begin{bmatrix} \lambda_y I_{n+1} & 0 \\ 0 & \lambda_u I_n \end{bmatrix} \tag{20}$$

in these normal equations will cause a bias in the least squares estimator

$$\hat{\theta}_{LS} = -\hat{R}_{\varphi\varphi}^{-1}\hat{r} . \tag{21}$$

Assuming the noise variances are known or estimated, a bias-compensated least squares scheme can be constructed as

$$\hat{\theta}_{BCLS} = \left(\hat{R}_{\varphi\varphi} - \begin{bmatrix} \hat{\lambda}_y I_{n+1} & 0 \\ 0 & \hat{\lambda}_u I_n \end{bmatrix}\right)^{-1} \hat{r} . \tag{22}$$

In the case of no input noise, it is fairly straightforward to devise an estimator for λ_y, using a TLS scheme, see [25].

The Frisch scheme can be viewed as a more advanced bias-compensating scheme. Note that the matrix

$$R_{\overline{\varphi}_o\overline{\varphi}_o} = \mathbf{E}\overline{\varphi}_o(t)\overline{\varphi}_o^T(t) \tag{23}$$

is positive semidefinite with one eigenvalue equal to zero. Let $\hat{\lambda}_u$ be some estimate of λ_u. The estimated output noise variance $\hat{\lambda}_y$ is then constructed so that

$$\hat{R}_{\overline{\varphi}\overline{\varphi}} - \begin{bmatrix} \hat{\lambda}_y I_{n+1} & 0 \\ 0 & \hat{\lambda}_u I_n \end{bmatrix} \tag{24}$$

is singular. Hence $\hat{\lambda}_y$ will be an implicit function of $\hat{\lambda}_u$, [5]. The procedure can be repeated with the model order n increased by one unit. The two functions $\hat{\lambda}_y(\hat{\lambda}_u)$ for model orders n and $n+1$, respectively, will share a common point corresponding to the true noise variances.

Comments on early work in the field: One of the first solutions for identification of dynamic errors-in-variables models using a bias-compensated scheme was presented in the works [12], [13], and [3]. In more recent years it has been re-proposed by several authors with slightly different variants [6], [8], [28]. One of the main drawbacks of the Koopmans-Levin method is constituted by the fact that the input and output noise are assumed white and the ratio of their variances is *a priori* known. These restrictive assumptions drastically reduce the practical use of the method.

5. Frequency domain estimators

The idea here is similar to that of the Frisch scheme, with the important difference that we use the full spectral information in the data and not only the covariance matrix (23). To be more specific, introduce the notation

$$z(t) = \begin{bmatrix} u(t) \\ y(t) \end{bmatrix}, \qquad z_o(t) = \begin{bmatrix} u_o(t) \\ y_o(t) \end{bmatrix} \tag{25}$$

for the input-output data. The spectral density of the input-output data satisfies

$$\phi_z(\omega) = \begin{bmatrix} 1 \\ G(e^{i\omega}) \end{bmatrix} \begin{bmatrix} 1 & G^*(e^{i\omega}) \end{bmatrix} \phi_{u_o}(\omega) + \begin{bmatrix} \lambda_y & 0 \\ 0 & \lambda_u \end{bmatrix}, \qquad \forall \omega \tag{26}$$

If the spectral density matrix ϕ_z is known, and an appropriate diagonal matrix is subtracted, then one would get a rank 1 matrix, corresponding to the first term of (26), for all frequencies ω. In case the decomposition as in (26) can be carried out, the first term would easily lead to estimates of the transfer function $G(e^{i\omega})$ and the true input spectrum $\phi_{u_o}(\omega)$. The above idea is developed in [4] using an FFT-based spectral estimator. The quality of the estimates is moderate. A more promising approach, based on a black-box modelling of the spectral density ϕ_z is described in [19].

Another frequency domain identification technique is proposed in [11]. It is essential though for the method to work that the signal $u_o(t)$ is periodic.

6. Maximum likelihood and prediction error approach

In this method the errors-in-variables model (1), (3) is regarded as a multivariable system with both $u(t)$ and $y(t)$ as outputs, [17]. Of crucial importance for this approach is the modified assumption **A2**, i.e. the noise-free input $u_o(t)$ is characterized by a rational spectrum. The signal $u_o(t)$ can thus be represented by its innovations form, described as an ARMA process of the type

$$u_o(t) = \frac{C(q^{-1})}{D(q^{-1})}\, e(t) = \frac{1 + c_1\, q^{-1} + \cdots + c_p\, q^{-p}}{1 + d_1\, q^{-1} + \cdots + d_p\, q^{-p}} e(t) \ , \qquad (27)$$

where $e(t)$ is a white noise with variance λ_e and the polynomials $C(q^{-1})$, $D(q^{-1})$ have known degree p; moreover, $C(z)$ and $D(z)$ are relatively prime and asymptotically stable polynomials. In this way, the whole errors-in-variables model can be considered as a system with a two-dimensional output vector $z(t) = [u(t)\, y(t)]^T$ and three mutually uncorrelated white noise sources $e(t)$, $\tilde{u}(t)$ and $\tilde{y}(t)$. Using well-known techniques the system can be converted into the innovations form

$$z(t) = S(q^{-1})\, \varepsilon(t) \ , \qquad (28)$$

where $S(z)$ is the square (2×2) canonical (stable and minimum-phase) spectral factor of $\phi_z(\omega)$ and $\varepsilon(t)$ is the two-dimensional innovations vector. Since the innovations $\varepsilon(t)$ of model (28) can also be interpreted as the one step ahead prediction errors, a prediction error method can be applied for the identification of the system parameters. The parameter vector θ includes the coefficients of polynomials $A(q^{-1})$, $B(q^{-1})$, $C(q^{-1})$, $D(q^{-1})$ and the variances λ_e, λ_y, λ_u. The solution of the identification problem in Section 2 is then determined by minimizing the loss function

$$J(\theta) = \det\left(\frac{1}{N}\sum_{t=1}^{N} \varepsilon(t)\, \varepsilon^T(t)\right) \ , \qquad (29)$$

where N is the number of input-output samples.

Comments and experiences: As it always happens for identification of multivariable systems, the choice of a suitable parameterization of the model is a crucial point of the approach. In [17] it was proved that this method yields consistent parameter estimates under quite mild conditions. Its main disadvantage is that the numerical optimization procedure is, in general, quite involved since at each iteration a Riccati equation must be solved in order to find the innovations $\varepsilon(t)$ used in (29). The procedure may fail to give good results when only poor initial parameter estimates are available. Even if this technique can work with rather arbitrarily correlated noises, an extended system representation by including also models for the noises is needed.

Acknowledgments

The research reported here was supported by Swedish Research Council for Engineering Sciences under contract 98-654, and by the Ministry for University and Scientific and Technologic Research, Rome, which is gratefully acknowledged.

References

[1] B. D. O. Anderson. Identification of scalar errors-in-variables models with dynamics. *Automatica*, 21:709–716, 1985.

[2] B. D. O. Anderson and M. Deistler. Identifiability of dynamic errors-in-variables models. *J. Time Series Analysis*, 5:1–13, 1984.

[3] M. Aoki and P. C. Yue. On *a priori* error estimate of some identification methods. *IEEE Transactions on Automatic Control*, AC-15:541–548, October 1970.

[4] S. Beghelli, P. Castaldi, and U. Soverini. A frequential approach for errors-in-variables models. In *Proc. European Control Conference, ECC '97*, Brussels, Belgium, July 1–4 1997.

[5] S. Beghelli, R.P. Guidorzi, and U. Soverini. The Frisch scheme in dynamic system identification. *Automatica*, 26:171–176, 1990.

[6] J. A. Cadzow and O. M. Solomon. Algebraic approach to system identification. *IEEE Transactions on Signal Processing*, ASSP-34(3):462–469, June 1986.

[7] C. T. Chou and M. Verhaegen. Subspace algorithms for the identification of multivariable dynamic errors-in-variables models. *Automatica*, 33(10):1857–1869, October 1997.

[8] K. V. Fernando and H. Nicholson. Identification of linear systems with input and output noise: the Koopmans–Levin method. *IEE Proceedings, Part D*, 132(1):30–36, January 1985.

[9] U. Forsell, F. Gustafsson, and T. McKelvey. Time-domain identification of dynamic errors-in-variables systems using periodic excitation signals. In *Proc. IFAC 14th World Congress*, Beijing, P.R.China, 1999.

[10] W. A. Fuller. *Measurement Error Models*. Wiley, New York, NY, 1987.

[11] P. Guillaume, R. Pintelon, and J. Schoukens. Robust parametric transfer function estimaton using complex logarithmic frequency response data. *IEEE Transactions on Automatic Control*, 40:1180–1190, 1995.

[12] T. J. Koopmans. *Linear Regression Analysis of Economic Time Series*. N. V. Haarlem, The Netherlands, 1937.

[13] M. J. Levin. Estimation of a system pulse transfer function in the presence of noise. *IEEE Transactions on Automatic. Control*, AC–9:229–235, July 1964.

[14] L. Ljung. *System Identification - Theory for the User, 2nd edition.* Prentice Hall, Upper Saddle River, NJ, USA, 1999.

[15] J. Schoukens, R. Pintelon, G. Vandersteen, and P. Guillaume. Frequency domain system identification using non-parametric noise models estimated from a small number of data sets. *Automatica*, 33:1073–1086, 1997.

[16] J. H. Van Schuppen. Stochastic realization problems. In N. Nijmeijer and J. M. Schumacher, editors, *Three Decades of Mathematical System Theory*, Berlin, 1989. Springer-Verlag.

[17] T. Söderström. Identification of stochastic linear systems in presence of input noise. *Automatica*, 17:713–725, 1981.

[18] T. Söderström and K. Mahata. On subspace fitting and instrumental variable approaches for identification of error-in-variables models. 2001. Submitted for publication.

[19] T. Söderström, K. Mahata, and U. Soverini. A frequency domain Frisch scheme for dynamic errors-in-variables identification. 2001. In preparation.

[20] T. Söderström, U. Soverini, and K. Mahata. Perspectives on errors-in-variables estimation for dynamic systems. Technical Report 2001-011, Systems and Control, Uppsala University, Uppsala, Sweden, 2001.

[21] T. Söderström and P. Stoica. *Instrumental Variable Methods for System Identification.* Springer-Verlag, Berlin, 1983.

[22] T. Söderström and P. Stoica. *System Identification.* Prentice Hall International, Hemel Hempstead, UK, 1989.

[23] U. Soverini and T. Söderström. Identification methods of dynamic systems in presence of input noise. In *Proc. SYSID 2000, IFAC 12th Symposium on System Identification*, Santa Barbara, California, June 21–23 2000.

[24] P. Stoica, M. Cedervall, and A. Eriksson. Combined instrumental variable and subspace fitting approach to parameter estimation of noisy input-output systems. *IEEE Transactions on Signal Processing*, 43:2386–2397, 1995.

[25] P. Stoica and T. Söderström. Bias correction in least–squares identification. *International Journal of Control*, 35(3):449–457, 1982.

[26] J. K. Tugnait. Stochastic system identification with noisy input using cumulant statistics. *IEEE Transactions on Automatic Control*, AC-37:476–485, 1992.

[27] S. Van Huffel, editor. *Recent Advances in Total Least Squares Techniques and Errors-in-Variables Modelling.* SIAM, Philadelphia, USA, 1997.

[28] S. Van Huffel and J. Vandewalle. *The Total Least Squares Problem: Computation Aspects and Analysis.* SIAM, Philadelphia, USA, 1991.

ERRORS-IN-VARIABLES FILTERING IN BEHAVIOURAL AND STATE-SPACE CONTEXTS

Roberto Guidorzi, Roberto Diversi and Umberto Soverini
Dipartimento di Elettronica, Informatica e Sistemistica, Università di Bologna
Viale del Risorgimento 2, 40136 Bologna, Italy
{rguidorzi, rdiversi, usoverini}@deis.unibo.it

Abstract This paper considers the problem of filtering data sequences generated by errors-in-variables processes where all measured signals, differently from the classical Kalman filtering context, are affected by additive noise.

The design of optimal (minimal variance) filters leading to estimates of the process inputs and outputs is first carried out in a behavioural context.

The state-space context where EIV filtering can be performed relying on modified Kalman filtering techniques is then considered and a Monte Carlo simulation is finally proposed.

Keywords: optimal filtering, dynamic errors-in-variables models, behavioural models, recursive filtering, Kalman filtering.

1. Introduction

Most stochastic contexts in science and engineering refer to models where the input is assumed as exactly known and uncertainties are introduced by means of unknown inputs with given statistical properties and of additive noise on the outputs. In contexts of this kind the filtering problem consists in determining the (minimal variance) estimate of the process state and of the noiseless component of the output.

The stochastic context of Kalman filtering is intrinsically asymmetric since while the output, $y(t)$, is considered as affected by additive noise, the input, $u(t)$, is assumed as exactly known. This assumption can be true in many control applications but is, in general, non realistic in other fields where errors are present on all signals and even a partition between inputs and outputs could be arbitrary. This limit is not present in errors-

S. Van Huffel and P. Lemmerling (eds.), Total Least Squares and Errors-in-Variables Modeling, 281–291.

in-variables (EIV) models that treat all variables in a symmetric way. State-space EIV models are given by relations of the following type where $\hat{x}(t) \in \mathbb{R}^n$, $\hat{y}(t) \in \mathbb{R}^m$, $\hat{u}(t) \in \mathbb{R}^r$ and $\tilde{y}(t)$ and $\tilde{u}(t)$ are white processes with covariance matrices R and Q respectively and cross-covariance S.

$$\hat{x}(t+1) = A\,\hat{x}(t) + B\,\hat{u}(t) \quad (1)$$
$$\hat{y}(t) = C\,\hat{x}(t) + D\,\hat{u}(t) \quad (2)$$
$$y(t) = \hat{y}(t) + \tilde{y}(t) \quad (3)$$
$$u(t) = \hat{u}(t) + \tilde{u}(t) \quad (4)$$

The many results available in the literature for the stochastic context of Kalman filtering [1, 2] cannot be directly applied to the EIV framework where research has been performed mainly in the field of identification [3, 4, 5, 6] and only seldom in other areas [7].

This paper describes and compares new solutions to the EIV minimal variance filtering problem in a behavioural framework and in the more traditional state-space one. More precisely, Section 2 defines the filtering problem considered in the paper starting from the formulation of the associated interpolation problem. Section 3 concerns least squares and minimal variance estimates for the inputs and outputs of EIV processes while Section 4 develops EIV filtering. A comparison with Kalman filtering is performed in Section 5 and the results of a Monte Carlo simulation are reported in Section 6. Some concluding remarks are finally proposed in Section 7.

2. Context and statement of the problem

This version of the paper considers SISO systems, described by models of the type

$$q(z)\,\hat{y}(t) = p(z)\,\hat{u}(t) \quad (5)$$

where $q(z)$ and $p(z)$ are order-n polynomials in the unitary advance operator z

$$p(z) = z^n - \alpha_n\,z^{n-1} - \cdots - \alpha_2\,z - \alpha_1 \quad (6)$$
$$q(z) = \beta_{n+1}\,z^n + \beta_n\,z^{n-1} + \cdots + \beta_2\,z + \beta_1\,. \quad (7)$$

These models can be also written in the form

$$\hat{y}(t+n) = \sum_{k=1}^{n} \alpha_k\,\hat{y}(t+k-1) + \sum_{k=1}^{n+1} \beta_k\,\hat{u}(t+k-1)\,. \quad (8)$$

In the following it will also be assumed, without loss of generality, that $\hat{u}(t)$ and $\hat{y}(t)$ have null mean values and that $\mathbf{E}[\tilde{u}(t)] = 0$, $\mathbf{E}[\tilde{y}(t)] = 0$.

Defining the Hankel structure

$$H_k(x) = \begin{bmatrix} x(1) & \cdots & x(k) \\ \vdots & \ddots & \vdots \\ x(N) & \cdots & x(k+N-1) \end{bmatrix}, \tag{9}$$

where N is linked to the number of samples L by the relation $N = L - k + 1$, and the matrix of input/output samples

$$H_k = \left[\, H_{k+1}(y)\ H_{k+1}(u) \,\right], \tag{10}$$

it is possible to rewrite model (8) in the form

$$\hat{H}_n\, \theta = 0, \tag{11}$$

where θ denotes the parameter vector

$$\theta = \left[\, \theta_\alpha\ \theta_\beta \,\right]^T = \left[\, \alpha_1, \ldots, \alpha_n, -1,\ \beta_1, \ldots, \beta_{n+1} \,\right]^T. \tag{12}$$

Relations (3) and (4) lead immediately to the condition

$$H_k = \hat{H}_k + \tilde{H}_k \tag{13}$$

with obvious meanings of $\hat{H}_k$ and $\tilde{H}_k$. From relation (11), it follows also that

$$H_n\, \theta = \tilde{H}_n\, \theta. \tag{14}$$

Some editing on equation (11) allows to obtain the equivalent form

$$G\,\hat{v} = 0 \tag{15}$$

where G is the $N \times 2(N+n)$ matrix

$$G = \left[\, G_\alpha\ G_\beta \,\right] = \tag{16}$$

$$\begin{bmatrix} \alpha_1 & \cdots & \alpha_n & -1 & 0 & \cdots & 0 & \beta_1 & \cdots & \beta_{n+1} & 0 & \cdots & 0 \\ 0 & \alpha_1 & \cdots & \alpha_n & -1 & \cdots & 0 & 0 & \beta_1 & \cdots & \beta_{n+1} & \cdots & 0 \\ \vdots & & \ddots & & \ddots & \ddots & \vdots & \vdots & & \ddots & & \ddots & \vdots \\ 0 & \cdots & 0 & \alpha_1 & \cdots & \alpha_n & -1 & 0 & \cdots & 0 & \beta_1 & \cdots & \beta_{n+1} \end{bmatrix},$$

and $\hat{v}$ is the $2(N+n)$-dimensional vector

$$\hat{v} = \left[\, \hat{v}_y^T\ \hat{v}_u^T \,\right]^T = \left[\, \hat{y}(1) \ldots \hat{y}(N+n)\ \hat{u}(1) \ldots \hat{u}(N+n) \,\right]^T. \tag{17}$$

Relation (14) can thus be rewritten in the equivalent form

$$G\,v = G\left[\, \hat{v} + \tilde{v} \,\right] = G\,\tilde{v} = \gamma \tag{18}$$

where v and $\tilde{v}$ have the same structure as $\hat{v}$.

Remark 1. Model (8) is non purely dynamic ($D \neq 0$ in (2)). In purely dynamic systems $\beta_{n+1} = 0$. The extension to MISO systems can be obtained by defining matrix G as $[G_\alpha \, G_{\beta 1} \ldots G_{\beta r}]$ where r denotes the number of inputs [7].

The model of the considered EIV system is constituted by the description of its deterministic part, i.e. by the quadruple (A, B, C, D), or equivalently by $p(z)$, $q(z)$ or by matrix G, by the variances of the additive noise $\tilde{u}(t)$ and $\tilde{y}(t)$, $\tilde{\sigma}_u^2$ and $\tilde{\sigma}_y^2$ respectively and by their cross-covariance $\tilde{\sigma}_{yu}$. In this context, according to the definition of Willems [8], the vector of input and output observations v defines a noisy behaviour while $\hat{v}$ is the *regular* behaviour of the process. The problems treated in this paper are the following.

Problem 1 *(Interpolation).* Given the model of the process and the noisy behaviour v, determine an unbiased estimate of its regular part $\hat{v}$.

Problem 2 *(Filtering).* Given the model of the process and an increasing sequence of observations, estimate, as soon as a new observation $u(t)$, $y(t)$, becomes available, an unbiased estimate of its regular part $\hat{u}(t), \hat{y}(t)$.

3. LS and minimal variance interpolations

In this section we will describe the family of solutions of the interpolation problem. Observing that rank G is N, and denoting with $G^\dagger$ its pseudoinverse, the family of all solutions of (18) is described by the linear variety

$$\tilde{v}^* = G^\dagger G \, v + \hat{v}^* \tag{19}$$

where $\hat{v}^* \in \ker G$ and $G^\dagger = G^T(GG^T)^{-1}$. Since $G^\dagger G$ is the orthogonal projection of v on $\operatorname{im} G^T$, the LS solution, i.e. the solution with minimal Euclidean norm is

$$\tilde{v}^*_{LS} = G^\dagger G \, v \, . \tag{20}$$

In this case the interpolation vector of the noisy observations, $\hat{v}^*_{LS} = \left[I - G^T(GG^T)^{-1} G \right] v$ is associated with the maximization of the Euclidean norm of $\hat{v}^*$. This solution has been defined as *Frisch interpolation* in [7]. The covariance matrix of the LS estimate error $\tilde{e}_{LS} = \tilde{v} - \tilde{v}^*_{LS}$ is given by [9]

$$\operatorname{cov}(\tilde{e}_{LS}) = \left(I - G^\dagger G\right) \tilde{\Sigma} \left(I - G^\dagger G\right)^T = \left(I - G^\dagger G\right) \tilde{\Sigma} \left(I - G^\dagger G\right) \tag{21}$$

where $\tilde{\Sigma}$ denotes the covariance matrix of the vector $\tilde{v}$

$$\tilde{\Sigma} = \mathbf{E}\left[\tilde{v}\,\tilde{v}^T\right] = \begin{bmatrix} \tilde{\sigma}_y^2\, I_{N+n} & \tilde{\sigma}_{yu}\, I_{N+n} \\ \tilde{\sigma}_{yu}\, I_{N+n} & \tilde{\sigma}_u^2\, I_{N+n} \end{bmatrix}. \tag{22}$$

Let us consider again relation (18); the minimal variance estimate of $\tilde{v}$ that can be obtained from γ is given by the mean value of $\tilde{v}$ conditioned by γ [2] $\tilde{v}^*_{MV} = \mathbf{E}\left[\tilde{v}\,|\,\gamma\right]$. Under the assumption of Gaussian distributions for $\tilde{u}(t)$ and $\tilde{y}(t)$, we have

$$\tilde{v}^*_{MV} = \mathbf{E}\left[\tilde{v}\right] + \mathbf{E}\left[\tilde{v}\,\gamma^T\right]\mathbf{E}\left[\gamma\gamma^T\right]^{-1}\left(\gamma - \mathbf{E}[\gamma]\right). \tag{23}$$

When this assumption is not satisfied, expression (23) constitutes, in any case, the best linear estimate of $\tilde{v}$ that can be obtained from the knowledge of γ. Recalling that $v = \hat{v} + \tilde{v}$ and $\mathbf{E}[\tilde{v}\,\hat{v}^T] = 0$, it is easy to obtain, from expression (23), the following estimate

$$\tilde{v}^*_{MV} = \tilde{\Sigma}\,G^T\left(G\,\tilde{\Sigma}\,G^T\right)^{-1} G\,v. \tag{24}$$

The covariance matrix of the estimation error $\tilde{e}_{MV} = \tilde{v} - \tilde{v}^*_{MV}$ is [9]

$$\operatorname{cov}(\tilde{e}_{MV}) = \tilde{\Sigma}\left(I - G^T(G\tilde{\Sigma}G^T)^{-1}G\tilde{\Sigma}\right)^2 = \tilde{\Sigma}\left(I - G^T(G\tilde{\Sigma}G^T)^{-1}G\tilde{\Sigma}\right). \tag{25}$$

The minimal variance estimate of the regular behaviour $\hat{v}$ is thus given by

$$\hat{v}^*_{MV} = v - \tilde{\Sigma}\,G^T\left(G\,\tilde{\Sigma}\,G^T\right)^{-1} G\,v. \tag{26}$$

It is immediate to verify that (24) coincides with (20) when $\tilde{\sigma}_y^2 = \tilde{\sigma}_u^2$ and $\tilde{\sigma}_{yu} = 0$ and that both belong to the family of admissible solutions (19).

Remark 2. It can be noted that the estimates (20) and (24) of $\tilde{v}$ have a structure of the type $\tilde{v}^* = SG\,v = R\,v$ so that

$$\mathbf{E}\left[\tilde{v}^*\right] = \mathbf{E}\left[SG\,\tilde{v} + SG\,\hat{v}\right] = SG\,\mathbf{E}\left[\tilde{v}\right] + SG\,\hat{v} = 0,$$

i.e. they are both unbiased.

4. EIV filtering

The samples $\hat{u}^*(t)$, $\hat{y}^*(t)$ obtained from (26) will depend, in general, on both previous and subsequent noisy samples $u(\cdot)$ and $y(\cdot)$; the only exception concerns the first sample $\hat{u}^*(1)$, $\hat{y}^*(1)$, that will depend only on subsequent samples and the last one $\hat{u}^*(N+n)$, $\hat{y}^*(N+n)$, that will depend only on previous samples. Relation (26) could thus be used in filtering applications by performing at every step an interpolation and selecting in the whole sequences $\hat{u}^*(\cdot)$, $\hat{y}^*(\cdot)$ only the last samples. A procedure of this kind would however exhibit poor efficiency because it does not rely on previous computations; moreover the amount of computations increases at every step with the dimension of G. It is however possible to take advantage of the specific structure of G to develop an on-line and finite-memory procedure; rewrite, to this purpose, G (16) and $\hat{v}$ (17) as follows

$$G(N) = \begin{bmatrix} \alpha_1 & \beta_1 & \alpha_2 & \beta_2 & \cdots & -1 & \beta_{n+1} & 0 & \cdots & \cdots & \cdots & 0 \\ 0 & 0 & \alpha_1 & \beta_1 & \cdots & \alpha_n & \beta_n & -1 & \beta_{n+1} & 0 & \cdots & 0 \\ \vdots & & & & & & & & & & & \vdots \\ 0 & \cdots & \cdots & \cdots & \cdots & \cdots & 0 & \alpha_1 & \beta_1 & \cdots & -1 & \beta_{n+1} \end{bmatrix} \tag{27}$$

$$\hat{v}(t) = \left[\, \hat{y}(1)\ \hat{u}(1) \mid \hat{y}(2)\ \hat{u}(2) \mid \cdots \mid \hat{y}(N+n)\ \hat{u}(N+n) \,\right]^T \tag{28}$$

where N is the number of rows of G. Passing from N to $N+1$ (i.e. from $t = N+n$ to $t = N+n+1$), the update of $\hat{v}(t)$ and $G(N)$ is given by

$$\hat{v}(t+1) = \left[\, \hat{v}(t)^T\ \hat{y}(t+1)\ \hat{u}(t+1) \,\right]^T \tag{29}$$

$$G(N+1) = \begin{bmatrix} & G(N) & & 0_{(N\times 2)} \\ 0 \cdots 0 & \alpha_1\,\beta_1 \cdots \alpha_n\,\beta_n & & -1\ \ \beta_{n+1} \end{bmatrix} = \begin{bmatrix} G(N) & 0_{(N\times 2)} \\ m_1 & m_2 \end{bmatrix}. \tag{30}$$

Making reference to the minimal variance estimate (26), and omitting the suffix MV, denote now with $R(N)$ the product

$$R(N) = \tilde{\Sigma}(N)\, G^T(N) \left(G(N)\tilde{\Sigma}(N)G^T(N) \right)^{-1} G(N) \tag{31}$$

where, according with the structure of $\hat{v}(t)$ (28), $\tilde{\Sigma}(N)$ will exhibit the following structure

$$\tilde{\Sigma}(N) = \begin{bmatrix} \tilde{\Sigma}_{yu} & \cdots & 0 \\ \vdots & \ddots & \vdots \\ 0 & \cdots & \tilde{\Sigma}_{yu} \end{bmatrix} \quad \text{with} \quad \tilde{\Sigma}_{yu} = \begin{bmatrix} \tilde{\sigma}_y^2 & \tilde{\sigma}_{yu} \\ \tilde{\sigma}_{yu} & \tilde{\sigma}_u^2 \end{bmatrix}. \tag{32}$$

Then the interpolation at time t is

$$\hat{v}^*(t) = \left[\, I - R(N) \,\right] v(t)\,. \tag{33}$$

In [9] it has been shown that the filtering problem can be solved by using a recursive filter with memory n allowing the update of the last $2n+2$ entries of $\hat{v}^*(t)$ by means of the relation

$$\hat{v}^*(t+1)_{2n+2} \;=\; \begin{bmatrix} \hat{v}^*_{2n}(t) \\ y(t+1) \\ u(t+1) \end{bmatrix} + k(t)\,\tilde{\Sigma}(1)\, S^T(t)\, \nu(t+1) \tag{34}$$

(the suffix $2n+2$ refers to the last $2n+2$ entries for vectors and to the lower right $(2n+2)\times(2n+2)$ submatrix for matrices) and

$$\begin{aligned}\nu(t+1) &= y(t+1) - \beta_{n+1}\, u(t+1) \\ &\quad -\sum_{k=1}^{n} \alpha_k\, \hat{y}^*(t-n+k-1) - \sum_{k=1}^{n} \beta_k\, \hat{u}^*(t-n+k-1)\end{aligned} \tag{35}$$

$$S(t) = \overline{m} \begin{bmatrix} \left(I - R(N)\right)_{2n} & 0 \\ 0 & I_2 \end{bmatrix} \tag{36}$$

where $\overline{m}$ denotes the constant vector $\overline{m} = [\,\alpha_1,\, \beta_1, \ldots, -1, \beta_{n+1}]$,

$$k(t) = 1/\big(S(t)\,\tilde{\Sigma}(1)\,\overline{m}^T\big)\,. \tag{37}$$

$S(t)$ and $k(t)$ can, finally, be updated by means of the recursion

$$R(N+1)_{2n} = \begin{bmatrix} R(N)_{2n} & 0 \\ 0 & 0 \end{bmatrix}_{2n} + k(t)\,\tilde{\Sigma}(0)\,\big(S^T(t)\,S(t)\big)_{2n}. \tag{38}$$

The steps of the whole algorithm [9] can thus be summarised as follows:

Algorithm 1.

1) At time $t = n+1$ compute, by means of (33), $\hat{v}^*(t)$ and, consequently, $\hat{y}^*(t)$ and $\hat{u}^*(t)$.

2) Compute $S(t)$ (36) and $k(t)$ (37).

3) At time $t = t+1$ compute the quantities $\nu(t)$ (35), $\hat{v}^*(t)_{2n+2}$ (34) and, consequently, $\hat{y}^*(t)$ and $\hat{u}^*(t)$.

4) Update now $R(N)_{2n}$ by means of (38).

5) Return to step 2.

5. A state-space approach

Kalman filtering cannot be directly applied to non-purely dynamic errors-in-variables models. To introduce a state-space approach in EIV filtering consider the process

$$\eta(t) = \frac{p(z)}{q(z)}\,u(t) - y(t)\,. \tag{39}$$

Because of (3), (4) and (8), relation (39) can also be rewritten in the form

$$\eta(t) = \frac{p(z)}{q(z)}\,\tilde{u}(t) - \tilde{y}(t) \tag{40}$$

that admits a state-space realization of the type

$$\begin{aligned} \xi(t+1) &= A\,\xi(t) + B\,\tilde{u}(t) && (41)\\ \eta(t) &= C\,\xi(t) + D\,\tilde{u}(t) - \tilde{y}(t) && (42) \end{aligned}$$

corresponding to the stochastic environment of Kalman filtering, so that [2]

$$\begin{aligned} \xi(t+1\,|\,t) &= \left(A - B\,S\,R^{-1}\,C\right)\xi(t\,|\,t) + B\,S\,R^{-1}\,\eta(t) && (43)\\ \xi(t\,|\,t) &= \xi(t\,|\,t-1) + \kappa(t)\left(\eta(t) - C\,\xi(t\,|\,t-1)\right) && (44)\\ \kappa(t) &= P(t\,|\,t-1)\,C^T\left(CP(t\,|\,t-1)\,C^T + R\right)^{-1} && (45)\\ P(t\,|\,t) &= \left(I - \kappa(t)\,C\right)P(t\,|\,t-1) && (46)\\ P(t+1\,|\,t) &= \left(A - B\,S\,R^{-1}\,C\right)P(t\,|\,t)\left(A - B\,S\,R^{-1}\,C\right)^T && (47)\\ &+\; B\left(Q - S\,R^{-1}\,S^T\right)B^T\,, \end{aligned}$$

where

$$Q = \tilde{\sigma}_u^2\,, \qquad R = D^2\,\tilde{\sigma}_u^2 + \tilde{\sigma}_y^2 - 2\,D\,\tilde{\sigma}_{yu}\,, \qquad S = D\,\tilde{\sigma}_u^2 - \tilde{\sigma}_{yu}\,. \tag{48}$$

Since $\eta(t)$, because of (39), is known for every t, the minimal variance estimates of $\tilde{y}(t)$ and $\tilde{u}(t)$ can be obtained by means of the relations

$$\begin{aligned} \eta(t) - C\,\xi(t\,|\,t) &= D\,\tilde{u}^*(t) - \tilde{y}^*(t) && (49)\\ \tilde{u}^*(t) &= \tilde{y}^*(t)\,\frac{D\,\tilde{\sigma}_u^2 - \tilde{\sigma}_{yu}}{D\,\tilde{\sigma}_{yu} - \tilde{\sigma}_y^2} = \tilde{y}^*(t)\Delta\,\cdot && (50) \end{aligned}$$

Note that (49) provides an estimate of the term $D\,\tilde{u}(t) - \tilde{y}(t)$ while (50) follows from the fact that $\tilde{y}^*(t)$ and $\tilde{u}^*(t)$ are estimates conditioned by the same process $\eta(t)$ [9]. Minimal variance estimates of $\hat{y}(t), \hat{u}(t)$ are finally given by

$$\hat{y}^*(t) = y(t) + \frac{\eta(t) - C\,\xi(t\,|\,t)}{1 - D\Delta}\,, \quad \hat{u}^*(t) = u(t) + \Delta\,\frac{\eta(t) - C\,\xi(t\,|\,t)}{1 - D\Delta}\,\cdot \tag{51}$$

6. A Monte Carlo simulation

Previous algorithms are equivalent since they lead, for every t, to the same minimal variance estimate of $\hat{u}(t)$ and $\hat{y}(t)$. They exhibit, however, different computational performances, as investigated in [10].

The effectiveness of EIV filtering will now be illustrated by means of a 100-runs Monte Carlo simulation based on the following order 3 model

$$q(z) = z^3 - 1.2\,z^2 + 0.81\,z - 0.27 \tag{52}$$

$$p(z) = 0.1510\,(z^3 + 0.9\,z^2 + 0.49\,z + 0.145)\ , \tag{53}$$

where the gain assures an output $\hat{y}(\cdot)$ with the same variance of the input $\hat{u}(\cdot)$. Sequences of mutually correlated white noise have then been added to obtain the noisy sequences $u(t) = \hat{u}(t) + \tilde{u}(t)$ and $y(t) = \hat{y}(t) + \tilde{y}(t)$. The length of the sequences is $L = 400$, the variances of the (normalized) input and output are $\hat{\sigma}_u^2 = 1$, $\hat{\sigma}_y^2 = 1$, while the additive noise on input and output has variances given by $\tilde{\sigma}_u^2 = 0.28$, $\tilde{\sigma}_y^2 = 0.64$ and their covariance is $\tilde{\sigma}_{yu} = 0.36$. Note that the percent amounts of noise defined as one hundred times the ratio between the standard deviations of the noise and those of the signals are equal to 53% on the input and 80% on the output; their correlation is equal to 62%. Figures 1 and 2 show the noiseless input and output (black line) compared with the noise-corrupted ones (dotted line) in a typical run. The mean values of the variances of the differences between the filtered input and output sequences, $\hat{u}^*(t)$, $\hat{y}^*(t)$, and the noiseless ones, as well as their standard deviations, are

$$\boxed{\sigma_{\epsilon u} = 0.0791 \pm 0.0047 \qquad \sigma_{\epsilon y} = 0.0431 \pm 0.0065}$$

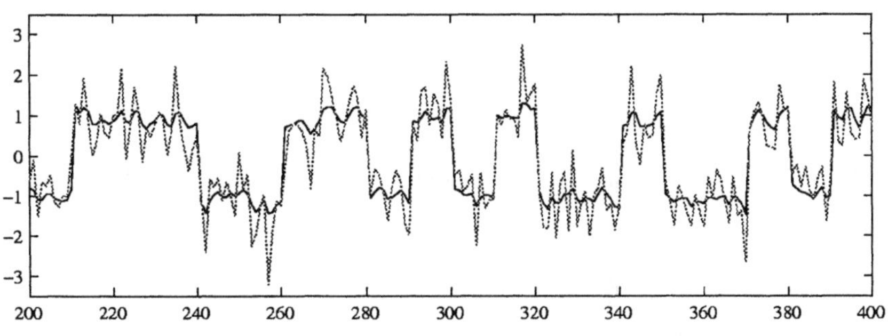

Figure 1. Comparison between the noiseless input and its observation

which correspond to 28.1% and 20.7% percent errors on the standard deviations. Figures 3 and 4 show the noiseless input and output (black line) compared with the filtered ones (dotted line) in a typical run.

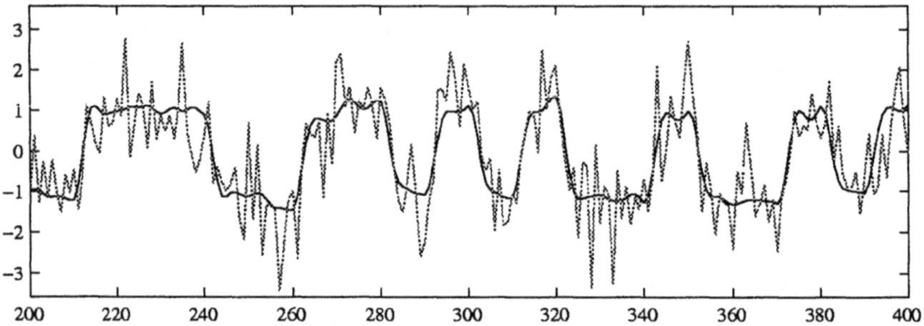

Figure 2. Comparison between the noiseless output and its observation

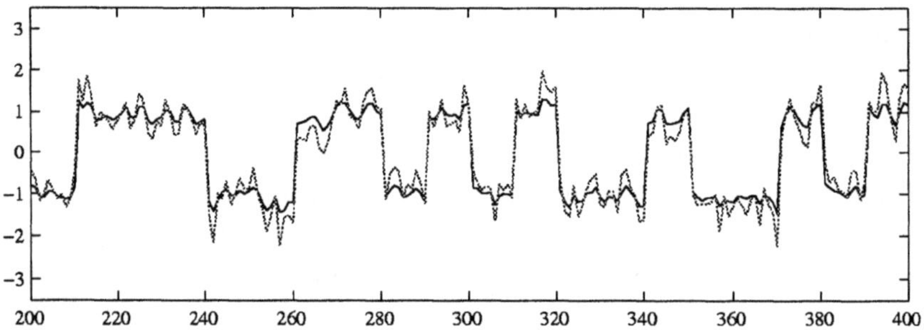

Figure 3. Comparison between the noiseless input and the filtered one

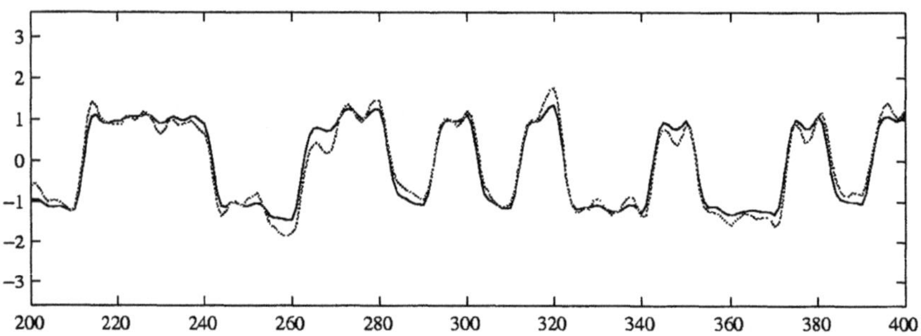

Figure 4. Comparison between the noiseless output and the filtered one

7. Concluding remarks

EIV filtering is a new area of research where the use of behavioural models looks as particularly attractive both on conceptual and implementation grounds. This paper has introduced the EIV optimal interpolation and filtering problems making reference to a geometrical interpretation of these problems in a behavioural context. It has, subsequently, reformulated the problem in a state-space context relying, for its solution, on a modified Kalman filter. The effectiveness of EIV filtering has then been shown by means of a numerical simulation.

References

[1] R.E. Kalman, A new approach to linear filtering and prediction problems, *Transactions of the ASME*, 82:35–45, 1960.

[2] B.D.O. Anderson and J.B. Moore, *Optimal filtering*, Prentice-Hall, Englewood Cliffs, New Jersey, 1979.

[3] T. Söderström, Identification of stochastic linear systems in presence of input noise, *Automatica*, 17:713–725, 1981.

[4] M. Deistler, Linear errors-in-variables models, *on Time Series and Linear Systems*, S. Bittanti ed., 37–68, Springer-Verlag, Berlin, 1986.

[5] S. Beghelli, R.P. Guidorzi and U. Soverini, The Frisch scheme in dynamic system identification, *Automatica*, 26(1):171–176, 1990.

[6] W.X. Zheng and C.B. Feng, Unbiased parameter estimation of linear systems in the presence of input and output noise, *Int. J. of Adaptive Control and Signal Processing*, 3:231–251, 1989.

[7] P. Guidorzi and R. Guidorzi, Frisch filtering of noisy signals, *Proceedings of EUSIPCO 98*, vol. IV, 2117-2120, Rhodes, September 1998.

[8] J.C. Willems, From time series to linear system, part I: Finite dimensional linear time invariant systems; part II: Exact modelling; part III: Approximate modelling, *Automatica*, 22/23:561–580, 675–694, 87–115, 1986, 1987.

[9] R. Guidorzi, R. Diversi and U. Soverini, Optimal Errors-in-Variables filtering, *under review.*

[10] R. Diversi, R. Guidorzi and U. Soverini, Algorithms for optimal Errors-in-Variables filtering, *under review.*

WEIGHTED TOTAL LEAST SQUARES, RANK DEFICIENCY AND LINEAR MATRIX STRUCTURES

Bart De Moor

ESAT-SISTA, Katholieke Universiteit Leuven, Kasteelpark 10, B-3001 Leuven, Belgium

bart.demoor@esat.kuleuven.ac.be web: www.esat.kuleuven.ac.be/~demoor

Abstract In this contribution we explain how the rank deficiency of a given data matrix, its linear matrix structure and its interpretation in terms of discrete-time dynamical systems, are intimately connected. This relation permits to formulate least squares dynamical system identification problems, the solution of which leads to structured total least squares. We also consider the insertion of given weights in the least squares objective function, leading to weighted TLS problems. The Riemannian SVD, which is a "nonlinear" generalized SVD, provides an elegant framework for these structured and/or weighted TLS problems.

Keywords: dynamic errors-in-variables, elementwise weighted total least squares, Hankel TLS, Riemannian SVD, system identification.

1. Rank deficiency and linear matrix structures

There is a trivial, yet important connection between the rank deficiency of a given matrix A and the existence of linear relations between its columns. Indeed, let $A \in \mathbb{R}^{p\times q}$, with $p \geq q$. When $\text{rank}(A) = r_A < q$, there exist $q - r_A$ linearly independent linear relations between the columns of A, but in this paper we will confine ourselves to one single linear relation, in which case $r_A = q - 1$. In a certain sense, the rank deficiency of a matrix is a "non-generic" property (a notion that can be made very precise as e.g. in [17]). This explains the fact that, in most applications, the observed data matrix $A \in \mathbb{R}^{p\times q}$, $p \geq q$ is of full column rank q. The central idea then of *total linear least squares* (TLLS) is to approximate the given data matrix A, by a matrix B of rank $q-1$, so that at least one linear relation exists between the columns of B:

$$\min_{B\in\mathbb{R}^{p\times q},\ v\in\mathbb{R}^q} \|A - B\|_F^2 \ \text{subject to} \ \begin{cases} B\,.\,v &= 0\,, \\ v^T\,.\,v &= 1\,. \end{cases} \tag{1}$$

S. Van Huffel and P. Lemmerling (eds.), Total Least Squares and Errors-in-Variables Modeling, 293–304.

A complete treatment of the TLS problem (1) and its solution via the singular value decomposition, including many references and comparisons with other fitting techniques, generic and non-generic cases, geometrical, statistical and algorithmic issues, may be found in the books [15] [16].

In this paper, we discuss so-called *structured and/or weighted total least squares (S/W-TLS)* problems, which are an extension of the TLS problem (1), in which the matrix A and its rank deficient approximant B are required to have a certain linear matrix structure, and/or the least squares objective function is modified by introducing weights. We call a matrix $A \in \mathbb{R}^{p\times q}$ linearly structured, when it can be written as a linear combination of given, known basis matrices A_i, $i = 1, 2, \dots, N \in \mathbb{R}^{p\times q}$ as $A = A_1\alpha_1 + \cdots + A_N\alpha_N$, with $\alpha_i \in \mathbb{R}$. N is the number of different coefficients α_i, needed to generate the matrix A. It is assumed throughout that the matrices A_i are linearly independent. Examples of such linear structures are symmetric, centro-symmetric, per-symmetric, row-rhomboidal, negacyclic matrices, Toeplitz, Hankel and circulant Brownian matrices and their block versions. These linear structures are ubiquitous in signal processing, systems and control theory, image processing and statistics, numerical linear algebra, mechanical and electrical engineering, physics, etc ... Observe that linear matrix structures are closed under addition and intersection: Examples are "symmetric Toeplitz" matrices (intersection of symmetric and Toeplitz) or "Hankel + Toeplitz" matrices, which are the sum of a Hankel and a Toeplitz matrix. While rank deficiency of the data matrix hints at an underlying linear model that "explains" the data, the fact that in addition the data matrix is also linearly structured, implies that these data are generated by a dynamical system, described by linear difference or differential equations. Let us give some examples.

Example 1 *(Rank deficient Hankel matrices).* Let $Z \in \mathbb{R}^{p\times q}$ be a Hankel matrix with $p \geq q$ and denote its $N = p + q - 1$ different elements by z_k, $k = 0, 1, \dots, N$. It is well known that $\text{rank}(Z) = n$ if and only if the elements z_k are generated by the output of a linear time-invariant discrete time system of the form

$$\begin{aligned} x_{k+1} &= Fx_k \,, \\ z_k &= Hx_k \,, \end{aligned} \tag{2}$$

with $F \in \mathbb{R}^{n\times n}$ the system matrix, $H \in \mathbb{R}^{1\times n}$ the output matrix, x_k the state and z_k the output at time instant k. In the case that $q = n + 1$, the vector $v \in \mathbb{R}^{n+1}$ that satisfies $Zv = 0$ contains the coefficients (up to within a scalar) of the characteristic polynomial $\det(F - \lambda I_n)$ (We assume here that the characteristic and minimal polynomial of F coincide). The problem of finding the model matrices F and H and the order n is called "the realization problem" and is, by now, a "classical" problem in system theory. The model (2) is used in many applications, ranging from biomedical signal processing (e.g. analysis of NMR spectra), vibrational analysis of dynamical structures (e.g. characterization of flutter of wings of air planes), industrial process system identification, stochastic identification and telecommunications, etc ... We refer to [1] [4] [5] [8] for examples and details.

Example 2 *(Rank deficient "double" Hankel matrices).* Next we extend the model (2) with scalar inputs $w_k \in \mathbb{R}$:

$$\begin{aligned} x_{k+1} &= Fx_k + Gw_k \,, \\ z_k &= Hx_k + Jw_k \,. \end{aligned} \tag{3}$$

Here, $G \in \mathbb{R}^{n\times 1}$ is the input matrix and $J \in \mathbb{R}^{1\times 1}$ is the so-called direct-feedthrough term. Let Z be a $p \times q$ Hankel matrix with the outputs ($p \geq q$) and W be a $p \times q$ Hankel matrix with the inputs. Then, under fairly general conditions on the input sequence ("persistancy of excitation", i.e. rank$(W) = q$) and the system (3) (equal number of poles and zeros, observability and controllability) we have

$$\text{rank}[\, Z \;\; W \,] = q + n \,.$$

In particular, with $q = n + 1$, rank$[\, Z\, W \,] = 2n + 1$ and

$$[\, Z\, W \,] \begin{bmatrix} a \\ -b \end{bmatrix} = 0 \,, \tag{4}$$

where a and b contain (up to within a scalar) the coeffients of the so-called transfer function $T(z)$ of the linear system (3), which is rational in z: $T(z) = J + H(zI_n - F)^{-1}G = b(z)/a(z)$. Here $a(z)$ and $b(z)$ are n-th degree polynomials in z with the coefficients of the vectors a and b. Eq. (4) is nothing else then a difference equation of the form

$$z_k\alpha_n + z_{k+1}\alpha_{n-1} + \cdots + z_{k+n}\alpha_0 = w_k\beta_n + w_{k+1}\beta_{n-1} + \cdots + w_{k+n}\beta_0 \,, \tag{5}$$

for $k = 0, 1, 2, \ldots$

Example 3 *(Rank deficient "bilinear" Hankel matrices).* Yet another extension of the linear model (3) is the bilinear model, in which products between inputs and states are included in the state equation by introducing an additional matrix $L \in \mathbb{R}^{n\times n}$:

$$\begin{aligned} x_{k+1} &= Fx_k + Gw_k + Lx_k w_k \,, \\ z_k &= Hx_k + Jw_k \,, \end{aligned} \tag{6}$$

The Volterra kernels of this system are of the form: 0-th order: $K_0 = HG$; 1-st order: $K_1 = (K_{11}; K_{12}) = (HFG; HLG)$; 2nd order $K_2 = (K_{21}; K_{22}; K_{23}; K_{24}) = (HF^2G; HFLG; HLFG; HL^2G)$, etc ... It can be shown (see e.g. [13]) that the following linearly structured, "generalized" Hankel matrix

$$H = \left[\begin{array}{c|cc|cccc|c} K_{00} & K_{11} & K_{12} & K_{21} & K_{22} & K_{23} & K_{24} & \cdots \\ \hline K_{11} & K_{21} & K_{22} & K_{31} & K_{32} & K_{33} & K_{34} & \cdots \\ K_{12} & K_{23} & K_{24} & K_{35} & K_{36} & K_{37} & K_{38} & \cdots \\ \hline K_{21} & K_{31} & K_{32} & \cdots & \cdots & \cdots & \cdots & \cdots \\ K_{23} & K_{35} & K_{36} & \cdots & \cdots & \cdots & \cdots & \cdots \\ K_{22} & K_{33} & K_{34} & \cdots & \cdots & \cdots & \cdots & \cdots \\ K_{24} & K_{37} & K_{38} & \cdots & \cdots & \cdots & \cdots & \cdots \\ \hline \cdots & \cdots & \cdots & \cdots & \cdots & \cdots & \cdots & \cdots \end{array}\right]$$

is rank deficient of rank n, if its elements are Volterra kernels of a bilinear system of order n of the form (6). Again, we see how the combination of a linear matrix structure and rank deficiency has a very precise dynamical system theoretic interpretation.

Observe that in all these examples, the rank of the structured matrices reveals the order (= number of states) of the "underlying" dynamical system!

2. A commuting Lemma for linear structures

Let $A \in \mathbb{R}^{p\times q}$ be a linearly structured matrix with N different elements collected in the vector $a \in \mathbb{R}^N$. For example, a $p \times q$ Hankel matrix has $N = p+q-1$ different elements and a $p \times p$ symmetric Toeplitz matrix has $N = p$. Let $v \in \mathbb{R}^q$ be a vector. The matrix-vector product Av is bilinear in the elements of A and the elements of v, and because of this bilinearity we can reverse the order:

$$A\,v = T_v\,a\,, \tag{7}$$

where $T_v \in \mathbb{R}^{p\times N}$ is linearly structured in the elements of v. Observe that the linear structure of A and the ordering of the different elements of A in the vector a, uniquely determine the structure of T_v. As an example, consider a $(p = 4) \times (q = 3)$ Hankel matrix A with $N = p + q - 1 = 6$ different elements. Then T_v follows from (7) as

$$\begin{bmatrix} a_0 & a_1 & a_2 \\ a_1 & a_2 & a_3 \\ a_2 & a_3 & a_4 \\ y_3 & a_4 & a_5 \end{bmatrix} \begin{bmatrix} v_0 \\ v_1 \\ v_2 \end{bmatrix} = \begin{bmatrix} v_0 & v_1 & v_2 & 0 & 0 & 0 \\ 0 & v_0 & v_1 & v_2 & 0 & 0 \\ 0 & 0 & v_0 & v_1 & v_2 & 0 \\ 0 & 0 & 0 & v_0 & v_1 & v_2 \end{bmatrix} \begin{bmatrix} a_0 \\ a_1 \\ a_2 \\ a_3 \\ a_4 \\ a_5 \end{bmatrix}.$$

Similarly we can define a linearly structured matrix $T_u \in \mathbb{R}^{N\times q}$ from $A^T u = T_u^T a$. These commuting results lead to simpler formula manipulation, in multilinear expressions of the form $\mathcal{L} = \cdots + u^T Av + \cdots$ in which they allow one to calculate expressions of the form $\frac{\partial \mathcal{L}}{\partial a} = \cdots + T_v^T u + \cdots$ The Commuting Lemma also allows to exchange the role of $v \in \mathbb{R}^q$ and $u \in \mathbb{R}^p$ in $T_v^T u = T_u^T v$.

3. Data and noise models

In practical engineering applications, data are never "exact", but they are always corrupted by so-called "measurement noise". More specifically, let $a \in \mathbb{R}^N$ be the vector of observed data. We will assume that the noise model is additive and Gaussian, i.e. $a = \hat{a} + \tilde{a}$, in which $\hat{a}$ contains the unknown exact (noisefree) data and $\tilde{a}$ the unobserved noise samples. As for the noise model, we assume that $\tilde{a} = M\tilde{\tilde{a}}$, where $M \in \mathbb{R}^{N\times N_1}$ with $N_1 \leq N$ is a given known matrix of full column rank N_1. The random vector $\tilde{\tilde{a}}$ is zero mean, normally distributed with covariance matrix equal to the identity matrix: $\tilde{\tilde{a}} \sim \mathcal{N}(0, I_{N_1})$. Hence, the noise random variable $\tilde{a}$ is zero mean normally distributed with covariance matrix $Q = MM^T$. Note that Q itself is singular whenever $N_1 < N$.

4. The Riemannian SVD

Let us now formulate structured and weighted TLS problems, the ingredients of which are a data vector $a \in \mathbb{R}^N$, a noise model $\tilde{a} = M\tilde{\tilde{a}}$ with a given specified matrix $M \in \mathbb{R}^{N\times N_1}$ with $\text{rank}(M) = N_1$, the specification of a linear matrix structure (Hankel, double Hankel, bilinear Hankel, Toeplitz, etc ...) (with corresponding linearly structured matrices T_u and T_v as defined above) and the specification of the number q of columns of the structured matrix.

Having specified these elements, every structured/weighted TLS problem can be phrased as

$$\min_{b\in\mathbb{R}^N,\, v\in\mathbb{R}^q,\, e\in\mathbb{R}^{N_1}} \frac{1}{2} e^T e \text{ subject to } \begin{cases} b = a - Me\,, \\ Bv = 0\,, \\ v^T v = 1\,. \end{cases} \tag{8}$$

In these equations, $B \in \mathbb{R}^{p\times q}$ is the linearly structured matrix, generated from the elements of the vector $b \in \mathbb{R}^N$, that collects the different elements of B. This structured matrix B is required to be rank deficient, a condition that is enforced via the vector v in its null space. The vector b contains the estimates of the "exact" data $\hat{a}$. The noise model is represented by the given matrix M. The least squares criterion (in which we introduce the factor 1/2 just for convenience) derives from the principle of maximum likelihood. It is easy to verify how the "ordinary" TLS problem (1) is just a special instance of this general formulation.

To solve the constrained optimization problem (8), we consider the Langrangean function $\mathcal{L}(m,e,v,l,\lambda,b) = \frac{1}{2}e^Te + l^TBv + \lambda(1-v^Tv) + m^T(b-a+Me)$. Setting all derivatives to zero (and repeatedly using the Commuting Lemma) results in

$$\begin{aligned}
\frac{\partial\mathcal{L}}{\partial m} &= 0 \Longrightarrow b = a - Me\,, & \frac{\partial\mathcal{L}}{\partial e} &= 0 \Longrightarrow e + M^Tm = 0\,,\\
\frac{\partial\mathcal{L}}{\partial v} &= 0 \Longrightarrow B^Tl = 2v\lambda\,, & \frac{\partial\mathcal{L}}{\partial l} &= 0 \Longrightarrow Bv = 0\,,\\
\frac{\partial\mathcal{L}}{\partial b} &= 0 \Longrightarrow T_v^Tl + m = 0\,, & \frac{\partial\mathcal{L}}{\partial \lambda} &= 0 \Longrightarrow v^Tv = 1\,.
\end{aligned}$$

From these equations, one can easily show that $\lambda = 0$, eliminate $m = -T_v^Tl$, eliminate $e = -M^Tm = M^TT_v^Tl$ and introduce $Q = MM^T$ to find that $b = a - QT_v^Tl$, $B^Tl = T_l^Tb = 0$ and $Bv = T_vb = 0$. Using these last two equations, we can eliminate the vector b and the matrix B, to find $T_l^Ta = (T_l^TQT_l)v$ and $T_va = (T_vQT_v^T)l$. Next we define the matrices D_{vv} and D_{ll} as $D_{vv} = (T_vQT_v^T)$ and $D_{ll} = (T_l^TQT_l)$, we normalize l as $l = u\sigma$ in which σ is such that $u^TD_{vv}u = 1$. Since now $D_{ll} = D_{uu}\sigma^2$, we then find (once again using the Commuting Lemma) that

$$\begin{aligned}
Av &= D_{vv}u\sigma\,, & u^TD_{vv}u &= 1\,,\\
A^Tu &= D_{uu}v\sigma\,, & v^Tv &= 1\,.
\end{aligned} \tag{9}$$

In these equations, which we have called the Riemannian SVD, the matrix A has the same affine structure as the matrix B we started from in (8). The value of σ that we need follows from the objective function as $e^Te = m^TMM^Tm = l^TT_vQT_v^Tl = u^TD_{vv}u\sigma^2 = \sigma^2$. Obviously, we need the minimal value of σ and corresponding vectors u and v that satisfy the equations (9).

Observe that the matrices D_{uu} and D_{vv} are defined from the given required linear matrix structure (as "represented" by the matrices T_u and T_v from the Commuting Lemma), and from the noise model represented by the covariance matrix $Q = MM^T$. By construction, the matrices D_{uu} and D_{vv} are symmetric, nonnegative definite matrix functions of the elements of u, resp. v. If they would be fixed matrices (i.e. independent of u and v), the Riemannian SVD would reduce to the so-called *Restricted Singular Value Decomposition* (see e.g. [3]). By analogy, we call u and v left resp. right singular vectors and σ the corresponding singular value of the Riemannian SVD. In the case of the ordinary TLS problem (1), we have $D_{uu} = I_p$ and $D_{vv} = I_q$. One can easily show that

$u^T D_{vv} u = v^T D_{uu} v$ always, as from the Commuting Lemma it follows that $u^T D_{vv} u = u^T T_v Q T_v^T u = v^T T_u^T Q T_u v = v^T D_{uu} v$. The "filtered" data (i.e. the least squares estimates of the unknown "exact" data) can be estimated from $b = a - QT_v^T u\sigma$. The linearly structured matrix B generated from the elements of the vector b is rank deficient because $Bv = T_v b = T_v(a - QT_v^T l) = Av - D_{vv} u\sigma = 0$; The left singular vector u is the normalized vector l of Lagrange multipliers corresponding to the rank deficiency constraint. From the Riemannian SVD (9) we easily find, assuming that D_{vv} is invertible:

$$\sigma^2 = v^T A^T D_{vv}^{-1} Av \; . \tag{10}$$

This equation can be interpreted as the weighted norm of the so-called equation error. Assume that, in the noiseless case $\hat{A}\hat{v} = 0$. From the noise model specified above, we find that $A\hat{v} = (\hat{A} + \tilde{A})\hat{v} = \tilde{A}\hat{v} = T_{\hat{v}}\tilde{a} = T_{\hat{v}} M \tilde{\tilde{a}}$. For obvious reasons, $A\hat{v}$ is called the equation error. Its covariance matrix is given by $\mathbf{E}(A\hat{v}\hat{v}^T A) = T_{\hat{v}} Q T_{\hat{v}}^T = D_{\hat{v}\hat{v}}$. Therefore, the objective function in (10) is the weighted norm of the equation error, in which the weight is the inverse covariance matrix of the equation error.

Finally, it can be shown that the vector of residuals $(b - a)$ is "$Q^\dagger$-orthogonal" to the vector with filtered data b as one can show that $b^T.Q^\dagger.(b - a) = 0$. For more details and additional properties, we refer to [4] [5].

5. Elementwise weighted TLS

So far, we have obtained the general formulas for solving a structured and/or weighted TLS problem, leading to the Riemannian SVD (9). In this section, we derive the weighting matrices D_{uu} and D_{vv} for an elementwise weighted TLS problem. Let $A \in \mathbb{R}^{p\times q}$ be a given data matrix; Assume that the noise is additive, $A = \hat{A} + \tilde{A}$, and that the elements of $\tilde{A}$ are zero mean, Gaussian, independently distributed scalar random variables with noise variance σ_{ij}^2 for element (i, j). When approximating in a least squares sense, this given data matrix A, by a rank deficient matrix B, the principle of maximum likelihood leads to the following constrained optimization problem:

$$\min_{b_{ij}\in\mathbb{R}\,,\, v\in\mathbb{R}^q} \; \frac{1}{2}\sum_{i=1}^{p}\sum_{j=1}^{2}(a_{ij} - b_{ij})^2/\sigma_{ij}^2 \text{ subject to } \left\{ \begin{array}{l} B\,.\,v = 0\,, \\ v^T\,.\,v = 1\,. \end{array} \right. \tag{11}$$

Let Σ be the $p \times q$ matrix that has σ_{ij}^2 as its (i, j)-th element. Let diag(.) be the operator that turns a vector into a diagonal matrix. By specializing the general derivation of Section 4 to this case, one obtains the Riemannian SVD for the weighted TLS problem with $D_{uu} =$

$\text{diag}(\Sigma^T \text{diag}(u)u)$, $D_{vv} = \text{diag}(\Sigma \text{diag}(v)v)$. The rank deficient matrix B can be reconstructed from $B = A - \text{diag}(u)\Sigma\text{diag}(v)\sigma$, where σ is the smallest singular value of (9) and u and v are the corresponding left and right singular vectors. There are several "extreme" cases that one can consider:

- Choosing $\sigma_{ij} \to 0$ implies that there is no noise on the corresponding element in A, hence that element is considered to be exact, and will not be modified, i.e. $a_{ij} = b_{ij}$; As an example, consider the matrix A

$$A = \begin{bmatrix} 1 & \boxed{2} & 3 \\ \boxed{-5} & 6 & \boxed{7} \\ 0 & 2 & -1 \\ \boxed{3} & 4 & 1 \end{bmatrix},$$

  where elements in a box can not be modified, when trying to least squares approximate A by a rank deficient matrix. The matrices D_{uu} and D_{vv} for this example are

$$D_{uu} = \text{diag}\begin{bmatrix} u_1^2 + u_3^2 \\ u_2^2 + u_3^2 + u_4^2 \\ u_1^2 + u_3^2 + u_4^2 \end{bmatrix}, D_{vv} = \text{diag}\begin{bmatrix} v_1^2 + v_3^2 \\ v_2^2 \\ v_1^2 + v_2^2 + v_3^2 (= 1) \\ v_2^2 + v_3^2 \end{bmatrix}.$$

- By taking $\sigma_{ij} \to \infty$, we take the noise variance on element (i,j) in A to be extremely large; In this way, one can tackle *missing observations*, i.e. b_{ij} will be filled in "automatically", without taking into account the corresponding element a_{ij} in A (Said in other words, because $\sigma_{ij} \to \infty$, the corresponding term does not contribute anything in the weighted objective function (11), leaving the choice for b_{ij} independent of a_{ij} (but constrained to make the matrix B rank deficient)).

- By taking $\sigma_{ij} = a_{ij}$, we get an objective function that minimizes the sum-of-relative-errors-squared, instead of "absolute" errors.

- One can also consider the case in which every noise element $\tilde{a}_{ij}$ is correlated with every other noise element $\tilde{a}_{kl}$, as expressed in a correlation matrix $Q \in \mathbb{R}^{pq \times pq}$. For details we refer to [4].

- When Σ is a rank one matrix, i.e. $\sigma_{ij} = \xi_i \cdot \eta_j$, for positive scalars $\xi_i, \eta_j \in \mathbf{R}_0^+$, the solution to the weighted TLS problem (11) can be derived from the SVD of the diagonally scaled matrix with elements $a_{ij}/(\xi_i \cdot \eta_j)$.

6. Double Hankel structured TLS

The "errors-in-variables" formulation of the linear dynamic system identification problem, for models of the form (3) is the following: Given noise corrupted scalar observations (u_k, y_k) of the input, resp. output, of a single-input single-output system. A least squares estimate w_k of u_k and z_k of y_k will be determined as follows. Let w, z, u, y be vectors with the scalar w_k, z_k, y_k, u_k. The structured TLS problem is then

$$\min_{w,z,a,b} \|y - z\|_2^2 + \|u - w\|_2^2 \text{ subject to } \begin{cases} [\,Z\ \ W\,]\begin{bmatrix} a \\ -b \end{bmatrix} = 0\,, \\ a^T a + b^T b = 1\,. \end{cases} \tag{12}$$

Here, the vectors a and b contain the coefficients of the difference equation that relates the "filtered" inputs w_k to the "filtered" outputs z_k as in (5). From a similar derivation as the general one in Section 4, one can show that one has to find the minimal singular triplet of the Riemannian SVD

$$\begin{aligned} [\,Y\ \ U\,]\begin{bmatrix} a \\ -b \end{bmatrix} &= (D_{aa} + D_{bb})\, u\, \sigma\,, & a^T a + b^T b = 1\,, \\ \begin{bmatrix} Y^T \\ U^T \end{bmatrix} u &= \begin{bmatrix} D_{uu} & 0 \\ 0 & D_{uu} \end{bmatrix}\begin{bmatrix} a \\ -b \end{bmatrix}\sigma\,, & u^T(D_{aa} + D_{bb})u = 1\,. \end{aligned}$$

Here $D_{aa} = T_a T_a^T$, $D_{bb} = T_b T_b^T$ and $D_{uu} = T_u T_u^T$, in which the matrices T_a, T_b and T_u are banded Toeplitz. More details, such as expressions for the filtered input and output sequences, the proof that the residuals and filtered data are orthogonal, and the fact that the residuals themselves can be described by a linear difference equation, can be found in [6] [7]. As an extension to these results, one might also include weights in the objective function (12). In this way, one can treat missing observations, time-varying observation noise variances, etc ... A general framework for doing so, including the incorporation of "unobserved latent" inputs, has been formulated in [12].

7. Rank 2 reduction

What if we want to reduce the rank of a linearly structured matrix A by 2, instead of by 1 ? Obviously, we now have to extend the constrained optimization problem (8) with some additional constraints:

$$\min_{b\in\mathbb{R}^N\,,\ v_1,v_2\in\mathbb{R}^q\,,\ e\in\mathbb{R}^{N_1}} \frac{1}{2}e^T e \text{ subject to } \begin{cases} b = a - Me\,, \\ B\,.\,(v_1\ v_2) = 0\,, \\ v_1^T\,.\,v_1 = 1\,, \\ v_2^T\,.\,v_2 = 1\,, \\ v_1^T\,.\,v_2 = 0\,. \end{cases}$$

We leave it to the reader to derive that the solution follows from the singular triplet corresponding to the smallest singular value of the "double sized" Riemannian SVD

$$\begin{bmatrix} A & 0 \\ 0 & A \end{bmatrix}\begin{bmatrix} v_1 \\ v_2 \end{bmatrix} = \begin{bmatrix} D_{v_1v_1} & D_{v_1v_2} \\ D^T_{v_1v_2} & D_{v_2v_2} \end{bmatrix}\begin{bmatrix} u_1 \\ u_2 \end{bmatrix}\sigma ,$$
$$\begin{bmatrix} A^T & 0 \\ 0 & A^T \end{bmatrix}\begin{bmatrix} u_1 \\ u_2 \end{bmatrix} = \begin{bmatrix} D_{u_1u_1} & D_{u_1u_2} \\ D^T_{u_1u_2} & D_{u_2u_2} \end{bmatrix}\begin{bmatrix} v_1 \\ v_2 \end{bmatrix}\sigma ,$$

with obvious definitions for the matrix $D_{v_iv_j}$ and $D_{u_iu_j}$ (e.g. $D_{v_1v_2} = T_{v_1}QT^T_{v_2}$), and with the normalization constraints

$$\begin{bmatrix} u_1^T \; u_2^T \end{bmatrix}\begin{bmatrix} D_{v_1v_1} & D_{v_1v_2} \\ D^T_{v_1v_2} & D_{v_2v_2} \end{bmatrix}\begin{bmatrix} u_1 \\ u_2 \end{bmatrix} = 1 , \begin{bmatrix} v_1^T \\ v_2^T \end{bmatrix}\begin{bmatrix} v_1 \; v_2 \end{bmatrix} = \begin{bmatrix} 1 & 0 \\ 0 & 1 \end{bmatrix} .$$

8. Conclusions and some research challenges

In this contribution, we have presented a short survey of recent results on the formulation of S/W-TLS problems via the so-called Riemannian SVD, which is a "nonlinear generalized" SVD. A lot of work remains to be done in constructing efficient algorithms to find the "smallest" singular triplet (u, σ, v) of (9), from which the optimal solution to the S/W-TLS can be calculated. These algorithms also will have to cope with the fact that the optimization problem (8) is non-convex, hence will typically have several local minima. In [4] [5] we have described an algorithm, the basic inspiration of which is the classical power method:

1. Initialize $u^{[0]}, v^{[0]}$;
2. At iteration step k :
   2.1. Calculate $D_{u^{[k]}u^{[k]}}$, $D_{v^{[k]}v^{[k]}}$;
   2.2. Solve sets of linear equations for $u^{[k+1]}$, $v^{[k+1]}$;
   2.3. Normalize $u^{[k+1]}$, $v^{[k+1]}$ properly;
3. Test for convergence .

When convergent, the convergence rate is linear, just like the power method. For sure, a lot of progress could be made here. For instance, the power method can be interpreted as an alternating weighted least squares optimization method, with "variable metrics" D_{uu} and D_{vv} in each iteration step. Maybe this interpretation helps in analysing algorithms and their convergence behavior. IQML-like algorithms have been analysed in [11]. One could also try to gain insight by analysing the so-called gradient flows that can be derived from the optimization formulation (8) (see e.g. [2]), in which ideas from differential geometry could play a role (the matrices D_{uu} and D_{vv} are positive definite metrics in the tangent spaces of certain manifolds). For large problems, one can also exploit the structure of the matrices D_{uu} and D_{vv} and of the data matrix A (e.g. by using displacement rank notions), in order to speeden up the calculations. Ideas in this direction and in this context have been proposed in [14]. Also algorithms to solve the rank-2 or higher reduction still need to be developed.

As for the statistical properties of the S/W-TLS estimates, we have to distinguish between the so-called "incidental parameters" (the elements of the structured least squares rank deficient matrix B) and the "model parameters", the elements of the vector v. We conjecture that, under the assumptions put forward in Section 3, the error $\tilde{v}$ on the estimate v, for fixed q and $p \to \infty$, asymptotically behaves like $\tilde{v} \sim \mathcal{N}(0, (A^T D_{vv}^{-1} A - D_{uu}\sigma^2)^\dagger)$. Some inspiration may be collected from [10]. As $p \to \infty$, it could also be interesting to investigate recursive algorithms to update the solutions of (9).

Finally, while in this paper, we have presented the "scalar" versions of the identification problems (i.e. for systems with scalar inputs and/or outputs), the derivation and analysis of the S/W-TLS problems corresponding to systems with vector inputs and outputs, remains to be done. There is a close relation with so-called global total least squares problems in the "behavioral" framework (see e.g. [9]).

Acknowledgments

Dr. Bart De Moor is a full professor at the Katholieke Universiteit Leuven. This work was supported by grants and funding from the Research Council of the K.U.Leuven (PhD/postdoc grants, IDO, GOA Mefisto-666), the Foundation for Scientific Research Flanders (FWO)(PhD and postdoc grants and projects (G.0115.01, G.0197.02, G.0407.02), research communities (ICCoS, ANMMM)), the Flemish regional government (Bilateral Coll., IWT-project Soft4s, Eureka projects Synopsis, Impact, FLiTE, STWW-project Genprom, GBOU-project McKnow, IWT PhD grants), the Belgian Federal Government (DWTC, IUAP IV-02, Sustainable Development MD/01/024), the European Commission (TMR Alapedes, Ernsi) and industry supported direct contract research (Electrabel, ELIA, ISMC, Data4s).

References

[1] A. Bultheel and B. De Moor. Rational approximations in linear systems and control. *Journal of Computational and Applied Mathematics*, Special Issue on Numerical Analysis in the 20-th Century, Vol. I: Approximation Theory, 121:355–

378, September 2000.

[2] J. Cambre and B. De Moor. Solving the structured total least squares problem using the Riemannian SVD. *Mathematical Theory of Networks and Systems*, A. Beghi, L. Finesso, G. Picci (Editors), MTNS, 715–718, 1998, Padua, Italy.

[3] B. De Moor and G.H. Golub. The restricted singular value decomposition: properties and applications. *SIAM Journal on Matrix Analysis and Applications*, 12(3)401–425, July 1991.

[4] B. De Moor. Structured total least squares and L_2 approximation problems, *Lin. Alg. Appl.*, Spec. issue on Numerical Linear Algebra in Systems & Control, (eds: Van Dooren, Ammar, Nichols, Mehrmann), 188/189:163–207 1993.

[5] B. De Moor. Total least squares for affinely structured matrices and the noisy realization problem, *IEEE Transactions in Signal Processing*, 42(11):3104–3113, 1994.

[6] B. De Moor and B. Roorda. L_2-optimal linear system identification: Dynamic total least squares for SISO systems. *Proc. of 33rd IEEE* CDC, 2874–2880, Florida, December 1994.

[7] B. De Moor. Linear system identification, structured total least squares and the Riemannian SVD. in *Recent Advances in Total Least Squares Techniques and Errors-in-Variables modeling*, ed. S. Van Huffel, SIAM, 225–239, 1997.

[8] B. De Moor. Structured total least squares for Hankel matrices. in *Communications, Computation, Control and Signal Processing*, eds: A. Paulraj, V. Roychowdhury, C. Schaper, Kluwer Academic Publishers, 243–258, 1997.

[9] C. Heij and W. Scherrer. Consistency of system identification by global total least squares. *Automatica*, 35:993–1008, 1999.

[10] E. Karlsson, T. Söderström and P. Stoica. The Cramer-Rao lower bound for noisy input-output systems. *Signal Processing*, 80:2421–2447, 2000.

[11] P. Lemmerling, L. Vanhamme, S. Van Huffel and B. De Moor. IQML-like algorithms for solving structured total least squares problems. *Signal Processing*, 81:1935–1945, 2001.

[12] P. Lemmerling and B. De Moor. Misfit versus latency. *Automatica*, 37:2057–2067, 2001.

[13] A. Isidori and A. Ruberti. Realization theory of bilinear systems, in *Geometric Methods in System Theory*, edited by D.Q.Mayne and R.W. Brockett, Nato Advanced Study Institute Series, 83–130, 1973.

[14] N. Mastronardi. *Fast Algorithms for Hankel/Toeplitz structured total least norm problems and applications in parameter estimation.* PhD Thesis, Dept. EE, K.U. Leuven, May 2001.

[15] S. Van Huffel and J. Vandewalle. *The Total Least Squares Problem: Computational Aspects and Analysis.* SIAM Philadelphia, 1991.

[16] S. Van Huffel (Ed.). *Recent Advances in Total Least Squares Techniques and Errors-in-Variables modeling*, SIAM, 225–239, 1997.

[17] M. Wonham. *Linear Multivariable Control: A Geometric Approach.* New York Springer, 1974.

FREQUENCY-DOMAIN TLS AND GTLS ALGORITHMS FOR MODAL ANALYSIS APPLICATIONS

Peter Verboven, Patrick Guillaume and Eli Parloo
Mechanical Engineering department (TW-WERK), Vrije Universiteit Brussel
Pleinlaan 2, B-1050 Brussel, Belgium
{ Peter.Verboven, Patrick.Guillaume } @vub.ac.be

Abstract This contribution focuses on the area of modal analysis and studies the applicability of Total Least Squares algorithms for the estimation of modal parameters in the frequency domain from Input-Output Fourier data. These algorithms provide an alternative for the frequency response function based estimators, especially in the case that the measurement data is characterized by low signal-to-noise ratios, short data sequences and leakage. In this paper it is shown how frequency-domain TLS and GTLS estimators can be numerically optimized to handle large amounts of modal data. Using an errors-in-variables noise model, a linear approximation is necessary in order to obtain a fast implementation of the GTLS algorithm. The validity of this approximation is a function of the signal-to-noise ratio of the input Fourier data and is evaluated by means of Monte Carlo simulations and experimental data.

Keywords: modal analysis, generalized TLS, system identification.

1. Introduction

Modal parameter estimation has become a common technique used by engineers to analyze complex mechanical and civil structures such as cars, aircraft, industrial machinery, bridges, buildings, ... The representation of the dynamical behavior of such structures by means of the modal parameters (*i.e* damped resonance frequencies, damping ratio and mode shapes) allows a better physical interpretation and a way to get condensed experimental data suitable for a variety of mechanical design purposes such as finite element model correlation, structural redesign as well as for monitoring and control applications.

S. Van Huffel and P. Lemmerling (eds.), Total Least Squares and Errors-in-Variables Modeling, 305–318.

The use of Total Least Squares (TLS) techniques for frequency-domain identification has already been extensively discussed in several publications (both for SISO and MIMO models) [11, 5]. However, modal analysis has its typical characteristics such as extensive data sets (e.g., systems with 4 inputs and 1000 outputs), high modal density with closely spaced modes, high dynamical range, and low signal-to-noise ratio's. Another important feature of modal analysis identification is the use of so-called stabilization charts, useful to separate the physical modes from the mathematical ones. To construct these charts, pole estimates for different model orders are required. Consequently, adapted TLS algorithms need to be developed.

The common approach for experimental modal identification starts from frequency response functions (FRF). Based on the inverse FFT of this FRF data, the time-domain Least Squares Complex Exponential (LSCE) parametric estimator [3] yields the poles and participation factors, while the mode shapes are found during a second step by means of a Least Squares estimation in the frequency domain (LSFD) [3]. In [2, 13] FRF-based TLS estimators optimized for modal analysis applications are presented. However, as is well known, FRF-based identification can have serious drawbacks. First of all, FRF data is obtained using an non-parametric FRF estimator, such as the H_1, H_2 or H_v estimator [3]. Restrictive assumptions about the noise model make these FRF estimators inconsistent. FRF-based errors-in-variables modelling is possible when using the H_{iv} for arbitrary excitation [12] or H_{ev} in case of periodic excitation [1]. Furthermore, it is not meaningful to use these average-based FRF estimators when only short data sequences are available, since a minimal frequency resolution is required for accurate system identification. This is often the case for flight flutter testing and structural health monitoring practices due to cost, safety, and time-invariance reasons.

In this contribution the FRF-based TLS approach is extended to Errors-in-Variables modelling based on Input/Output Fourier modal data. In case of the GTLS, the measurement noise can be incorporated in the estimation process, which improves the accuracy of the parameter estimates. A similar approach that was used for numerical optimization of FRF-based TLS algorithms [2, 13] is applied (Section 3). Nevertheless, some important differences compared to these FRF-based algorithms can be noticed. Once also input noise sources are taken into account, these algorithms are not directly applicable. In the next section it is shown how this can be overcome and the validation of the TLS and GTLS algorithms is done by means of Monte Carlo simulations (Section 4) and experimental data from an aircraft slat track (Section 5).

2. Frequency domain identification - MIMO systems

2.1 Parametric model

In this contribution, the frequency response function between output o and input i (for $o = 1, 2, \ldots, N_o$ and $i = 1, 2, \ldots, N_i$) is modelled using a scalar matrix fraction description model [4]

$$\hat{H}_{oi}(\omega_k, \theta) = \frac{n_{oi}(\omega_k, \theta)}{d(\omega_k, \theta)} \tag{1}$$

with $n_{oi}(\omega_k, \theta) = \sum_{j=0}^{n} \Omega_j(\omega_k) b_{oij}$ the numerator polynomial for output o and input i and $d(\omega_k, \theta) = \sum_{j=0}^{n} \Omega_j(\omega_k) a_j$ the common-denominator polynomial. Using a discrete-time domain model, the polynomial basis functions are given by $\Omega_j(\omega_k) = e^{-\mathrm{i}\omega_k T_s j}$ with T_s the sampling period, $k = 1, \ldots, N_f$ (N_f the number of spectral lines) and $j = 1, \ldots, n$ (n order of the polynomials). The polynomial coefficients a_j and b_{oij} are the (unknown) parameters θ to be estimated.

2.2 Errors-in-variables framework

In practice, the measured responses (velocities and/or accelerations) as well as forces are affected by *"errors"*. Consequently, what is observed are not the *"true"* input-output (I/O) Fourier data, $Z_0 = [F_0^H, X_0^H]^H$, but Z_0 together with some random perturbations, $E_Z = [E_F^H, E_X^H]^H$. The measured I/O Fourier data, $Z = [F^H, X^H]^H$, are defined by the following *"errors-in-variables"* stochastic model with $Z(\omega_k)$, $Z_0(\omega_k)$, $E_Z(\omega_k) \in \mathbb{C}^{N_i+N_o}$.

$$\left.\begin{array}{l} Z(\omega_k) = Z_0(\omega_k) + E_Z(\omega_k) \\ {[H_0(\omega_k), -I_{N_o}]}\, Z_0(\omega_k) = 0 \end{array}\right\} \quad k = 1, \ldots, N_f \tag{2}$$

The errors $E_Z(\omega_k)$ are assumed to be complex normally distributed with *a-priori* known covariance matrix

$$C_{E_Z}(\omega_k) = \mathbf{E}\{E_Z(\omega_k) E_Z(\omega_k)^H\} = \left[\begin{array}{cc} C_{E_F}(\omega_k) & C_{E_F E_X}(\omega_k) \\ C_{E_X E_F}(\omega_k) & C_{E_X}(\omega_k) \end{array} \right] \tag{3}$$

Taking, besides the noise variances, also the cross-correlations of the errors on the I/O Fourier data into account, can further decrease the uncertainty of the estimates. However, in the case that the output correlations are considered, the time efficient implementation, presented in Section 3, cannot be applied. Moreover, since the covariance matrix $C_{E_Z}(\omega_k)$ is usually obtained by means of measurements, obtaining a consistent estimate of the full output covariance matrix $C_{E_X}(\omega_k)$ for a large number of responses (N_o) requires at each response location at least N_o measurements [8] and as a result only the output variances are used in practice. On the other hand, since the number of inputs is typically small ($N_i \leq 5$), input correlations as well as input-output correlations can be experimentally determined, resulting in $C_{E_F}(\omega_k)$ and $C_{E_X E_F}(\omega_k)$ being full matrices. In the next sections, correlations are only considered between the inputs, although a fast implementation is still possible when also the input-output correlations are used.

2.3 Model equations

Using (1) and (2) under the noise assumptions made in Section 2.2, the model equations are written for output o and spectral line k as (Levi's method)

$$\mathrm{w}_o(\omega_k)\left(n_o(\omega_k,\theta)\{F(\omega_k)\} - d(\omega_k,\theta)X_o(\omega_k)\right) \approx 0 \tag{4}$$

where $\{F(\omega_k)\}$ represents the input Fourier vector and $n_o(\omega_k,\theta)$ row o of the $(N_o \times N_i)$ numerator polynomial matrix $N(\omega_k,\theta)$. By introducing an adequate (frequency-dependent) weighting function $\mathrm{w}_o(\omega_k)$ in the equations (4), the quality of the parameter estimates can often be improved [5]. By estimating simultaneously initial conditions and the system model parameters, it is possible to deal with arbitrary signals in the frequency-domain without any approximation and under the same assumptions as in the time-domain [6]. Using this approach, the proposed I/O GTLS estimator can be made robust against errors due to leakage effects, while this approach is not applicable for FRF-based estimators. As the equations (4) are linear-in-the-parameters and in the Fourier data, they can be reformulated as $J\theta \approx 0$

$$\begin{bmatrix} P_1 & 0 & 0 & \cdots & Q_1 \\ 0 & P_2 & 0 & \cdots & Q_2 \\ \vdots & \vdots & \ddots & \vdots & \vdots \\ 0 & 0 & \cdots & P_{N_o} & Q_{N_o} \end{bmatrix} \begin{bmatrix} \theta_{n_1} \\ \theta_{n_2} \\ \vdots \\ \theta_{n_{N_o}} \\ \theta_d \end{bmatrix} \approx 0 \tag{5}$$

with J the Jacobian matrix having $N_f.N_o$ rows and $(n+1)(N_o.N_i+1)$ columns. A similar formulation can be obtained for real coefficients for which the number of equations is doubled. For $o = 1,\ldots,N_o$ and $i = 1,\ldots,N_i$, the entries of the sub-matrices in (5) $P_o = [P_{o1},\ldots,P_{oN_i}]$ $(N_f \times N_i.(n+1))$ and Q_o $(N_f \times (n+1))$ are given by $(k = 1,\ldots,N_f$ and $j = 1,\ldots,n)$

$$\begin{aligned} P_{oi}(k,j) &= \mathrm{w}_o(\omega_k)\Omega_j(\omega_k)F_i(\omega_k) \\ Q_o(k,j) &= -\mathrm{w}_o(\omega_k)\Omega_j(\omega_k)X_o(\omega_k) \end{aligned} \tag{6}$$

while the parameter vector entries contain the (unknown) coefficients

$$\theta_{n_o} = [b_{o10}, b_{o11}, \ldots, b_{oN_in}]^T \quad , \quad \theta_d = [a_0, a_1, \ldots, a_n]^T \tag{7}$$

3. Fast weighted generalized TLS algorithm

This paragraph, discusses a fast implementation for solving the model equations (5) in a Weighted Generalized Total Least Squares (WGTLS) approach.

Given that $N_f \gg n$, formulating the normal equations $J^H J\theta \approx 0$ results in a more compact formulation of the identification problem ($J^H J$ having $(n+1)(N_o.N_i)$ rows and columns)

$$\begin{bmatrix} R_1 & 0 & 0 & \cdots & S_1 \\ 0 & R_2 & 0 & \cdots & S_2 \\ \vdots & \vdots & \ddots & \vdots & \vdots \\ 0 & 0 & \cdots & R_{N_o} & S_{N_o} \\ S_1^H & S_2^H & \cdots & S_{N_o}^H & T \end{bmatrix} \begin{bmatrix} \theta_{n_1} \\ \theta_{n_2} \\ \vdots \\ \theta_{n_{N_o}} \\ \theta_d \end{bmatrix} \approx 0 \tag{8}$$

with matrices R_o having $(n+1).N_i$ rows and columns, S_o having $(n+1).N_i$ rows and $(n+1)$ columns, and $T = \sum_o T_o$ having $(n+1)$ rows and columns.

Using a discrete time-domain model often leads to a well-conditioned Jacobian matrix, which justifies the explicit calculation of the normal equations. In the case of a continuous time-domain model, orthogonal polynomials (*e.g.* Forsythe or Chebychev) can be used to improve the numerical conditioning [7]. Moreover, if the frequencies are uniformly distributed, a closer look at equations (5) and (8) learns that the Fast Fourier Transform (FFT) algorithm can be used to compute the entries of matrices R_o, S_o and T_o [9, 2], with these matrices also having Toeplitz structure, which is beneficial for both computation time and memory usage.

Considering a Generalized Total Least Squares approach, equation (8) can also be formulated under the form of a (generalized) eigenvalue problem

$$J^H J\theta = \lambda C\theta \tag{9}$$

with C being the covariance matrix containing a measure for the errors on the matrix $J^H J$ as a result from measurement noise on the I/O Fourier data. Considering the errors-in-variables model with the noise assumptions made in Section 2.2, the covariance matrix C is given by (with $C_T = \sum_o C_{T_o}$)

$$C = \text{blockdiag}(C_{R_1}, ..., C_{R_{N_o}}, C_T) \tag{10}$$

for which the matrix formulation of C is similar as for the normal equations (8), taking benefit of the Fast Fourier Transform algorithm and the Toeplitz structure of the sub-matrices.

In [2], it is shown how the computation time can be further significantly reduced by eliminating the numerator coefficients θ_{n_o}, since only the denominator coefficients θ_d are required to compute the system poles. In the case of I/O Fourier data, elimination of the numerator coefficients θ_{n_o} in (9) yields

$$\theta_{n_o} = -\left[I - \lambda R_o^{-1} C_{R_o}\right]^{-1} R_o^{-1} S_o \theta_d \quad , \; o = 1, \ldots, N_o \tag{11}$$

from which it can be seen that an additional term in λ (with λ unknown) appears in the case that also the input noise sources are taken into account ($C_{R_o} \neq 0$). Notice that for FRF-based GTLS identification this problem does not occur since errors are only present on the FRF data, only present in the last $(n+1)$ columns/rows of the normal equations [2]. Consequently, it is not possible to formulate a generalized eigenvalue problem in the denominator coefficients θ_d by eliminating θ_{n_o} from the last n equations of (9), *i.e*

$$\sum_{o=1}^{N_o} S_o^H \theta_{n_o} + T\theta_d = \lambda C_T \theta_d \tag{12}$$

and as a result, the exact solution of the GTLS identification problem (9) is only given by solving (9). However, the computational load for solving this eigenvalue problem , in the case of typical modal analysis applications (*i.e.* $N_o = 500$, $N_i = 3$, $n = 50$), is not acceptable in practice.

One obvious way to benefit from the elimination approach is by considering only errors on the measured output sequences (*i.e.* H_1 noise-model assumptions [3]), with all matrices C_{R_o} equal to zero in (10) and (11) resulting after eliminating the numerator coefficients in the following generalized eigenvalue problem in θ_d

$$\left[T - \sum_{o=1}^{N_o} S_o^H R_o^{-1} S_o\right] \theta_d = \lambda C_T \theta_d \tag{13}$$

Once the θ_d coefficients are known, equation (11) (with $C_{R_o} = 0$) can be used to derive all numerator coefficients. However, ignoring the input noise sources has the important disadvantage of loosing the EV noise-model characteristics together with the consistency properties of the GTLS approach.

Table 1. Poles and residues of theoretical SISO transfer function.

mode	*pole*	*residue*
1	-1.5234E-3 + 0.7500E+1i	3.9282E-8 + 2.5015E-3i
2	-1.3441E-3 + 1.2167E+1i	1.2874E-7 + 1.4997E-3i
3	-0.7553E-3 + 2.2164E+1i	-4.9174E-6 + 1.1966E-3i
4	-1.2464E-3 + 2.3837E+1i	1.0541E-5 + 1.3087E-3i
5	-0.5506E-3 + 2.5336E+1i	-7.9402E-7 + 1.7064E-3i
6	-1.1167E-3 + 4.4167E+1i	1.2672E-9 + 1.9000E-3i

Therefore, another option was considered in order to combine the time-efficient elimination process with the advantages of EV (GTLS) identification. Using a Taylor expansion, the matrix $\left[I - \lambda R_o^{-1} C_{R_o}\right]^{-1}$ in equation (11) can be approximated as

$$\left[I - \lambda R_o^{-1} C_{R_o}\right]^{-1} \cong I + \lambda R_o^{-1} C_{R_o} + (\lambda R_o^{-1} C_{R_o})^2 + \ldots \qquad (14)$$

Introducing this approximation for the first order in (11) and substitution in (12) results now again in a generalized eigenvalue problem in θ_d

$$\left[T - \sum_{o=1}^{N_o} S_o^H R_o^{-1} S_o\right] \theta_d = \lambda \left[C_T + \sum_{o=1}^{N_o} S_o^H R_o^{-1} C_{R_o} R_o^{-1} S_o\right] \theta_d \qquad (15)$$

Compared to (13), an additional term in λ appears after linearizing, containing a measure of the errors on the input Fourier data, *i.e.* C_{R_o}. Instead of solving (9) directly, this elimination approach results in an important reduction of the computation time since the GSVD algorithm is applied to much smaller matrices (dimensions reduced by $N_o^2.N_i^2$). From equation (14) it follows that the linear approximation will be valid in the case that $\text{eig}(R_o^{-1} C_{R_o}) \ll \lambda$, where $R_o^{-1} C_{R_o}$ is a measure for the signal-to-noise ratio (SNR) of the input Fourier data. In the next section the validity of this linear approximation is studied by means of Monte Carlo simulations.

4. Monte Carlo simulation results

A Single Input Single Output system having 6 modes with poles and residues given in Table 1 was used to generate an I/O data set of 1200 equally distributed frequencies in the band of [0.6Hz;7.9Hz] : *i.e.* $F_0(\omega_k) = 1$, $X_0(\omega_k) = H_0(\omega_k)$ $(k = 1, \ldots, 1200)$. Independent zero-mean Gaussian noise was added to both the input and output data. In order to evaluate the validity of approximation (14) in the generalized eigenvalue problem (15), the variance of the input noise was increased over 21 logarithmically distributed steps in the range of [1E-8,1E+0] while the output noise level remained constant with a variance of 1E-6. For each I/O noise combination, 1000 disturbed data sets were generated and for each set the modal parameters of 20 modes (poles, residues) were estimated using the following estimators: GTLS eqn.(9), FGTLS eqn.(15), FGTLSX eqn.(13), TLS eqn.(9) with $C = I$ and FTLS eqn.(13) with $C_T = I$, while the weighting function $w_o(\omega_k) = 1$. The accuracy of the different estimators can be compared with respect to their squared bias, total variance and mean squared error (MSE) ([10], p.244) for both the estimated poles and residues. The Monte Carlo simulation results can be compared for the different estimators as shown in Figure 1. The important errors in case of both TLS estimators are due to the high bias errors, originating from the inconsistency properties of the TLS approach, since no noise covariance information is taken into account. Considering only output noise results in the FGTLSX, improving the accuracy as long as the input variance is small. However, once the effect of the input noise outweighs the output noise, the errors for the FGTLSX increase significantly. Since input noise is disregarded, the FGTLSX is still inconsistent, explaining the increase in bias and the agreement with the TLS results for high input noise variances. The gain in accuracy by taking input noise into account is clearly illustrated by comparing the FGTLSX and the GTLS where the MSE is 5dB to 10dB smaller for the GTLS. The good agreement of the errors for the GTLS and FGTLS indicates the validity of the approximation (14). Only for very high variances of the input noise (SNR around 0dB), the errors are slightly higher for the FGTLS. This result proves that no loss in accuracy is encountered when using the approximation (within practical noise levels), making the fast implementation of the fast algorithm of the (W)GTLS suitable for an accurate analysis of large modal data sets.

5. Experimental results

The device under test is a slat track (Figure 2), which is a safety critical component in the wing of an aircraft, used to enlarge the wing surface during takeoff and landing when airspeed is reduced. By means of a scanning Laser Doppler Vibrometer setup (Figures 3, 4) the response to a single shaker input was measured in 352 points. A periodic signal excitation in a frequency band [0Hz;4kHz] with 3200 spectral lines was applied. Notice that this setup completely satisfies the errors-in-variables noise model defined in Section 2.2.

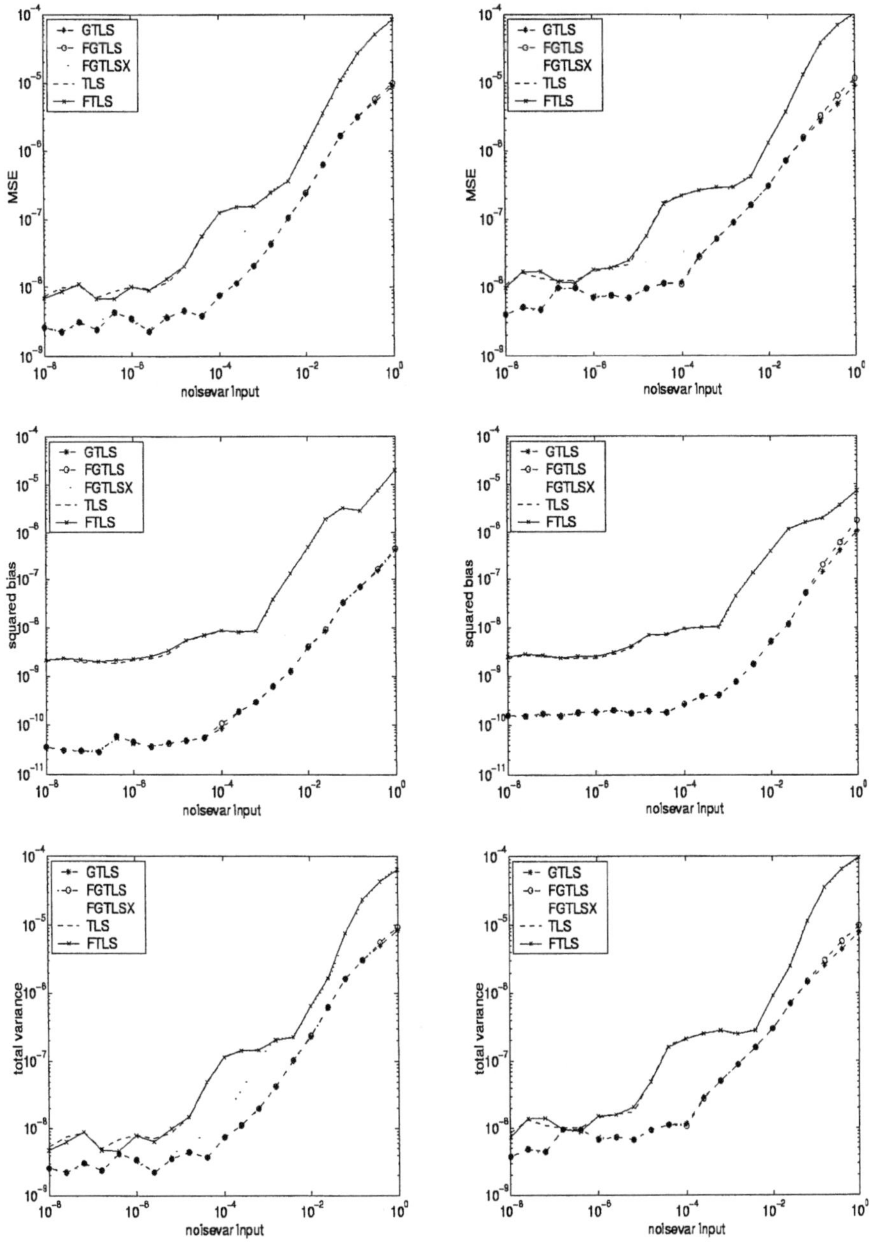

Figure 1. Monte Carlo simulation results: mean squared error (MSE), squared bias and total variance for poles (left) and residues (right).

Figure 2. Slat tracks mounted in wing of Airbus A320.

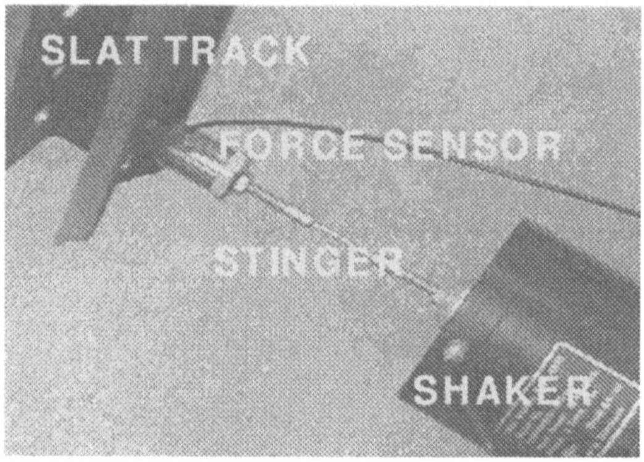

Figure 3. Force (input) measurement with shaker, stinger and force sensor attached to the slat track.

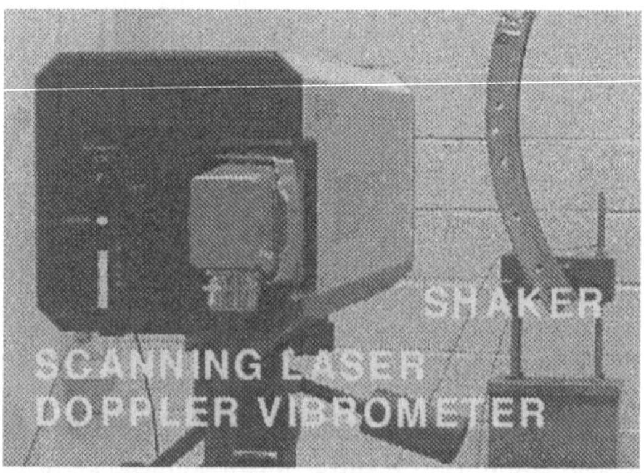

Figure 4. Velocity (output) measurement with scanning laser Doppler vibrometer.

Figure 5 (top) shows the ratio of a measured velocity and force and the estimated transfer function model by means of the FTLS and FGTLS for a model with 22 modes. The FGTLS has a better performance, explained by the use of noise covariance matrix information during the estimation process, making the FGTLS consistent and less sensitive to measurement noise. For modal analysis, a stabilization chart is an im-

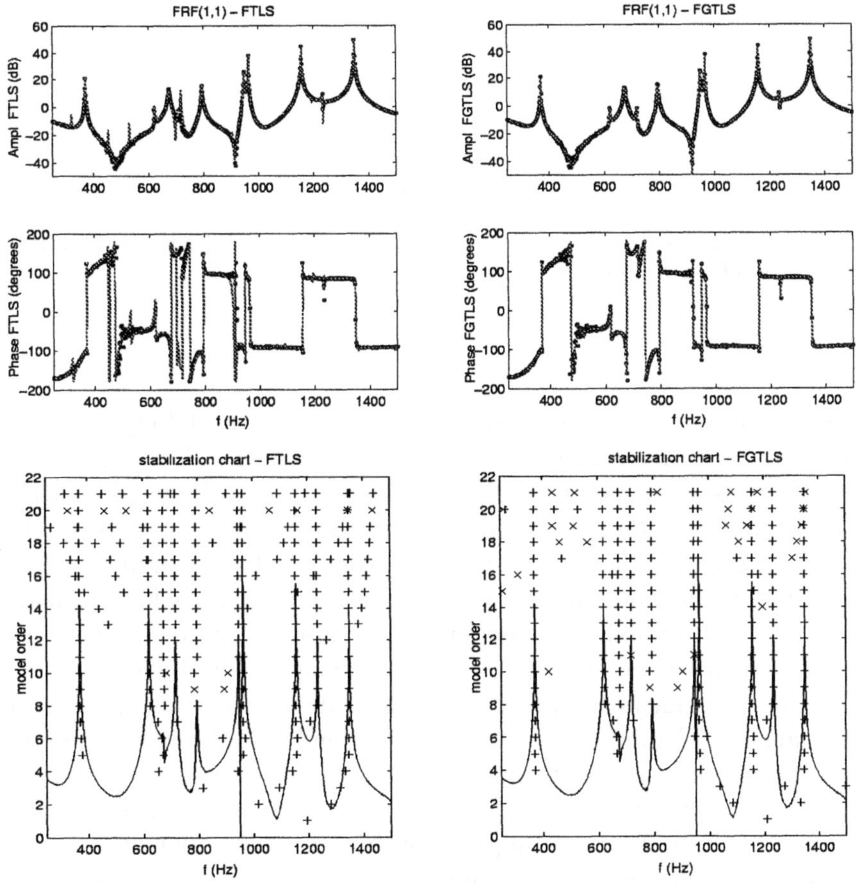

Figure 5. Parametric results for the slat track using the FTLS (left), FGTLS (right) algorithms. (top) Amplitude and phase of the measurements (dotted line) and the estimated transfer function model (solid line). (bottom) Stabilization chart with the stable (+) and unstable (x) poles.

portant tool that is often used to assist the user in separating physical poles from mathematical ones in order to select the optimal model order. To construct this chart, the poles have to be estimated for increasing model orders as shown in Figure 5 (bottom), with (+) indicating stable poles and (x) unstable poles (*i.e.* positive real part), while the line is

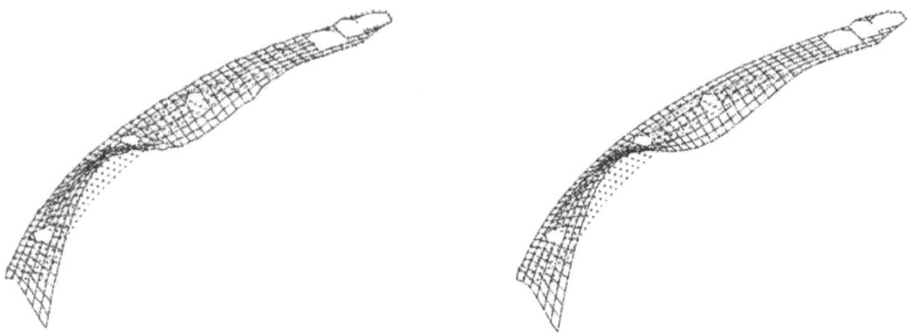

Figure 6. Parametric results for the slat track using the FTLS (left), FGTLS (right) algorithms. Mode shape for the mode of 1234Hz.

the averaged sum of all measurements. Since the model equations are formulated for a given maximum model order, the FGTLS problem can be re-solved for all smaller model orders by simply taking sub-matrices of the full matrices given in equation (15). By doing so, a set of FGTLS solutions is obtained for an increasing order of the denominator polynomial, while the order of the numerator polynomial is kept constant and equal to the maximum specified (polynomial) model order. This approach is very time efficient for constructing the stabilization chart. Comparing the charts for both estimators, it is clear that the use of a proper noise (EV) model results in an improved stabilization behavior of the physical modes. In addition, the computational poles are estimated as unstable in the case of the FGTLS, improving the user-friendliness of the chart, since in general, only the stable poles are plotted. Another important part of the modal model is the spatial information present in the estimated mode shapes. Figure 6 shows the mode shape for the mode at 1234Hz. Again the results obtained by the FGTLS are better since an accurate noise model is taken into account. The mode shape for the FTLS is clearly corrupted by the measurement noise.

Finally, the numerical efficiency is illustrated in Figure 7 by comparing the GTLS and FGTLS algorithms in terms of (Matlab) flops count for a varying number of outputs N_o (with n=20, N_i=1) and model order n (with N_o=10, N_i=1). Considering the output dimension N_o, the results indicate that the GTLS problem is solved in $\mathcal{O}(N_o^3)$ flops, whereas $\mathcal{O}(N_o)$ flops for the FGTLS, resulting in a gain of $\mathcal{O}(N_o)^2$ flops, explained by the elimination approach. However, for an increasing model order, the number of flops for the GTLS and FGTLS is similar, *i.e.* $\mathcal{O}(n^3)$.

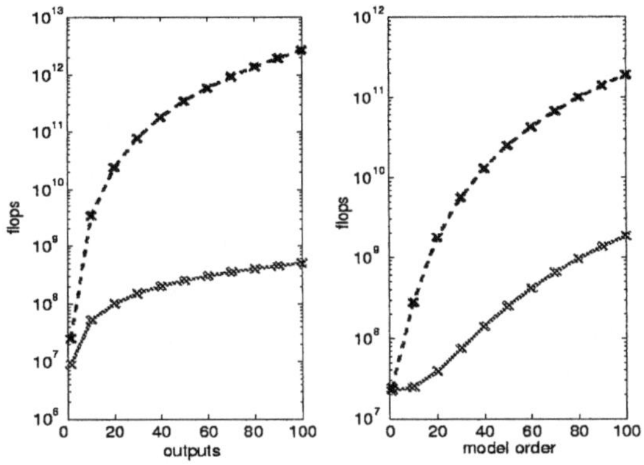

Figure 7. Flops count for GTLS (dashed line) and FGTLS (solid line) for varying number of outputs and model order.

6. Conclusions

A fast implementation of the frequency-domain Input-Output GTLS estimator makes it suitable to analyze large amounts of modal data within an Errors-in-Variables framework. Although a linear approximation is necessary, Monte Carlo simulations show that only a very small loss of accuracy appears for very high noise levels on the inputs and as a result the fast GTLS algorithm has the same statistical properties as the exact GTLS. Both accuracy as well as computation speed and memory usage are important issues in the field of modal analysis, such as for modal-based design and comfort improvement, damage detection and structural health monitoring, and finite element model updating. The proposed approach also provides an alternative for FRF-based modal analysis in case of poor measurement signal quality, short data sequences and leakage.

Acknowledgments

This research has been supported by the Institute for the Promotion of Innovation by Science and Technology in Flanders (IWT), the Fund for Scientific Research - Flanders (Belgium) (FWO) and by the Research Council (OZR) of the Vrije Universiteit Brussel (VUB).

References

[1] P. Guillaume, R. Pintelon, and J. Schoukens. Accurate estimation of multivariable frequency response functions. *13th Triennial World Congress*, pages 423–428, 1996.

[2] P. Guillaume, P. Verboven, and S. Vanlanduit. Frequency-domain maximum likelihood identification of modal parameters with confidence intervals. *23rd International Seminar on Modal Analysis*, pages 359–366, September 1998.

[3] W. Heylen, S. Lammens, and P. Sas. *Modal Analysis Theory and Testing.* KULeuven, Fac. of Engineering, PMA, Leuven, Belgium, 1998.

[4] T. Kailath. *Linear Systems.* Prentice-Hall, 1980.

[5] R. Pintelon, P. Guillaume, G. Vandersteen, and Y. Rolain. Analyses, development, and applications of TLS algorithms in frequency-domain system identification. *SIAM Journal of Matrix Analyses and Applications*, 19(4):983–1004, October 1998.

[6] R. Pintelon, J. Schoukens, and G. Vandersteen. Frequency domain system identification using arbitrary signals. *IEEE Transactions on Automatic Control*, 42(12):1717–1720, December 1997.

[7] Y. Rolain, R. Pintelon, K. Q. Xu, and H. Vold. Best conditioned parametric identification of transfer function models in the frequency domain. *IEEE Transactions on Automatic Control*, 40(11):1954–1960, November 1995.

[8] J. Schoukens, R. Pintellon, G. Vandersteen, and P. Guillaume. Frequency-domain system identification using nonparametric noise models estimated from a small number of data sets. *Automatica*, 33(6):1073–1986, 1997.

[9] J. Schoukens, Y. Rolain, F. Gustafsson, and R. Pintelon. Fast calculation of least squares estimates for system identification. *37th IEEE Conference on Decision and Control*, 3:3408–3410, December 16-18 1998.

[10] S. Van Huffel and J. Vandewalle. *The Total Least Squares Problem: Computational Aspects and Analysis.* SIAM, Philadelphia, PA, 1991.

[11] S. Van Huffel, editor. *Recent Advances in Total Least Squares Techniques and Errors-in-Variables Modeling.* SIAM, Philadelphia,PA, USA, 1997.

[12] P. Verboven, P. Guillaume, and M. Van Overmeire. Improved modal parameter identification by non-parametric modeling of the measurement noise. *17th International Modal Analysis Conference*, pages 1984–1990, February 1999.

[13] P. Verboven, P. Guillaume, M. Van Overmeire, and S. Vanlanduit. Improved frequency-domain total least squares identification for modal analysis applications. *5th National Congress on Theoretical and Applied Mechanics*, pages 211–214, May 2000.

Part VIII

ERRORS-IN-VARIABLES ESTIMATION IN SIGNAL PROCESSING

A NEW DECIMATIVE SPECTRAL ESTIMATION METHOD WITH UNCONSTRAINED MODEL ORDER AND DECIMATION FACTOR

Stavroula-Evita Fotinea, Ioannis Dologlou and George Carayannis
Institute for Language and Speech Processing
6, Artemidos str. & Epidavrou, Paradissos Amaroussiou 151 25, Athens, Greece
{evita,ydol,gcara}@ilsp.gr

Abstract This paper presents a new state-space method for spectral estimation that performs decimation by any factor D while it imposes no constraints to the model order with respect to D. The new method, called DESED, as well as its Total Least Squares version called DESED_TLS, makes use of the full data set available and is based on SVD in order to estimate frequency, damping factor, amplitude and phase of exponential sinusoids.

The new methods are tested in the field of Nuclear Magnetic Resonance (NMR) spectroscopy, where accuracy of parameter estimation is of utmost importance. They are compared against HTLS -an existing method for spectral estimation- its decimative version HTLSD and against a purely decimative method (CONDED). Monte-Carlo based experiments performed on a typical simulated NMR signal prove the new methods to be more robust, especially for low signal to noise ratio.

Keywords: Spectral estimation, singular value decomposition, decimation, subspace estimation, total least squares, Hankel TLS, Hankel/Toeplitz matrix, damped sinusoids, signal processing, regularisation.

1. Introduction

Various applications of digital signal processing, including speech processing and NMR spectroscopy, are employing exponential sinusoidal models in order to represent a signal segment as a sum of exponentially damped complex-valued sinusoids ([6, 8]).

S. Van Huffel and P. Lemmerling (eds.), Total Least Squares and Errors-in-Variables Modeling, 321–330.

The generalised sinusoidal model we use is given by

$$s(n) = \sum_{i=1}^{p}(a_i e^{j\phi_i})e^{(-d_i+j2\pi f_i)n} = \sum_{i=1}^{p} g_i\, z_i^n, n = 0, ..., N-1 \qquad (1)$$

where p is the number of sinusoids that comprise the measured signal. The objective is to estimate the frequencies f_i, damping factors d_i, amplitudes a_i and phases ϕ_i, $i = 1, \cdots, p$.

The new method proposed here is called DESED (DEcimative Spectral Estimation by factor D), can perform decimation by any factor, it exploits the full data set whereas it is not obliged to reduce the model order (size of the Hankel matrix) as D increases, allowing the use of model order $N/2$ approximately. Moreover, the new method makes use of decimated Hankel derived matrices and Singular Value Decomposition. The DESED is a generalisation of the DESE2 method proposed in [2], which performs decimation by factor 2.

2. The DESED method

2.1 Derivation

Let S be the $L \times M$ Hankel signal observation matrix of our deterministic signal of p exponentials $s(n), n = 0, \cdots, N-1$.

$$S = [\tilde{s}_0 \tilde{s}_1 \cdots \tilde{s}_{M-1}] \qquad (2)$$

with $L - D \leq M$, $p < L - D$ and $L + M - 1 = N$. Note that $\tilde{s}_n$ are the column vectors of S, for $n = 0, 1, \cdots, M-1$. Let the $(L-D) \times M$ matrices $S{\downarrow}_D$ and $S{\uparrow}_D$ be the D order lower shift (top D rows deleted) and the D order upper shift (bottom D rows deleted) equivalents of S.

Theorem 1. *There is an (L-D) order matrix X, such that,*

$$XS{\uparrow}_D = S{\downarrow}_D \qquad (3)$$

and all the signal's decimated poles are contained in the eigenvalues of X. A solution of Eq.(3), is given by $X = S{\downarrow}_D(S{\uparrow}_D)^{\dagger}$ and contains the decimated poles of the signal.

Proof. The first claim of the theorem is true because S is constructed of p sinusoids and therefore any row of $S{\downarrow}_D$ can be written as a linear combination of the rows of $S{\uparrow}_D$.

We want then to prove that p of the eigenvalues of the matrix

$$X = S{\downarrow}_D(S{\uparrow}_D)^{\dagger} \qquad (4)$$

are equal to the decimated signal pole estimates z_i^D, $i = 1, \cdots, p$.

Let us consider the well known Vandermonde decomposition of S:

$$S = AGB^T \tag{5}$$

where the $L \times p$ matrix A, $p \times p$ matrix G and $M \times p$ matrix B are defined as follows:

$$\begin{array}{c} A = (a_H(z_1) \cdots a_H(z_p)), a_H(z) = (z^0 z^1 \cdots z^{L-1})^T \\ B = (b_H(z_1) \cdots b_H(z_p)), b_H(z) = (z^0 z^1 \cdots z^{M-1})^T \\ G = diag(z_1, ..., z_p) \end{array}$$

We can then easily write $S{\downarrow}_D = A{\downarrow}_D GB^T$ and $S{\uparrow}_D = A{\uparrow}_D GB^T$ where $A{\downarrow}_D$ and $A{\uparrow}_D$ are defined from A, similarly to the way $S{\downarrow}_D$ and $S{\uparrow}_D$ are defined from S.

Hence, Eq. (3) can be written as:

$$(XA{\uparrow}_D - A{\downarrow}_D)GB^T = 0 <=> XA{\uparrow}_D = A{\downarrow}_D \tag{6}$$

The latter system of linear equations has an infinite number of solutions given by $X = A{\downarrow}_D(A^H{\uparrow}_D A{\uparrow}_D)^{-1}A^H{\uparrow}_D + \Delta^H$, where $\Delta^H A{\uparrow}_D = 0$. Let us consider now the $p \times p$ diagonal matrix Φ_D containing the decimated signal poles z_i^D, $i = 1, \cdots, p$. It is easy then to see that $A{\downarrow}_D = A{\uparrow}_D \Phi_D$. Moreover, $A{\uparrow}_D = A{\downarrow}_D \Phi_D^{-1}$ resulting in $\Delta^H A{\downarrow}_D = 0$.

Hence, since $\|X\|_F^2 = \|X_0\|_F^2 + \|\Delta^H\|_F^2$, the minimum-norm solution to Eq. (6) which we compute is

$$X_0 = A{\downarrow}_D(A^H{\uparrow}_D A{\uparrow}_D)^{-1}A^H{\uparrow}_D \tag{7}$$

X_0 has only p nonzero eigenvalues ($rank(A{\uparrow}_D) = p$), that are equal to the eigenvalues of $(A^H{\uparrow}_D A{\uparrow}_D)^{-1}A^H{\uparrow}_D A{\downarrow}_D = \Phi_D$, i.e. the decimated poles. □

2.2 Least squares DESED

Let S be the $L \times M$ Hankel signal observation matrix of our deterministic signal of p exponentials $s(n), n = 0, \cdots, N-1$, with $L - D \leq M$, $p < L - D$ and $L + M - 1 = N$, where D denotes the decimation factor.

The proposed algorithm, involves the following five steps:

Step 1. We compute the $L \times M$ Hankel matrix S from the N data points $s(n)$ of Eq.(1).

Step 2. We compute the $S{\downarrow}_D$ and $S{\uparrow}_D$ as the D order lower shift (top D rows deleted) and the D order upper shift (bottom D rows deleted) equivalents of S. Best results are obtained when we use the $(L-D) \times M$ matrices $S{\downarrow}_D$ and $S{\uparrow}_D$ as square as possible ([1, 4]).

Step 3. We compute the enhanced version $S{\uparrow}_D e$ of $S{\uparrow}_D$ by employing the SVD of $S{\uparrow}_D$, $S{\uparrow}_D = U{\uparrow}_D \Sigma{\uparrow}_D V{\uparrow}_D^H$ and by truncating to order p (we retain only the largest p singular values).

Step 4. We compute matrix $X = S{\downarrow}_D (S{\uparrow}_D e)^{\dagger}$, also known as truncated SVD solution of $XS{\uparrow}_D = S{\downarrow}_D$. The eigenvalues $\hat{\lambda}_i$ of X give the decimated signal pole estimates, which in turn give the estimates for the damping factors and frequencies.

Step 5. The last step is to compute the phases and the amplitudes. This is done by finding a least squares solution to Eq. (1), with z_i replaced by the estimates and $s(n)$ given by the signal data points.

2.3 Total least squares DESED (DESED_TLS)

Matrix X of *Step 4* in the above described version of DESED, is computed in a least squares sense using the pseudo-inverse of matrix $S{\uparrow}_D e$ ($(S{\uparrow}_D e)^{\dagger}$). We could however, compute matrix X in a total least squares sense using the theorem 3.10 presented in [5]. We can, hence, obtain the DESED_TLS method, which is also used in the experimentation reported below.

3. Decimative version of HTLS (HTLSD)

The HTLS method ([7, 4]) consists of using the Hankel matrix, performing an SVD decomposition and reducing the size of matrices to order p. Damping factors and the frequencies are estimated in a total least squares sense. Phases and amplitudes are estimated using the least squares method.

By using the concept introduced by DESED that implies minor reduction of the Hankel matrix size with respect to decimation factor D, the decimative version of HTLS can be easily derived. The HTLS/HTLSD algorithmic presentation is presented below.

Step 1. We compute the SVD of the $L \times M$ Hankel matrix S from the N data points $s(n)$ of Eq.(1): $S = U_{L\times L}\Sigma_{L\times M}V^H_{M\times M}$, where $L \leq M$. The best results are obtained when we use $L = M(+1) = N/2$.

Step 2. We truncate U, Σ, V to order p and compute: $S_p{=}U_p\Sigma_p V_p^H$ where U_p, Σ_p, V_p are the first p columns of $U_{L\times L}$, $\Sigma_{L\times M}$, $V^H_{M\times M}$.

Step 3. We compute the solution Q of $U{\uparrow}_D Q = U{\downarrow}_D$, in a total least squares sense. For HTLS, $U{\downarrow}_D (U{\uparrow}_D)$ are derived from U_p by deleting its first (last) row respectively, while for HTLSD, $U{\downarrow}_D (U{\uparrow}_D)$ are derived from U_p by deleting its top D (bottom D) rows respectively.

The eigenvalues $\hat{\lambda}_i$ of Q give the decimated signal pole estimates, which in turn give the estimates for the damping factors and frequencies of Eq. (1).

Step 4. The last step is to compute the phases and the amplitudes. This is done by finding a least squares solution to Eq. (1), with z_i replaced by the estimates and $s(n)$ given by the signal data points.

4. The CONDED decimative method

The drawbacks of conventional decimative methods are related to the size of the data set and to the overdetermined model order that can be used. Already proposed decimative methods even though make use of the full data set available, are obliged to reduce the maximum possible model order as D increases. Hence, they relate the model order n (size of the Hankel matrix) with D, according to $n = N/(2D)$. This implies that the efficiency of the overdetermined model is reduced. On the contrary, DESED does not present this drawback and allows the use of model order $n = (N+1)/2 - D/2$, that changes very slowly with respect to decimation factor D.

The DESED has been tested against an existing decimation method proposed in [9, 3], which we call below CONDED (CONventional DEcimative method for decimation factor D). The method makes use of the auto/cross-covariance matrices, decimated sequences of the input signal, SVD and TLS in order to estimate the frequency and damping factor of the sinusoids. The method's algorithmic presentation for decimation factor D can be found below.

Step 1. We compute the $L \times M$ $(L = M = N/(2D))$ Hankel matrix C_k that corresponds to the k-th decimated signal, $c_k(n)$, from the N data points $s(n)$ of Eq.(1). $c_k(n) = s(Dn + D - k), n = 0, 1, \cdots, (N-D)/D$.

Step 2. We compute a global matrix C by concatenating matrices C_k, $k = 1, \cdots, D$ as shown below: $C = [C_1 : C_2 : \cdots : C_D]$. We then compute a global covariance matrix $R = CC^H$.

Step 3. We compute the eigenanalysis of $R = U\Lambda U^H$ to deduce U, which in turn is truncated to order p, thus, yielding U_p.

Step 4. We compute the solution Q of $U{\uparrow}_p Q = U{\downarrow}_p$, in a total least squares sense, where $U{\downarrow}_p (U{\uparrow}_p)$ are derived from U_p by deleting its top (bottom) row. The eigenvalues $\hat{\lambda}_i$ of Q give the decimated signal pole estimates, which in turn give the estimates for the damping factors and frequencies of Eq. (1).

Step 5. The last step is to compute the phases and the amplitudes. This is done by finding a least squares solution to Eq. (1), with z_i replaced by the estimates and $s(n)$ given by the signal data points.

Note that CONDED in Steps 4 and 5 proceeds exactly as HTLS (D=1) in Steps 3 and 4.

5. Experimental results

All methods have been tested via simulations on a typical ^{31}P NMR signal, in order to evaluate both robustness as well as the improvement in accuracy of parameter estimation, when using them in the modelling problem defined by Eq.(1). The signal is a representative example simulating a typical ^{31}P NMR signal of perfused rat liver and it comprises a fifth-order model function given in Table 1 by which N data points uniformly sampled at 10KHz are exactly modelled. In Figure 1, the fast fourier transform (magnitude spectrum) of the signal is depicted.

Table 1. Exact parameter values of the simulated ^{31}P NMR signal.

peak i	f_i (Hz)	d_i (rad/s)	a_i	$\psi_i = \phi_i 180\pi$
1	-1379	208	6.1	15
2	-685	256	9.9	15
3	-271	197	6.0	15
4	353	117	2.8	15
5	478	808	17.0	15

The signal's data points are perturbed by Gaussian noise whose real and imaginary components have standard deviation σ_v. For all methods, N=128, $M = N/2 = 64$ have been used (except for CONDED, for which $M = 32(21)$ for $D = 2(3)$ respectively). Root mean-squared errors of the estimates of all signal parameters are computed using 500 noise realizations (excluding *failures*) for different noise levels. A failure occurs when *not all peaks* are resolved within specified intervals lying symmetrically around the exact frequencies. The halfwidths of the intervals are respectively 82, 82, 82, 43 and 82 Hz, the values being derived from the Cramer-Rao bounds of peaks 4 and 5 at the noise standard deviation where these intervals are touching each other. The estimated minimum model order, namely the number of sinusoids being present in the signal, is set to 5. The Cramer-Rao lower bounds are derived from the exact parameter values and σ_v.

Comparative results between all methods are presented below for different noise standard deviations. In Figure 2 failure rates (bad runs) in 500 realisations are depicted as a function of noise standard deviation ($\sigma_v \in (0, 2.6]$) for the five peak simulated ^{31}P NMR signal. In this graphical representation, results are presented for methods DESE2,

DESE2_TLS, HTLS, HTLS2 and CONDE2. Note that the DESE family presents considerably fewer bad runs than the other methods under consideration. This is even more evident as the noise increases.

Root mean square errors of frequency and damping factor estimates between DESE2, DESE2_TLS, DESE3, DESE3_TLS, HTLS, HTLS2, HTLS3, CONDE2 and CONDE3, are presented in Table 2, for different noise standard deviations σ_v, for peaks 1 and 5 of the simulated ^{31}P NMR signal. The results are presented for $\sigma_v \in [1, 2.6]$, where the number of bad runs for all methods is different than zero. Peak 5 of this signal is considered the most difficult to estimate since it is relatively close to peak 4. Clearly DESED and DESED_TLS are more robust than all the other methods since a decimative approach is expected to outperform any non decimative one. This is due to the fact that decimation brings peaks further apart, thus, increasing the discriminative capacity of spectral estimation methods. Note that comparison between DESE3 versus DESE2 proves DESE3 more robust in all cases, as expected. Moreover, the DESED technique also outperforms the CONDED approach, due to the limitations of the latter method in model order selection. The results suggest in all cases that the proposed decimative approach performs similarly for high S/N ratio. However, for low S/N ratio, despite the similarity of the root mean-squared errors of all parameters estimated, the DESED and DESED_TLS methods perform better due to their lower failure rate and are, thus, more robust. Moreover, the obtained results suggest that DESED and DESED_TLS perform rather similarly for small noise standard deviations, whereas DESED seems slightly more robust for large noise standard deviations than its total least squares counterpart.

6. Conclusion

In this paper a new state-space decimative method for spectral estimation, called DESED, was presented. DESED makes use of decimation and SVD, not only exploiting all data available but also using unconstrained model order and decimation factor, in order to estimate frequencies, damping factors, amplitudes and phases of exponential sinusoids. The version of DESED, called DESED_TLS that uses Total Least Squares was presented. Both DESED (solved in a least squares sense) and DESED_TLS were tested in NMR spectroscopy, where robustness and accuracy of parameter estimation is crucial. The methods were compared to a state-of-the-art non decimative method in spectroscopy, the HTLS method, its decimative version HTLSD as well as a purely decimative method already proposed in literature. Examples on a typical ^{31}P NMR signal were presented and the superior performance of the new decimative approaches was shown, especially for low signal to noise ratio. Different decimation factors were also used and it was shown that higher rates perform better when the peaks under consideration are very close.

Acknowledgments

This work was partially supported by the European TMR Network Project ERBFMRXCT970160.

References

[1] I.Dologlou and G.Carayannis, LPC/SVD analysis of signals with zero modeling error, *Signal Procesing*, 23(3):293-298, 1991.

[2] S-E.Fotinea, I.Dologlou and G.Carayannis, Decimation and SVD to estimate exponentially damped sinusoids in the presence of noise, Proceed. *ICASSP2001*, Vol. V, 3073–3076, Utah, USA, 2001.

[3] B. Halder and T. Kailath, Efficient estimation of closely spaced sinusoidal frequencies using subspace-based methods, *IEEE Signal Proc. Letters*, 4(2):49–51, February 1997.

[4] S. Van Huffel, H. Chen, C. Decanniere and P. Van Hecke, Algorithm for Time-Domain NMR Data Fitting Based on Total Least Squares, *J.Magn.Res.*, Series A 110, 228–237, 1994.

[5] S. Van Huffel and J. Vandewalle, The Total Least Squares Problem. Computational Aspects and Analysis, Frontiers in Applied Mathematics, Vol.9, *SIAM*, Philadelphia, 1991.

[6] S.Y.Kung, K.S.Arun and D.V. Bhaskar Rao, State space and singular-value decomposition-based approximation methods for the harmonic retrieval problem, *J.Amer.Opt.Soc.*, 73(12):1799–1811, 1983.

[7] R. Roy, A. Paulraj, T. Kailath, ESPRIT - A subspace rotation approach to estimation of parameters of cisoids in noise, *IEEE Trans. on Acoustics, Speech and Signal Proc.*, ASSP 34:1340–1342, 1986.

[8] P. Stoica and R. Moses, Introduction to spectral analysis, *Prentice Hall*, Uppers Saddle River, New Jersey, 1997.

[9] P. Stoica, A. Nordsjo, Subspace-based frequency estimation in the presence of moving-average noise using decimation, *Signal Proc.*, 63:211–220, 1997.

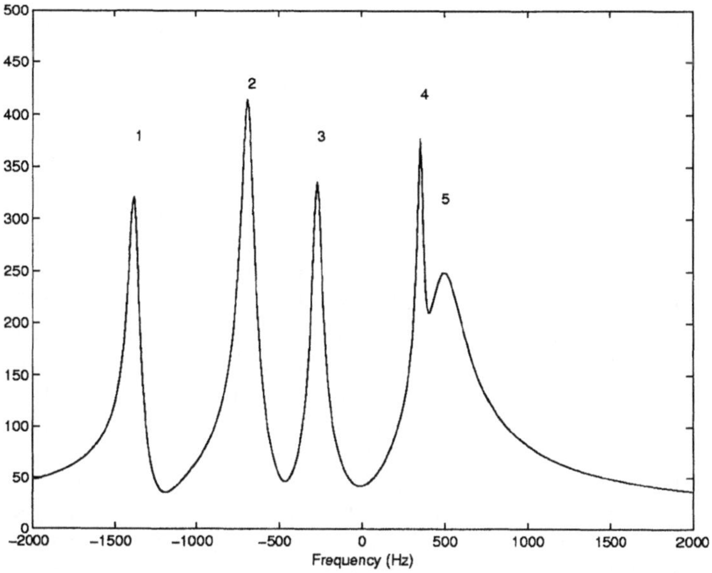

Figure 1. Fast Fourier transform (magnitude) of the simulated ^{31}P NMR signal.

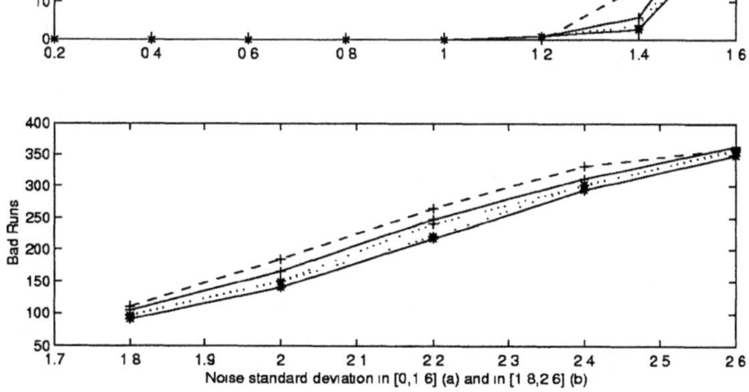

Figure 2. Number of bad runs for methods DESE2, DESE2_TLS, HTLS, HTLS2 and CONDE2.

Table 2. Number of bad runs (BR) and root mean-squared errors of frequency and damping factor estimates for peaks 1 and 5 of the simulated ^{31}P NMR signal as a function of noise standard deviation σ_v for all methods.

σ_v	Method	BR	f_1	d_1	f_5	d_5
1.0	DESE2	0	2.7056	0.0799	11.1992	0.0885
1.0	DESE2_TLS	0	2.7046	0.0784	12.9509	0.1001
1.0	DESE3	0	2.7240	0.0793	11.4102	0.0905
1.0	DESE3_TLS	0	2.7239	0.0787	13.2059	0.1015
1.0	HTLS	0	2.6817	0.0781	10.7178	0.0815
1.0	HTLS2	0	2.7045	0.0784	11.0223	0.08320
1.0	HTLS3	1	2.7260	0.0788	11.2129	0.08540
1.0	CONDE2	0	2.6872	0.0788	11.1000	0.08677
1.0	CONDE3	0	2.7483	0.0790	11.3619	0.08572
1.4	DESE2	3	3.9089	0.1169	17.2046	0.1315
1.4	DESE2_TLS	3	3.9456	0.1137	22.3152	0.1587
1.4	DESE3	3	3.9126	0.1175	17.1838	0.1398
1.4	DESE3_TLS	3	3.9360	0.1136	22.3889	0.1612
1.4	HTLS	6	3.9166	0.1133	15.8401	0.1073
1.4	HTLS2	4	3.9084	0.1138	15.9559	0.1105
1.4	HTLS3	3	3.9104	0.1144	16.1404	0.1141
1.4	CONDE2	14	3.9341	0.1147	16.9414	0.1214
1.4	CONDE3	11	4.0011	0.1137	17.9091	0.1234
1.8	DESE2	91	5.0751	0.1503	26.8908	0.2445
1.8	DESE2_TLS	97	5.1451	0.1453	31.4701	0.2306
1.8	DESE3	86	5.1006	0.1496	25.8218	0.2120
1.8	DESE3_TLS	88	5.0917	0.1445	31.7961	0.3168
1.8	HTLS	105	5.0719	0.1428	24.6343	0.1667
1.8	HTLS2	98	5.1184	0.1457	25.2973	0.1633
1.8	HTLS3	100	5.1375	0.1443	24.0197	0.1623
1.8	CONDE2	111	5.1835	0.1475	25.9503	0.2086
1.8	CONDE3	126	5.1266	0.1487	27.3524	0.2045
2.2	DESE2	217	6.2700	0.1835	31.5661	0.3023
2.2	DESE2_TLS	221	6.2695	0.1770	34.7310	0.2652
2.2	DESE3	214	6.2195	0.1838	31.7100	0.3121
2.2	DESE3_TLS	220	6.3110	0.1758	35.7617	0.3234
2.2	HTLS	248	6.1574	0.1754	28.6350	0.2190
2.2	HTLS2	241	6.4242	0.1775	28.6868	0.2303
2.2	HTLS3	233	6.4119	0.1829	30.0438	0.2324
2.2	CONDE2	265	6.1825	0.1808	30.1581	0.2505
2.2	CONDE3	252	6.5679	0.1733	33.0274	0.2651
2.6	DESE2	349	7.7193	0.2319	33.2003	0.3536
2.6	DESE2_TLS	356	7.7419	0.2078	35.3180	0.2638
2.6	DESE3	339	8.0044	0.2217	35.9888	0.3740
2.6	DESE3_TLS	351	8.1125	0.2086	36.1001	0.3039
2.6	HTLS	363	7.7699	0.2031	34.0230	0.2526
2.6	HTLS2	359	7.7300	0.2030	32.0033	0.2670
2.6	HTLS3	349	8.1864	0.2044	34.1685	0.3124
2.6	CONDE2	360	8.4645	0.2210	32.5606	0.3245
2.6	CONDE3	363	8.1582	0.2076	31.5678	0.4602

MODELING AUDIO WITH DAMPED SINUSOIDS USING TOTAL LEAST SQUARES ALGORITHMS

Werner Verhelst, K. Hermus, P. Lemmerling, P. Wambacq, S. Van Huffel
ESAT-PSI, K.U.Leuven, Kasteelpark Arenberg 10, B-3001 Leuven-Heverlee, Belgium
{Werner.Verhelst, Kris.Hermus, Philippe.Lemmerling, Patrick.Wambacq}@esat.kuleuven.ac.be

Abstract We demonstrate that damped sinusoidal modeling can be used to improve the modeling accuracy of current perceptual audio coders. We show that the model parameter estimation can be performed with TLS algorithms, and that a subband modeling approach results in TLS problems that are computationally much more tractable than the fullband approach. Experimental results indicate that subband TLS modeling can be effectively controlled using perceptual criteria.

Keywords: Audio coding, damped sinusoids, transients, total least squares, signal processing, Hankel TLS, subspace estimation, Hankel/Toeplitz matrix

1. Introduction

A fairly general model that is often used to represent speech and audio signals is based on an AM/FM representation, as shown in eq. (1).

$$s(n) \approx \sum_{k=1}^{K} a_k(n) \sin\left(2\pi f_k(n)\, n + \phi_k(n)\right) \tag{1}$$

In this model, the signal is represented as a sum of sinusoidal components with time-varying amplitude, phase and frequency. While this model can obviously generate any desired signal $s(n)$, it is required for effective modeling and efficient encoding that amplitude, frequency and phase are only slowly time varying functions of n. In practice these parameters are therefore often considered constant for the duration of an analysis frame (depending on the signal the length of quasi-stationary segments can vary from few tens of ms to several hundreds of ms).

Audio signals often contain noise-like segments and transient sounds that are not efficiently modeled by (1) [1]. Therefore, model (1) is often

S. Van Huffel and P. Lemmerling (eds.), Total Least Squares and Errors-in-Variables Modeling, 331–340.

extended with an explicit noise term $\eta(n)$ and sometimes also with a transient segments term $\tau(n)$ as shown in eq. (2). The transient term can be used to deal with fast transients (such as sharp attacks) in specialized ways [2].

$$s(n) \approx \sum_{k=1}^{K} a_k(n) \sin\left(2\pi f_k(n)\, n + \phi_k(n)\right) + \eta(n) + \tau(n) \tag{2}$$

Current industry standards for audio coding, such as MPEG1 Layer III [3], do not explicitly use signal model (2). Instead, they use a subband coding structure with perceptual bit allocation [4]. Subband coding [5] is efficient for coding wideband stationary noise, but less so for tones and transient sounds. Several researchers have therefore investigated schemes that attempted to extract tonal components from the subband signals and to model them separately with the sinusoidal model of eq. (1) [4]. Only the residual signal, which is obtained after subtracting the quantized tones from the subband signal, is then encoded using the traditional subband scheme.

Transient sounds are particularly difficult to encode well with a subband coder or with a sinusoidal coder. In particular, transient sounds can cause a type of distortion that is known as pre-echo. Pre-echo originates from the fact that, in order to obtain a high coding gain, subband and sinusoidal coders normally produce signals with a fairly high degree of pseudo-stationarity. Thus, the sharpness of transient attack segments will actually be building up gradually, before the attack. Many coding schemes detect the presence of transient sounds and switch to an encoding mode with a better time resolution to avoid pre-echo [4]. This, however, goes at the expense of coding efficiency.

As a possible way to overcome these problems, researchers have considered modeling and encoding audio signals using a superposition of slowly time-varying *exponentially weighted* sinusoids and quasi-stationary noise:

$$s(n) \approx \sum_{k=1}^{K} a_k(n) e^{-d_k(n)\, n} \sin\left(2\pi f_k(n)\, n + \phi_k(n)\right) + \eta(n) \tag{3}$$

In the next section we briefly recall how the parameters of such an exponential sinusoidal model (ESM) can be estimated using TLS algorithms and we study the time domain and frequency domain aspects of TLS-ESM modeling for audio. The relation and differences between sinusoidal modeling (SM) and exponential sinusoidal modeling (ESM) of sounds are also considered. The advantages of TLS modeling in the subband domain are discussed in Section 3, and some experiments on controlling TLS-ESM with a psycho-acoustic model are described in Section 4. Finally, a concluding discussion is given in Section 5.

2. Modeling audio with damped sinusoids

2.1 Damped sinusoidal modeling and TLS

Total Least Squares (TLS) algorithms determine the parameters of an Auto-Regressive (AR) model that exactly matches a (slightly) perturbed version $\hat{s}(n)$ of the input segment $s(n)$, $n = 1,\ldots,N$. The TLS modeling problem of order L is to find the model parameters $b(l)$, $l = 1,\ldots,L$ and $\Delta s(n)$, $n = 1,\ldots,N$ that minimize

$$\sum_{n=1}^{N} (\Delta s(n))^2 \tag{4}$$

subject to

$$\hat{s}(n) = s(n) + \Delta s(n) = \sum_{l=1}^{L} b(l)\,(s(n-l) + \Delta s(n-l)) \tag{5}$$

for $n = (L+1),\ldots,N$.

Since $\hat{s}(n)$ is purely autoregressive, it can be written as a sum of damped sinusoids:

$$\hat{s}(n) = \sum_{k=1}^{K} a_k e^{-d_k\,n} \sin(2\pi f_k\, n + \phi_k) \tag{6}$$

where the damping factors d_k can be positive, negative, or zero. Thus, TLS algorithms can be readily applied to audio frames in order to estimate the parameter values of the ESM model (3). The proper arrangement of the equations (5) into structured matrices (e.g. Hankel, Toeplitz) leads to a Structured TLS (STLS) problem [6, 7]. Several approaches exist for solving the STLS problem. Since the latter is a nonlinear optimization problem, all optimal approaches are implemented using iterative algorithms. As a consequence, it is hard to predict the CPU time necessary to process a particular speech frame. Therefore, the suboptimal Hankel TLS (HTLS) [8] approach will be used throughout this paper. HTLS is a subspace based approach in which the N data points are stored in a mxn data matrix with $m + n - 1 = N$. A straightforward implementation of HTLS requires $\mathcal{O}(mn^2)$ flops. Using the implicitly restarted Arnoldi[9] method or the Lanczos method with partial reorthogonalization[10] it is possible to reduce the complexity to $\mathcal{O}(K(m+n)\log_2(m+n))$ flops if $K << m, n$.

Using the notation from (5) we define the signal to modeling noise ratio (SNR) as

$$SNR = 10\log_{10}\frac{\sum_{n=1}^{N} s^2(n)}{\sum_{n=1}^{N}(s(n)-\hat{s}(n))^2} \tag{7}$$

This SNR can be used to measure the modeling accuracy. It can be noted from (4) that SNR is actually the quality measure that is maximized in TLS problems. Unfortunately the SNR measure is not always successful in minimizing the audibility of distortions.

2.2 Time domain aspects of TLS-ESM

It is customary in the field of music synthesis to consider that the amplitude envelope of sounds produced by musical instruments has an attack phase, followed by a decay and a sustain phase and dies out during the release phase (this is usually referred to as the ADSR model [11]). In audio signals we can therefore expect to find many transient segments besides sustained (quasi-stationary) segments. As mentioned earlier, segments with a transient nature can not be efficiently modeled with the sinusoidal model (1). In contrast, the exponential amplitude terms in ESM provide the capability for modeling fast amplitude fluctuations inside a frame. Thus, with ESM, we can consider the amplitude, frequency, phase, and damping parameters to be slowly time-varying during all the segment-types of a typical ADSR envelope, and not only in the quasi-stationary sustain regions. The modeling residual of ESM could therefore be better suited for subband coding than that of SM.

While it is clear that transient segments can be produced using ESM, the location of the transient in the frame matters as the modeled transients are superpositions of weighted sinusoids: the exponential weights $a_k e^{-d_k n}$ are restricted to have their maxima either in the beginning or at the end of the frame. Also, efficient modeling with ESM would require that all relevant segment types can be accurately located in the audio signal. In current ESM systems, however, the segmentation is usually done asynchronously for practical reasons, leading to suboptimal modeling, as illustrated in Figs. 1a and 1b. Fig. 1a shows a short transitory speech segment. This segment and all its circular rotations have been modeled using fixed order TLS-ESM. The result in Fig. 1b shows the dependence of the modeling accuracy on the position of the segment in the analysis frame (see also [12]).

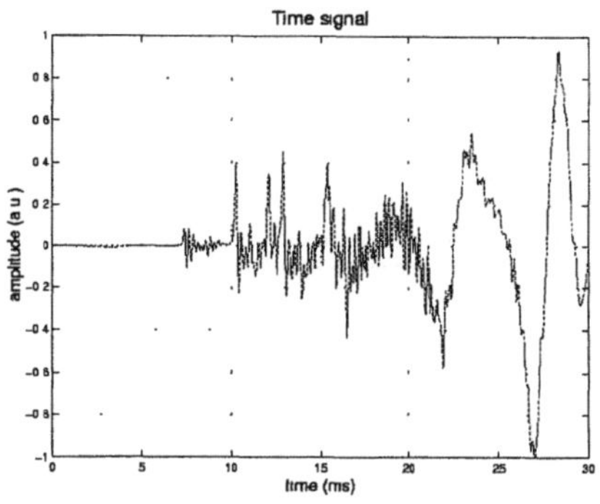

Figure 1a. Original speech frame

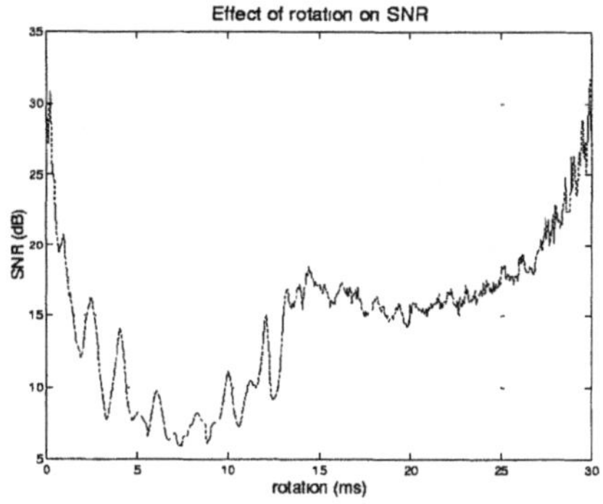

Figure 1b. Signal to modeling error energy ratio (SNR) in dB as a function of frame rotation

2.3 Frequency domain aspects of TLS-ESM

Fig. 2 illustrates how TLS operates in the spectral domain. From an example frame, we plot the original spectrum (dashed) and the spectra of the TLS modeled versions (solid), for 32 (top) and 48 (bottom) damped sinusoids in the model.

Observe that TLS models the original spectrum with spectrally localized contributions. TLS keeps spending a lot of components to model a high energetic region, until this local fit is 'good enough', before it goes on to model another high energetic spectral region. Up to a significant number of damped sinusoids, spectral gaps are present in the modeled speech and this is often incompatible with the criterion of optimal perceptual quality. As we already mentioned, the optimization of the SNR is clearly not a perceptual criterion.

3. Fullband versus subband TLS

As indicated in Section 2.1, the solution of a TLS problem requires solving a non-linear optimization problem of computational complexity $\mathcal{O}(N^3)$, with N the number of samples in the frame. For a frame of audio, sampled at typical sampling frequencies, e.g., 44.1 kHz, a large number of unknowns is involved, leading to a high computational load and a poor numerical conditioning. These problems are alleviated when applying the TLS-ESM modeling procedure to the downsampled channel signals of the subband coder, instead of to the fullband audio signal [13]. An additional advantage of subband TLS over fullband TLS is that better

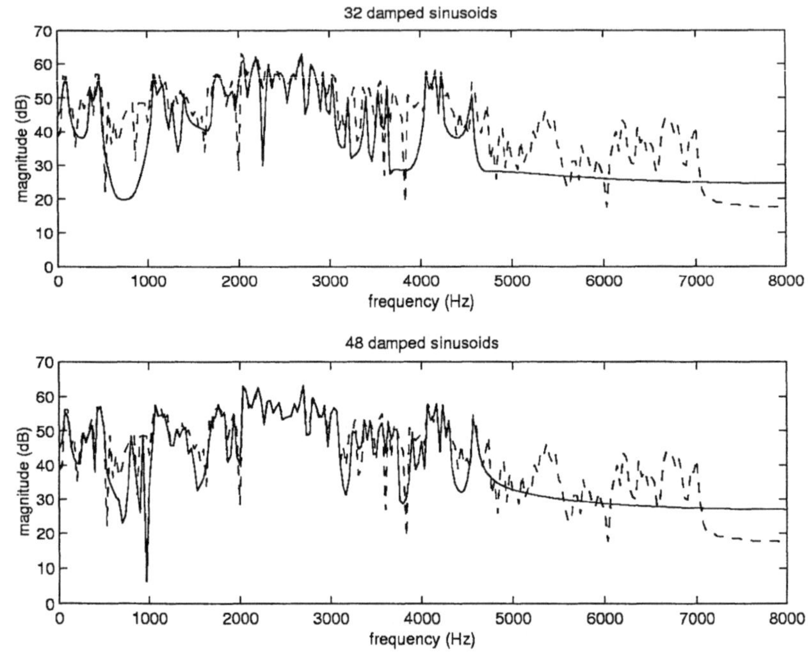

Figure 2. Illustration of TLS in the spectral domain. Original spectra (dashed) and TLS modeled spectra (solid) for phoneme /k/.

coverage of the whole spectral range can be achieved with a given model order.

Thus, for an M-channel fully decimated system, M TLS problems in N/M unknowns must be solved, which requires only

$$M \,.\, \mathcal{O}(\frac{m}{M}\frac{n^2}{M^2}) \approx \mathcal{O}(\frac{mn^2}{M^2}) \qquad (8)$$

calculations. This means a reduction by a factor M^2 compared to fullband TLS-ESM. Furthermore, the M subband problems will generally be numerically more tractable than the fullband problem. Fig. 3 shows the computation time required to model a test utterance with an accuracy of 30 dB SNR using SB-TLS and uniform QMF filterbanks with varying number of channels [13].

Since SB-TLS cannot model across the frequency range of the different channels (each component is restricted to model only its own subband), it requires more damped sinusoids to achieve the same SNR as a corresponding fullband model. The number of damped sinusoids needed to obtain a global SNR of 30 dB is indicated with 'x' in fig. 3. While the total number of required model components increases with the number

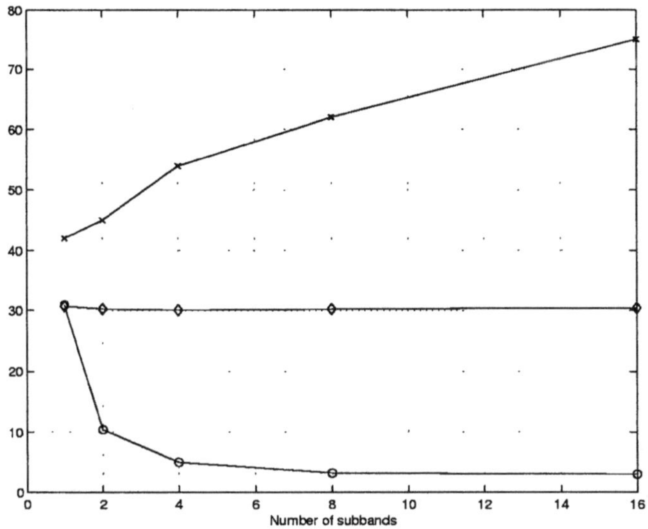

Figure 3. Computational complexity of SB-TLS. As a function of the number of channels: computation time measured in 100 ms (∘), total number of damped sinusoids (x), and global SNR of the utterance (⋄).

of channels in the filter bank, the overall decrease in computation time is very significant.

4. Towards perceptual TLS-ESM

While fullband TLS-ESM is optimal from the viewpoint of the SNR criterion, using subbands enables us to control the distribution of the damped sinusoids over the whole frequency range. This added flexibility can be exploited to improve the perceptual quality of the modeled audio.

In the most basic implementation of subband TLS-ESM, we would assign to each subband an equal number of damped sinusoids. The impact on the frequency spectrum is illustrated in Figure 4. We used the same speech frame as in Figure 2, but in this case the signal was first sent through a fully decimated uniform QMF analysis filter bank with 16 channels. Each subband was modeled with 2 (top) or 3 (bottom) damped sinusoids and reconstructed using the synthesis filter bank. The difference in modeling behavior compared to Figure 2 is apparent: the subband approach leads to a more uniform coverage of the spectral range.

A *fixed* distribution of the components is, however, far from optimal due to the changing spectral contents of consecutive frames: one could only rely on the (statistically modeled) *average* spectral contents of audio signals. A more accurate approach is to change the distribution

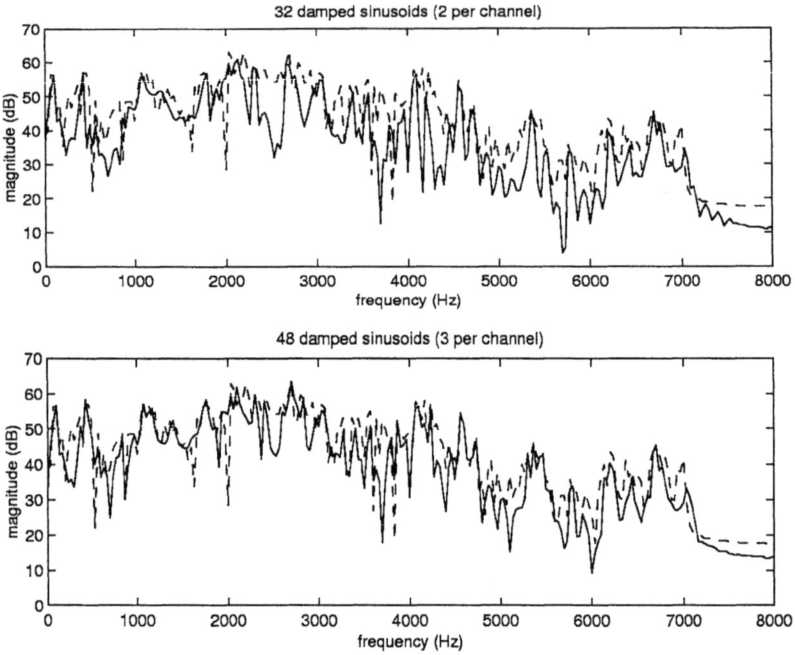

Figure 4. Spectral representation of basic subband TLS-ESM.

from frame to frame to model only those parts of the spectrum that are *perceptually* relevant. The necessary information to identify the relevant parts can be derived from a psycho-acoustic analysis. Based on time and frequency masking phenomena of our hearing system, psycho-acoustic models calculate a frame-specific (and hence time varying) frequency dependent masking curve. Spectral components of modeling distortions that fall below a masking threshold are perceptually irrelevant. In our current implementation we use MPEG 1, psycho-acoustic model I [3].

Of course, the masking curve by itself is not sufficient to decide *how many* damped sinusoids are needed for each subband in order to obtain a desired perceptual quality per subband. In a first experiment we used the following strategy. Assign an extra component to the subband with the largest ratio of modeling error noise to masking level (Noise to Mask Ratio (NMR)). During initialization every subband has zero components assigned to it and the noise is equal to the original subband signal. In each iteration step, the NMR in the subbands is updated before a new component can be assigned. The iteration stops wher the total number of assigned components reaches a predefined number.

However, this strategy runs a serious risk. If a subband is perceptually important, but can not be well modeled by ESM, all available components would be assigned to that subband in vein. We therefore first check the impact of an extra component, before assigning it to the subband that makes most effective use of it (i.e., such that the overall audio quality is maximally improved). In a first approximation, we assign the extra components such that the total NMR is maximally reduced in each step.

5. Concluding discussion

We applied TLS algorithms for modeling tonal and transitional audio segments. Because TLS solutions are superpositions of exponentially weighted sinusoids, they are well suited for this purpose. Using a subband approach makes the TLS modeling problem more tractable (less computationally complex and numerically better conditioned). The subband TLS-ESM modeling residuals can be further processed using standard perceptual coding strategies. Experiments have shown that the TLS subband modeling procedure itself can also be controlled in accordance with perceptual criteria.

Further work should be done before evaluating the quality that can be achieved at different bitrates with TLS-ESM strategies. One particularly interesting issue concerns the design of model order criteria: adding more modeling components requires a higher bitrate for the model parameters, but produces a residual that can be encoded with a lower bitrate. Thus, order selection criteria could be designed to estimate a frame dependent optimal model order. These order selection criteria should also take into account the necessary trade-off between model order and average number of bits that can be assigned to each model parameter.

Once a base coder is constructed, many optimizations could be considered to further increase the coding gain. In particular, it was shown that the modeling efficiency of TLS-ESM depends on the position of the audio signal in the frame. An intelligent non-uniform signal segmentation strategy could be used to locate sound fragments of different types, especially those resembling attack, decay, sustain and release phases. It can be expected that such a segmentation would boost overall modeling accuracy and coding gain. Finally, it can be noted that while perceptual TLS-ESM requires high computational power for real-time encoding, the decoding complexity is comparable to that of the current industry standard decoders.

Acknowledgments

Support is acknowledged from the Flemish Community - IWT. Werner Verhelst is also with the Vrije Universiteit Brussel - VUB. Philippe Lemmerling is supported by a post doctoral scholarship of FWO - Flanders.

References

[1] J. Jensen, S.H. Jensen and E. Hansen. Exponential Sinusoidal Modeling of Transitional Speech Segments, *Proc. Intl. Conference on Acoustics, Speech and Signal Processing (ICASSP)*, vol. I, 473-476, Phoenix, U.S.A., March 1999.

[2] S.N. Levine. *Audio representations for data compression and compressed domain processing.* PhD dissertation, Stanford University, December 1998.

[3] ISO/IEC 11172-3. Information technology — Coding of moving pictures and associated audio for digital storage media at up to about 1.5 Mbit/s — Part 3: Audio, ISO/IEC 1993.

[4] T. Painter and A. Spanias. Perceptual Coding of Digital Audio, *Proceedings of the IEEE*, 88(4):451–513, April 2000.

[5] N.S. Jayant and P. Noll. *Digital Coding of Waveforms*, Prentice-Hall, 1984.

[6] P. Lemmerling. *Structured total least squares: analysis, algorithms and applications*, PhD thesis, Faculty of Applied Sciences, K.U. Leuven, Belgium, 1999.

[7] S. Van Huffel, H. Park and J.B. Rosen. Formulation and Solution of Structured Total Least Norm Problems for Parameter Estimation, *IEEE Trans. Signal Processing*, 44(10):2464–2474, 1996.

[8] S. Van Huffel, H. Chen, C. Decanniere, and P. Van Hecke. Algorithm for Time-Domain NMR Data Fitting Based on Total Least Squares *J. Magn. Reson.*, A1(10):228–237, 1994.

[9] D. Calvetti, L. Reichel, and D.C. Sorensen. An implicitly restarted Lanczos method for large symmetric eigenvalue problems, *Electron. Trans. Numer. Anal.*, 2:1–21, March 1994.

[10] H.D. Simon. The lanczos algorithm with partial reorthogonalization *Math. of Computation*, 42:115–142, 1984.

[11] C. Roads. *The computer music tutorial*, third printing, MIT Press, 1998.

[12] J. Nieuwenhuijse, R. Heusdens and Ed F. Deprettere. Robust Exponential Modeling of Audio Signals, *Proc. of the International Conference on Acoustics, Speech and Signal Processing (ICASSP98)*, VI:3581–3584, Seattle, U.S.A., May 1998.

[13] K. Hermus, W. Verhelst, P. Wambacq and P. Lemmerling. Total Least Squares based Subband Modelling for Scalable Speech Representations with Damped Sinusoids, *Proc. of the International Conference on Spoken Language Processing (ICSLP2000)*, vol III, pp. 1129-1132, Beijing, China, October 2000.

[14] P. Lemmerling, I. Dologlou and S. Van Huffel. Speech Compression based on exact modeling and Structured Total Least Norm optimization, *Proc. of the International Conference on Acoustics, Speech and Signal Processing (ICASSP98)*, 353–356, Seattle, U.S.A., May 1998.

[15] J. Jensen. *Sinusoidal Models for Speech Signal Processing*, PhD dissertation, CPK — Aalborg University, Denmark, August 2000.

[16] R.J. McAuley, and T.F. Quatieri. Sinusoidal Coding, In W.B. Kleijn, K.K. Paliwal, *Speech Coding and Synthesis*, CH4, Elsevier Science BV, 1995.

REAL-TIME TLS ALGORITHMS IN GAUSSIAN AND IMPULSE NOISE ENVIRONMENTS

Da-Zheng Feng, Zheng Bao and Xian-Da Zhang

Key Lab. for Radar Signal Processing, Xidian University, Xi'an, 710071, China

dzfeng@rsp.xidian.edu.cn

Abstract On the basis of the total minimum mean square error or the minimum Raleigh quotient, we propose a modified total least mean squares (TLMS) algorithm with the computational complexity close to the well-known LMS algorithm. A new fast RTLS algorithm is developed by using the recursive computation of the TLS solution for adaptive finite impulse response (FIR) filters. Using the shift structure of the augmented data vector, a fast algorithm for computing the new gain vector is given. The new fast algorithm is numerically stable and of computational complexity $\mathcal{O}(n)$. A recursive total instrumental-variable (RTIV) algorithm is given for finding the TLS solution to the over-determined normal equations, and its applications to adaptive IIR filtering are presented. Its arithmetic operation complexity is $\mathcal{O}(mn)$ (where m is the number of instrumental variables and n the dimension of the input-vector) per iteration. We give the L_p-norm distance from a point to a hyper-plane. A real-time algorithm is introduced for finding the robust TLS solution in impulse noise, and has computational complexity $\mathcal{O}(n)$.

Keywords: total least mean squares, recursive least squares, instrumental variables, total L_p-norm approximation, impulse noise, signal processing.

1. Introduction

If interference only exists at the output of the analyzed system, some well-known LS algorithms [2] can obtain the optimal solution of a least-squares (LS) problem. However if there is interference at both the input and output of the analyzed system, some TLS algorithms [23] should be proposed in order to get the optimal solution for signal processing problems.

S. Van Huffel and P. Lemmerling (eds.), Total Least Squares and Errors-in-Variables Modeling, 341–350.

For computing the generalized eigenvector associated with the smallest eigenvalue of the auto-correlation matrix adaptively, a number of algorithms have been proposed and fall into two broad classes. A large variety of the first class of algorithms [16, 4] are often called recursive total least squares (RTLS) algorithms and have usually computational complexity $\mathcal{O}(N^2)$ per iteration. For solving TLS problems in adaptive FIR filtering on-line, Davila proposed a fast RTLS algorithm based on gradient search for the generalized Rayleigh quotient along the Kalman gain vector [11]. Its computational complexity is $\mathcal{O}(n)$ per iteration. Davila's RTLS algorithms are dependent on the fast computation of the Kalman gain vector. However, the computation of the Kalman gain vector is numerically unstable [12], although some efficient solution approaches have been presented [19].

The second class involves stochastic-type adaptive algorithms [16, 15, 24, 13]. Notice that Gao et al [9] proposed a constrained anti-Hebbian learning algorithm for solving TLS problems in adaptive FIR and IIR filtering. In [7], a total least mean squares (TLMS) algorithm was developed from the point of view of total adaptive signal processing.

The instrumental-variable (IV) approaches have been applied extensively for system identification [20, 8] and array processing [14, 21]. Applications of the IV techniques in array processing were presented in [14, 21]. A more detailed reference list can be found there.

In many real world problems, however the noise encountered is impulsive in nature and is more accurately represented by a heavy-tailed non-Gaussian distribution. Systems optimized under the Gaussian assumption can suffer from severe performance degradation when subjected to impulsive, non-Gaussian noise [18]. There exists a class of distributions called alpha-stable distributions [18, 3, 10], which have been shown to model these types of noise accurately. When introducing α-stable distributions into the signal processing community, the Least L_p-Norm (LPN) estimation [18, 10] is proposed. Later, Arikan et al. developed the Normalized LMP (NLMP) algorithm [3]. Especially, a new problem, known as structured total least norm (STLN), and an algorithm for computing the STLN solution, has been developed in [22, 17]. Moreover, the application of the STLN method to various parameter estimation problems is studied.

2. A modified total least mean squares algorithm

For a multi-input single-output (MISO) simultaneous system, the input-output relation can be represented as

$$y(t) = x^T(t)h \tag{1}$$

where $h = [h_1, \ldots, h_n] \in \mathbb{R}^{n\times 1}$; $x(t) = [x_1(t), \ldots, x_n(t)]^T \in \mathbb{R}^{n\times 1}$. Define an augmented observed data vector sequence as

$$\hat{z}(k) = \begin{bmatrix} \hat{x}(k) \\ \hat{y}(k) \end{bmatrix} = \begin{bmatrix} x(k) + n_x(k) \\ y(k) + n_y(k) \end{bmatrix} \in \mathbb{R}^{(n+1)\times 1} \tag{2}$$

where $n_x(k)$ represents the noise in the observed input, $n_y(k)$ represents the noise in the observed output. In order to find the TLS solution in the above adaptive parameter-identification problem, we should first track the smallest left singular vector associated with the smallest singular value of the augmented correlation matrix $R_z = E\{\hat{z}(k)\hat{z}^T(k)\}$. All the TLMS algorithms take the $\hat{z}(k)\hat{z}^T(k)$ as the approximate estimate of R_z [5,6,7,8,9], i.e.

$$R_z \approx \hat{z}(k)\hat{z}^T(k) \tag{3}$$

An iterative search procedure for the smallest non-normalized singular vector of R_z can be represented algebraically as follows [8]

$$\hat{o}(l+1) = \hat{z}^T(l+1)\,\hat{w}(l) \tag{4}$$

$$\tilde{w}(l+1) = w(l) - \mu\,\hat{o}(l+1)\,\hat{z}(l+1) \tag{5}$$

$$\hat{w}(l+1) = \frac{-\tilde{w}(l+1)}{\tilde{w}_{n+1}(l+1)} \tag{6}$$

This algorithm is a random search algorithm. The learning rate should be selected as a very small positive constant or a very small positive time-varying function with sampling time. In neural network theory [5,6,7,8], the term $\hat{o}(l+1)\,\hat{z}(l+1)$ in (5) is called the anti-Hebbian rule or iteration, and plays an important role in real-time extraction of the smallest singular vector. In order to avoid the extra division in (6), Gao et al developed an adaptive TLMS algorithm that only requires $3(n+1)$ multiplications per iteration. Here, a modified TLS algorithm with low computational complexity is described by

$$u(l) = \hat{w}(l) \tag{7}$$

$$\begin{cases} \hat{o}(l+k) = \hat{z}^T(l+k)\,u(l+k) \\ u(l+k+1) = u(l+k) - \mu\,\hat{o}(l+k)\,\hat{z}(l+k), \quad k = 0, 1, \ldots, q-1 \end{cases} \tag{8}$$

$$\hat{w}(l+q) = -u(l+q)/u_{n+1}(l+q) \tag{9}$$

On average, this algorithm requires $(2 + 1/q)(n+1)$ multiplications, $2(n+1)$ additions, $1/q$ extra divisions per iteration, and it is close to the computational complexity of the LMS algorithm [1]. In the above equations, (7) initiates weights before weights are updated q times, (8) updates the weights q times by the anti-Hebbian rule, (9) scales the weights, after updating the weights q times. As before, μ is the learning rate or the gain constant that regulates the speed and stability of adaptation. Once a stable $\hat{w}$ is found, the TLS solution of the above adaptive signal-processing problem is

$$w_{\text{TLS}} = [\hat{w}_1, \dots, \hat{w}_n]^T \tag{10}$$

3. A fast recursive total least squares algorithm for FIR filtering

Here, we search the optimal solution of the generalized Rayleigh quotient along the augmented data vector. The computational complexity of the new algorithm is $\mathcal{O}(N)$ and lower than that of Davila's algorithm per iteration. Moreover, it is independent of the recursive computation of the inversion of the correlation matrix, and numerically stable. The FIR of an unknown system is described by $h^* = [h_0^*, h_1^*, \dots, h_{M-1}^*]^T$. The so-called "expectation" output is given by

$$d(t) = x^T(t)h^* + n_0(t) \tag{11}$$

where the measured noise $n_0(t)$ the Gaussian white noise with variance σ_0^2, and independent of the input vector. Let the M-order linear regression input vector be as follows :

$$r(t) = [x(t)+n_1(t), x(t-1)+n_1(t-1), \dots, x(t-M+1)+n_1(t-M+1)]^T \tag{12}$$

where the noise $n_1(t)$ has variance σ_1^2. At t, the augmented data vector is given by

$$\bar{r}(t) = [r^T(t), d(t)]^T \in \mathbb{R}^{(M+1)\times 1} \tag{13}$$

The classical least-squares solution is described by the optimal problem :

$$\min_{w} E\{\|d(t) - r^T(t)w(t)\|^2\} \tag{14}$$

where $w(t)$ is the M-dimensional parameter vector. The TLS solution can be obtained by solving the following optimal problem :

$$\min_{\bar{w}} \left\{ \bar{w}^T \bar{r}(t)\, \bar{r}^T(t)\, \bar{w} / \left(\bar{w}^T \bar{D}\, \bar{w}\right) \right\} \tag{15}$$

$$\begin{bmatrix} w_{\text{TLS}} \\ -1 \end{bmatrix} = -\frac{\bar{w}}{w_{N+1}} \tag{16}$$

where $\bar{w}$ is an $(M+1)$-dimensional parameter vector. The weighting matrix is represented by $\bar{D} = \left[\begin{array}{c|c} I & 0 \\ \hline 0 & \beta \end{array}\right]$, where I is the M-dimensional identity matrix, $\beta = \sigma_0^2/\sigma_1^2$. The parameter vector is updated by

$$\bar{w}(t) = \bar{w}(t-1) + \alpha(t)\bar{r}(t) \tag{17}$$

where $\alpha(t)$ can be efficiently computed in $\mathcal{O}(M)$ multiplications by

$$\min_{\alpha(t)} \mathbf{E}\{\bar{w}(t)\} = \min_{\alpha(t)} \left\{ \bar{w}^T(t)\bar{R}(t)\,\bar{w}(t) / \left[\bar{w}^T(t)\bar{D}\,\bar{w}(t)\right] \right\} \tag{18}$$

Using an approach similar to [10] and the shift structure in the data [11], we get a new fast RTLS algorithm, Algorithm 1, further outlined in [25].

Algorithm 1 *(Fast recursive TLS algorithm).* Initialize $\bar{w}(0) = [0, 0, \ldots, -1]$, $\lambda_{\min}(0) = 0$, $N = M + 1$, $\mu \in [0.99, 1.0]$. For $t = 1, 2, \ldots$

1)	update extended data vector	
2)	update gain $k(t)$ using the approach in [25]	$14N + 12$
3)	$\lambda^0(t) = \mu\lambda_{\min}(t)\boldsymbol{TB}w(t-1)\boldsymbol{BDB}w(t-1) + (\boldsymbol{TB}w(t-1)\boldsymbol{B}r(t))^2$	$N + 5$
4)	$a = [\boldsymbol{T}k(t)\boldsymbol{B}r(t)][\boldsymbol{TB}w(t-1)\boldsymbol{BDB}r(t)] - [\boldsymbol{T}k(t)\boldsymbol{B}w(t)][\boldsymbol{TB}r(t)\boldsymbol{BDB}r(t)]$	$3N + 2$
5)	$b = [\boldsymbol{T}k(t)\boldsymbol{B}r(t)][\boldsymbol{TB}w(t-1)\boldsymbol{BDB}w(t-1)] - \lambda^0(t)[\boldsymbol{TB}r(t)\boldsymbol{BDB}r(t)]$	2
6)	$c = [\boldsymbol{T}k(t)\boldsymbol{B}w(t-1)][\boldsymbol{TB}w(t-1)\boldsymbol{BDB}w(t-1)] - \lambda^0(t)[\boldsymbol{TB}w(t-1)\boldsymbol{BDB}r(t)]$	2
7)	$\alpha(t) = (-b + \sqrt{b^2 - 4ac})/2a$	6
8)	$\lambda_{\min}(t) = \left(\lambda^0(t) + 2\alpha(t)\boldsymbol{T}k(t)\boldsymbol{B}w(t-1) + \alpha^2(t)\boldsymbol{T}k(t)\boldsymbol{B}r(t)\right) / \left(\boldsymbol{TB}w(t-1)\boldsymbol{BDB}w(t-1) + 2\alpha(t)\boldsymbol{TB}w(t)\boldsymbol{BDB}r(t) + \alpha^2(t)\boldsymbol{TB}r(t)\boldsymbol{BDB}r(t)\right)$	7
9)	$\tilde{w}(t) = \tilde{w}(t-1) + \alpha(t)\boldsymbol{B}r(t)$	N
10)	$\boldsymbol{B}w(t) = \tilde{w}(t)/\|\tilde{w}(t)\|$	$2N + 2$
11)	$h = -[\boldsymbol{B}w]_{1,N-1}/[w]_N$	$N - 1$
	Total real multiplications, divisions, and square roots	$22N + 40$

4. RTLS algorithm in instrumental variable methods and gaussian colored noise

Suppose that there is a variable $y(t)$ that is related linearly to a set of n variables $x(t) = [x_1(t), \ldots, x_n(t)]^T$, that is

$$y(t) = \sum_{i=1}^{n} w_i x_i(t) + n(t) \tag{19}$$

in which $w = [w_1, \ldots, w_n]^T$ is a set of constant parameters, and $y(t)$ and $n(t)$ are the output variables and the noise variables, respectively. The unknown parameter vector w will be estimated by minimizing a quadratic cost function, involving the data vector $x(t)$ and the m-dimensional IV vector $z(t)$ [15,16], where $m > n$:

$$w = \arg\min_{w} \left\| \sum_{t=1}^{k} z(t)\boldsymbol{T}x(t)w - \sum_{t=1}^{k} z(t)y(t) \right\|^2 , \tag{20}$$

where k is the number of data points. It is straightforward to show that the solution to (20) can be obtained by a least-squares solution of the following over-determined set of linear equations [15,16] :

$$\left[\sum_{t=1}^{k} z(t)\boldsymbol{T}x(t) \right] w \approx \left[\sum_{t=1}^{k} z(t)y(t) \right] . \tag{21}$$

These are called the generalized Yule-Walker equations [15]. Consider the case where the input vector $x(t)$ is corrupted by noise including the additive noise and the measurement error. Define the augmented input data vector as

$$\tilde{x}(t) = [x_1(t), \ldots, x_n(t), y(t)]^T . \tag{22}$$

Let the augmented cross-correlation matrix be:

$$\tilde{R}_k = \frac{1}{k} \sum_{t=1}^{k} z(t)\boldsymbol{T}\tilde{x}(t) . \tag{23}$$

The TLS solution of the over-determined IV problem (21) can be obtained by finding the single right singular vector associated with the smallest singular value of $\tilde{R}_k$. We refer to such a solution as the total instrumental variable (TIV) solution, denoted by w_{TIV}. Thus, we have

$$\begin{bmatrix} w_{\mathrm{TIV}} \\ -1 \end{bmatrix} = -\frac{v_{n+1}}{v_{n+1,n+1}} , \tag{24}$$

where is the right singular vector corresponding to the smallest singular value of $\tilde{R}_k$, whereas $v_{n+1,n+1}$ is the last component of v_{n+1}. The rank-one modification of the augmented cross-correlation matrix can be written as

$$\tilde{R}_{k+1} = \lambda_k \tilde{R}_k + (1 - \lambda_k)\big(z\,(k+1)\,\boldsymbol{T}\tilde{x}\,(k+1)\big)\ , \tag{25}$$

where $\lambda_k = k/(k+1)$. Based on the well-known power algorithm [2], the RTIV algorithm is developed as follows :

$$\tilde{R}_{k+1} = \lambda_k \tilde{R}_k + (1 - \lambda_k)\big(z\,(k+1)\,\boldsymbol{T}\tilde{x}\,(k+1)\big) \tag{26}$$

$$\tilde{R}_{k+1}q(k) = \tilde{p}(k+1)\ ,\ \boldsymbol{T}\tilde{R}_{k+1}\tilde{p}(k+1) = \tilde{q}(k+1) \quad \text{Power method} \tag{27}$$

$$\sigma^2_{\max}(k+1) = \|\tilde{q}(k+1)\| \quad \text{Square of the largest singular value} \tag{28}$$

$$q(k+1) = \sigma^{-2}_{\max}(k+1)\tilde{q}(k+1) \quad \text{Normalization} \tag{29}$$

$$\tilde{R}_{k+1}v(k) = \tilde{u}(k+1)\ ,\ \boldsymbol{T}\tilde{R}_{k+1}\tilde{u}(k+1) = \tilde{v}(k+1) \quad \text{Power method} \tag{30}$$

$$\sigma^2_{\min}(k+1) = \|\tilde{v}(k+1)\| \quad \text{Square of the smallest singular value} \tag{31}$$

$$\eta(k+1) = 0.5\,\sigma^2_{\max}(k+1) + \sigma^2_{\min}(k+1) \quad \text{Shift parameter} \tag{32}$$

$$\hat{v}(k+1) = \eta(k+1)v(k) - \tilde{v}(k+1) \quad \text{Shift operation} \tag{33}$$

$$v(k+1) = \hat{v}(k+1)/\,\|\hat{v}(k+1)\| \quad \text{Normalization} \tag{34}$$

$$\begin{bmatrix} w_{\text{TIV}}(k+1) \\ -1 \end{bmatrix} = \frac{-v(k+1)}{v_{m+1}(k+1)} \tag{35}$$

where $v_{m+1}(k+1)$ is the last component of $v(k+1)$, $\eta(k+1)$ is called the shift parameter. Equations (27) to (29) and (30) to (34) represent the recursive estimates for the singular vectors associated with the largest and the smallest singular values of the augmented cross-correlation matrix, respectively. Equations (27) and (30) are slightly different from the standard power method, in the sense that the matrix is non-square. Rearranging (27),(28),(29) yields the following equivalent equation :

$$p\,(k+1) = \tilde{R}^T_{k+1}\tilde{R}_{k+1}\,q(k)/\left\|\tilde{R}^T_{k+1}\tilde{R}_{k+1}\,q(k)\right\| \tag{36}$$

Equation (36) is the power iteration for tracking the right singular vector associated with the largest singular value of the augmented cross-correlation matrix and is applicable even if $\tilde{R}_{k+1}$ is singular. Note that the converging speed of (36) is proportional to $(\tilde{\sigma}_1(k+1)/\tilde{\sigma}_2(k+1))^2$ [2], where $\tilde{\sigma}_1(k+1)$ and $\tilde{\sigma}_2(k+1)$ are the first two largest distinct singular values of $\tilde{R}_{k+1}$. Similarly, rearranging (30) and (33) yields

$$B_{k+1} = \eta(k+1)I - \tilde{R}^T_{k+1}R_{k+1}\ , \tag{37}$$

$$v(k+1) = B_{k+1}\,v(k)/\|B_{k+1}v(k)\|\ , \tag{38}$$

where I is the $(n+1)\times(n+1)$ identity matrix. Equations (37-38) represent the power iteration for tracking the right singular vector associated with the largest singular value of the matrix B_{k+1}. In order to enforce the convergence of $v(k+1)$ in (37-38) to the singular vector associated with the smallest singular value of $\tilde{R}_{k+1}$, we should appropriately select the shift parameter $\eta(k+1)$ so that the largest singular value of B_{k+1} approaches the smallest singular value of $\tilde{R}_{k+1}$.
Since $B_{k+1} = \eta(k+1)I - \sum_{i=1}^{n+1}\sigma_i^2(k+1)\tilde{v}_i(k+1)\boldsymbol{T}\tilde{v}_i(k+1)$ (where $\tilde{v}_i(k+1)$ is the right singular vector associated with the ith singular value of $\tilde{R}_{k+1}$), the eigen-value distribution of the symmetric matrix B_{k+1} satisfies : $\eta(k+1)-\sigma_{n+1}^2(k+1) > \eta(k+1)-\sigma_n^2(k+1)\geq\cdots\geq \eta(k+1)-\sigma_1^2(k+1)$. When $\eta(k+1)\leq(\sigma_{n+1}^2+\sigma_1^2)/2$, the largest singular value of B_{k+1} is $\sigma_1^2(k+1)-\eta(k+1)$; when $\eta(k+1)>(\sigma_{n+1}^2+\sigma_1^2)/2$, the largest singular value of B_{k+1} is $\eta(k+1)-\sigma_{n+1}^2(k+1)$ associated with the smallest singular value of $\tilde{R}_{k+1}$. When $\eta(k+1)>(\sigma_{n+1}^2+\sigma_1^2)/2$, (37-38) can track the right singular vector associated with the smallest singular value of $\tilde{R}_{k+1}$. It can be shown that the computational complexity of the RTIV algorithm is $\mathcal{O}(mn)$ per iteration.

5. Robust TLS solution and TLS algorithms in impulse noise environments

In this section, a new adaptation is proposed. We consider the matrix equation

$$[A|b][\boldsymbol{T}x,-1]^T = 0 \tag{39}$$

where $A\in\mathbb{R}^{m\times n}$ is a data matrix; $b\in\mathbb{R}^m$ the data vector; $x\in\mathbb{R}^n$ a parameter vector. Notice that the classical TLS estimate is far from optimal when the data is contaminated by impulsive noise. For α-stable processes, we develop the total least L_p-norm (TLPN) estimate by minimizing the L_p-norm of the error matrix $D\in\mathbb{R}^{m\times(n+1)}$ of $[A|b]$. Given a constant $p\geq 1$, the TLPN estimate $\hat{x}_{\text{TLPN}}$ is defined by

$$\hat{x}_{\text{TLPN}} = \arg\min_x \|D\|_p^p \tag{40}$$

where $\|D\|_p$ is denoted by $\|D\|_p = \left(\sum_{i=1}^m\sum_{j=1}^n d_{ij}^p\right)^{1/p}$. Given a hyper-plane $a^Tx=0$ and an observed point (a_i,b_i), the L_p-norm distance from this point to the hyper-plane is given by

$$\|d_i\|_p^p = \left\|a_i^Tx-b_i\right\|_p^p / \left(\|x\|_p^{p/(p-1)}+1\right)^{p-1} \tag{41}$$

Then $\hat{x}_{\text{TLPN}}$ is given by

$$\hat{x}_{\text{TLPN}} = \arg\min_x \sum \left\|a_i^Tx-b_i\right\|_p^p / \left(\|x\|_p^{p/(p-1)}+1\right)^{p-1} \tag{42}$$

Let $x(k) \in \mathbb{R}^n$ be the input signal, $d(k)$ the desired output, $w(k)$ the weight vector. Both the input and output signal samples are corrupted by additive alpha-stable noise [19]. The TLPN solution of the above problem can be obtained by solving

$$\min E\{J(w)\} = E\left\{\left\|w^T(k)x(k) - d(k)\right\|^p / \left(\|w\|_p^{p/(p-1)} + 1\right)^{p-1}\right\} \tag{43}$$

The objective of an adaptation algorithm is to minimize the cost function by adjusting the coefficients of the filter. In the TLMP algorithm we take $J(w)$ itself as an estimate of $E\{J(w)\}$. Then, in the adaptive process, we have a gradient estimate of the form

$$\begin{aligned}\tilde{\nabla}(k) = &\; p\,\|e\|^{p-1}\operatorname{sign}(e)x\left(\|w\|^{p/(p-1)} + 1\right) \\ &- p\,\|e\|^{p}\,\|w\|^{1/(p-1)}\operatorname{sign}(w)/\left(\|w\|^{p/(p-1)} + 1\right)\end{aligned} \tag{44}$$

where $e(k) = \boldsymbol{T}x(k)w(k) - d(k)$. Using this estimate of the gradient, we can specify a steepest-descent-type algorithm

$$w(k+1) = w(k) - \mu\tilde{\nabla}(k) \tag{45}$$

This is the TLMP algorithm.

References

[1] G.H. Golub and C.F. Van Loan. *Matrix Computations.* The Johns Hopkins University Press, M.D., Baltimore, 1989.

[2] S.S. Haykin. *Adaptive Filter Theory.* Prentice-Hall, New York, 2nd edition, 1991.

[3] O. Arikan, M. Belge, E. Cetin, and E. Erzin. Adaptive Filtering Approaches for Non-Gaussian Stable Processes. In *Proceeding International Conference on Acoustics, Speech and Signal Processing'95*, volume 2, pages 1400–1403, 1995.

[4] C.E. Davila. An Efficient Recursive Total Least Mean Squares Algorithm for FIR Adaptive Filtering. *IEEE Transactions on Signal Processing*, 42(2):268–280, 1994.

[5] D-Z. Feng et al. An Adaptive Total Least Mean L_p-Norm Algorithm applied in α-stable Noise Environments. *Acta Electronica Sinica (Accepted, in Chinese).*

[6] D-Z. Feng et al. A Fast Recursive Total Least Squares Algorithm for Finite Impulse Response Filters. *J. Xi'an Jiaotong University*, 33(4):44–51, 1999 (in Chinese).

[7] D.Z. Feng, Z. Bao, and L.C. Jiao. Total Least Mean Squares Algorithm. *IEEE Transactions on Signal Processing*, 46(8):2212–2220, 1998.

[8] B. Friedlander. The Overdetermined Recursive Intrumental Variable Method. *IEEE Transactions on Automatic Control*, 29:353–356, February 1984.

[9] K.Q. Gao, M.O. Ahmad, and M.N.S. Swamy. A Constrained Anti-Hebbian Learning Algorithm for Total Least-Squares Estimation with Applications to Adaptive FIR and IIR Filtering. *IEEE Transactions on Circuits and Systems II : Analog Digital Signal Processing*, 41(11):718–729, 1994.

[10] S. Kalluri and G.R. Arce. Adaptive Weighted Myriad Filter Algorithms for Robust Signal Processing in α-stable Noise Environments. *IEEE Transactions on Signal Processing*, 46(2):322–334, February 1998.

[11] L. Ljung, M. Morf, and D. Falconer. Fast Calculation of Gain Matrices for Recursive Estimation Schemes. *International Journal of Control*, 27(1):1–19, 1978.

[12] S. Ljung and L. Ljung. Error Propagation Properties of Recursive Least-Squares Adaptation Algorithms. *Automatica*, 21(2):157–167, 1985.

[13] F-L. Luo, R. Unbehauen, and Cichocki A. A Minor Component Analysis Algorithm. *Neural Networks*, 10(2):291–297, 1997.

[14] R.L. Moses and A.A. (Luois) Beex. Instrumental Variable Adaptive Array Processing. *IEEE Transactions on Aerospace Electronics Systems*, 24:192–202, March 1988.

[15] E. Oja. Principal Components, Minor Components, and Linear Neural Networks. *Neural Networks*, 5:927–935, 1992.

[16] N.L. Owsley. Adaptive Data Orthogonalization. In *IEEE Conference on Acoustics Speech and Signal Processing*, pages 109–112, 1978.

[17] J.B. Rosen, H. Park, and J. Glick. Total Least Norm Formulation and Solution for Structured Problems. *SIAM Journal on Matrix Analysis and Applications*, 17(1):110–128, 1996.

[18] M. Shao and C.L. Nikias. Signal Processing with Fractional Lower-Order Moments : Stable Processess and their Applications. In *Proceedings of the IEEE*, volume 81, pages 986–1010, 1993.

[19] D.T.M. Slock and T. Kailath. Numerically Stable Fast Transversal Filters for Recursive Least Squares Adaptive Filtering. *IEEE Transactions on Signal Processing*, 39(1):92–114, 1991.

[20] T. Soderstrom and P. Stoica. *System Identification.* Prentice-Hall, London, U.K., 1989.

[21] P. Stoica, M. Ottersten, and B. Viberg. An Instrumental Variable Approach to Array Processing in Spatially Correlated Noise Fields. *IEEE Transactions on Signal Processing*, 42(1):121–133, January 1994.

[22] S. Van Huffel, H. Park, and J.B. Rosen. Formulation and Solution of Structured Total Least Norm Problems for Parameter Estimation. *IEEE Transactions on Signal Processing*, 44(10):2464–2474, 1996.

[23] S. Van Huffel and J. Vandewalle. *The Total Least Squares Problem, Computational Aspects and Analysis.* Frontiers in Applied Mathematics. SIAM, Philadelphia, 1991.

[24] L. Xu, E. Oja, and C.Y. Suen. Modified Hebbian Learning for Curve and Surface Fitting. *Neural Networks*, 5:441–457, 1992.

Part IX

ERRORS-IN-VARIABLES APPLICATIONS IN OTHER FIELDS

EFFICIENT COMPUTATION OF THE RIEMANNIAN SVD IN TOTAL LEAST SQUARES PROBLEMS IN INFORMATION RETRIEVAL

Ricardo D. Fierro

Department of Mathematics, California State University, San Marcos, CA 92096 USA

fierro@csusm.edu

Michael W. Berry

Department of Computer Science, University of Tennessee, Knoxville TN 37996-3450

berry@cs.utk.edu

Abstract Recently, a nonlinear generalization of the singular value decomposition (SVD), called the Riemannian-SVD (R-SVD), for solving full rank total least squares problems was extended to low rank matrices within the context of latent semantic indexing (LSI) in information retrieval. This new approach, called RSVD-LSI, is based on the full SVD of an $m \times n$ term-by-document matrix A and requires the dense $m \times m$ left singular matrix U and the $n \times n$ right singular matrix V. Here, m corresponds to the size of the dictionary and n corresponds to the number of documents. We dicuss this method along with an efficient implementation of the method that takes into account the sparsity of A.

Keywords: information retrieval, Riemannian singular value decomposition.

1. Introduction

In Section 1.1 we briefly discuss vector space models in information retrieval. In Section 1.2 we present our notation for orthogonal decompositions that can be used in vector space models. In Section 1.3 we discuss how the total least squares problem arises in information retrieval.

S. Van Huffel and P. Lemmerling (eds.), Total Least Squares and Errors-in-Variables Modeling, 353–364.

1.1 Vector space models for information retrieval

As digital libraries and the Web continue to grow in textual content, the ability to index concepts and retrieve relevant information to users' needs becomes more and more problematic. The ability to process both static and dynamic text collections (e.g., newsfeeds, electronic mail, and legislative proceedings) in an efficient manner requires automated technologies (supervised by humans, of course) for a truly sustainable information processing environment. Among the earliest computational frameworks for automated indexing is the *vector space model.* Salton and his colleagues [20] demonstrated early success in automating text retrieval using this model in the SMART [19] system.

The classical vector space model processed queries using a term-by-document matrix $A \in \mathbb{R}^{m \times n}$ in which each column vector of the matrix defines the m coordinates of the corresponding document. Each row of $A = (a_{ij})$ reflects a term from the dictionary of the collection (usually sorted alphanumerically) so that each matrix element, a_{ij}, denotes the (weighted) frequency of occurrence for term i in document j. In one sense, the matrix A is simply an incidence matrix which is normally quite sparse since most documents will use a small fraction of words from the dictionary of all terms parsed from the text collection. As the element a_{ij} can reflect the importance of term i in the semantic representation of document j, numerous statistically-based weighting schemes have been developed [1, 3] to improve the conceptual relationships between terms and documents. For example, one can define

$$a_{ij} = l_{ij} g_i d_j, \tag{1}$$

where l_{ij} is the local weight for term i occurring in document j, g_i is the global weight for term i in the text collection, and d_j is a document normalization factor which specifies whether or not the columns of A should be normalized. Local weights (l_{ij}) would be used to specify importance within a documents and global weights (g_i) are typically used to define the overall importance of the term across the entire collection. One simple but commonly used (local, global) weighting pair is (*tf*,*idf*), where *tf* is the term frequency within the document and *idf* is the inverse document frequency or number of documents from the collection that contain the term. Many other choices for term weighting are certainly possible.

As with each column of the matrix A, a query can also be represented by an m-vector of weighted frequencies. Most of the elements of such a vector will be zero, of course. By assumption, documents closest to the query in $\mathbb{R}^m$ (in a geometric sense) are considered *similar* or related to the query [4]. In fact, a simple measure such as the cosine can be used to rank documents in similarity to a given query $q = [q_1, q_2, \ldots, q_m]^T$. If we label the columns of matrix A as A_j, $j = 1, \ldots, n$, then we can compute the corresponding n cosines via

$$\cos\theta_j = \frac{A_j^T q}{\|A_j\|\|q\|} \qquad \text{for } j = 1, \ldots, n . \tag{2}$$

Clearly, as the size of the dictionary (m) grows so also does the dimension of each document vector in the vector space model. In addition, similarity (as computed by measures such as the cosine) greatly depends on the co-occurrence of terms in a document and the query. In natural language, there are usually many possible words (terms) that can be used to describe a given concept or idea (synonymy), and many words (nouns and verbs) can have more than one meaning (polysemy) such as *bank*, *mouse*, and *ship*. Such phenomena pose difficulties for the scoring of documents by formulas such as (2). High scoring documents will simply have a higher literal match of terms from a user's query. As the terms used by readers and authors will continue to differ in just about every social aspect, the literal matching offered by the vector space model can be rather limited for retrieving relevant information in a conceptual sense. A more effective approach would perhaps allow a user to retrieve information by conceptual topic or the meaning of a particular document.

As reported in the precursor to this work, Latent Semantic Indexing (LSI) [10] is one attempt to overcome the problems of lexical matching by computing a low-rank approximation of the term-by-document matrix A. Using the truncated SVD of A to produce k-dimensional term and document vectors (as opposed to m-dimensional vectors using the classical vector space model) can force the geometric alignment of document vectors having similar (but not necessarily equivalent) term usage patterns. As discussed in [3, 5], the truncated SVD computation can be viewed as a principal component analysis of the original term-to-document associations and thereby filter variations across the collection by representing terms and documents using only the k dominant principal components.

We mention that we will use Matlab notation in this paper (e.g., $M(:, c_1 : c_2)$ denotes the submatrix of M consisting of columns c_1 through c_2 and $M(r, :)$ denotes row r of M).

1.2 The SVD and ULV decomposition

Recall that for a matrix $A \in \mathbb{R}^{m\times n}$, with $m \geq n$, its singular value decomposition (SVD) [13] can be defined as

$$A = U\Sigma V^T = [U_k\ U_0] \begin{bmatrix} \Sigma_k & 0 \\ 0 & \Sigma_0 \end{bmatrix} \begin{bmatrix} V_k^T \\ V_0^T \end{bmatrix} \tag{3}$$

where

$$U \in \mathbb{R}^{m\times m},\ U^TU = I_m, \quad V \in \mathbb{R}^{n\times n},\ V^TV = I_n\ ,$$
$$\Sigma \in \mathbb{R}^{m\times n},\ \Sigma_k = \mathrm{diag}(\sigma_1 \cdots \sigma_k) \text{ and } \Sigma_0 = \mathrm{diag}(\sigma_{k+1} \cdots \sigma_n)\ .$$

Truncation of the lower portion of Σ yields the reduced-rank representation

$$A \approx A_k = [U_k\ U_0] \begin{bmatrix} \Sigma_k & 0 \\ 0 & 0 \end{bmatrix} \begin{bmatrix} V_k^T \\ V_0^T \end{bmatrix} = U_k\Sigma_k V_k^T, \tag{4}$$

where $U_k \in \mathbb{R}^{m\times k}, \Sigma_k \in \mathbb{R}^{k\times k}$, and $V_k \in \mathbb{R}^{n\times k}$. Normally, $k \ll n$ so that the LSI model achieves considerable compression of the vector representations used in the classic vector space model. The rows of U_k and V_k now encode the terms and documents, respectively. As shown in [3, 18], the $m \times 1$ query vector q denoting weighted frequencies of the appropriate dictionary terms can be projected into $\mathbb{R}^k$ via

$$\hat{q} = \Sigma_k^{-1} U_k^T q\ . \tag{5}$$

The LSI method computes the corresponding n cosines via

$$\cos\theta_j = \frac{V_k(j, :)\, \hat{q}}{\|V_k(j, :)\| \|\hat{q}\|} \qquad \text{for } j = 1, \ldots, n\ . \tag{6}$$

As first noted in [2], alternative matrix decompositions such as the ULV can be quite effective for approximating dominant singular values and singular vectors to suitable accuracy vector encodings for LSI. The ULV decomposition of A is given by

$$A = U_L L V_L^T = [U_{Lk}\ U_{L0}] \begin{bmatrix} L_k & 0 \\ H & E \end{bmatrix} \begin{bmatrix} V_{Lk}^T \\ V_{L0}^T \end{bmatrix} \tag{7}$$

where $U_L^T U_L = I_m$ and $V_L^T V_L = I_n$, and L_k is a $k \times k$ block lower triangular matrix whose singular values approximate the first k singular values of A.

As an alternative to the SVD, the rows of U_{Lk} and V_{Lk} now encode the terms and documents, respectively, and the $m \times 1$ query vector q can be projected into $\mathbb{R}^k$ via

$$\hat{q}_L = L_k^{-1} U_{Lk}^T q\ . \tag{8}$$

Similar to (6), the corresponding n cosines are computed via

$$\cos\theta_j = \frac{V_{Lk}(j,:)\,\hat{q}_L}{\|V_{Lk}(j,:)\|\|\hat{q}_L\|} \qquad \text{for } j = 1, \ldots, n \,. \tag{9}$$

1.3 Connection between total least squares and information retrieval

The total least problem can be reduced to an equivalent minimization problem of finding a rank-deficient matrix approximation to a given full-rank matrix. De Moor [8] developed the Riemannian SVD (RSVD) method for solving *structured* total least squares problems, where the rank-deficient matrix must also take into account some restrictions related to the structure of the full-rank matrix.

The RSVD method is modified in [17] for low-rank matrices, in the sense $k \ll \min(m,n)$, that arise in information retrieval, as follows. Suppose A_k defined in (4) has the splitting $A_k = A_k^F + A_k^C$, where A_k^F is the portion of A_k that is not allowed to change in the structured approximation. Then the following minimization problem

$$\begin{array}{ll} \min_{B_k \in \mathbb{R}^{m\times n},\, y\in\mathbb{R}^n} & \|A_k - B_k\|_F \\ \text{subject to} & B_k y = 0,\ y^T y = 1,\ B_k = A_k^F + B_k^C \end{array} \tag{10}$$

can be formulated as the equivalent RSVD problem

$$A_k v = \tau D_v u,\ u^T D_v u = 1,\ A_k^T u = \tau D_u v,\ v^T D_u v = 1 \,. \tag{11}$$

The generalized triplet (τ, u, v) corresponds to the minimal τ that satisfies (11). The solution B_k to (10) is then given by [8, 16]

$$B_k = A_k - \operatorname{diag}(u)\, P \operatorname{diag}(v)\, \tau \,.$$

P is a perturbation matrix that can be automatically generated having zeros at all entries of A_k^F and nonzeros elsewhere, $D_u = \operatorname{diag}(P^T \operatorname{diag}(u) u)$ and $D_v = \operatorname{diag}(P \operatorname{diag}(v) v)$ are positive-definite diagonal matrices, and finally, $\operatorname{rank}(B_k) \approx k$ [17]. The dominant SVD factors of B_k must then be computed and applied using the LSI approach as in (8) and (9), resulting in the RSVD-LSI method.

The following inverse iteration algorithm (RINVIT [17]) for computing the generalized triplet (τ, u, v) of A_k is based on the full SVD of A in (3) and forms the core of the RSVD-LSI method:

RINVIT requires $\mathcal{O}(m^2+n^2)$ flops per iteration, and the $\mathcal{O}(m^2+n^2)$ storage requirements are due to the explicit SVD factors U_k, U_0, V_k, and V_0. Numerical experiments in [17] based on the LATIMES [14] document collection demonstrate that RSVD-LSI is superior to the LSI model [10] based on the SVD as well as the LSI model with relevance feedback [9] in terms "precision" and "recall."

Table 1. RINVIT algorithm

0. Compute the full SVD of A as in (3).
1. Set $i = 0$
2. Initialize $\tau^{(0)}$, $v^{(0)}$, and $u^{(0)}$.
3. Compute matrices $D_u^{(0)}$ and $D_v^{(0)}$.
4. While (not converged) do
    5. Set $i = i + 1$.
    6. $x^{(i)} = \Sigma_k^{-1} V_k^T D_u^{(i-1)} v^{(i-1)} \tau^{(i-1)}$
    7. $y^{(i)} = U_0^T u^{(i-1)}$
    8. $u^{(i)} = U_k x^{(i)} + U_0 y^{(i)}$
    9. $s^{(i)} = \Sigma_k^{-1} U_k^T D_v^{(i-1)} u^{(i-1)} \tau^{(i-1)}$
    10. $t^{(i)} = V_0^T v^{(i-1)}$
    11. $v^{(i)} = V_k s^{(i)} + V_0 t^{(i)}$
    12. $\tau^{(i)} = u^{(i)\,T} A v^{(i)}$
    13. $D_u^{(i)} = \text{diag}(P \text{diag}(v^{(i)}) v^{(i)})$
    14. $D_v^{(i)} = \text{diag}(P^T \text{diag}(u^{(i)}) u^{(i)})$
    15. Convergence test

2. Efficient implementation of RSVD-LSI

In Section 2.1 we motivate a new method for computing the SVD factors, within a prescribed tolerance, of the sparse matrix A required in RSVD-LSI, and then outline the first and second steps of the new method in Sections 2.2 and 2.3, respectively.

2.1 Motivation

Our efficient implementation of RINVIT and RSVD-LSI relies on three main considerations. First, RINVIT does not need the SVD factors U_k, U_0, V_k, and V_0 explicitly, but rather matrix-vector products involving these factors (e.g., $U_k x^{(i)}$ and $V_0^T v^{(i-1)}$). Second, it is well known the SVD factors (e.g., U_k and V_0) can be substituted with the corresponding ULV factors (e.g., U_{Lk} and V_{L0}) in problems that involve truncated SVD solutions, reduced-rank matrix approximations, and projection matrices, such as in RINVIT, as long as the off-diagonal block $\|H\|$ is "sufficiently" small [2, 12]. More recently, Jessup and Martin [15] affirm that both QR and ULV decompositions can produce vector encodings for LSI that are equally competitive (in retrieval accuracy) to those produced by the

SVD. As demonstrated in [2], updating the ULV decomposition can yield at least an order of magnitude reduction in floating-point computations compared to the SVD. Third, the heavy computational and storage demands of the full SVD in large-scale problems must be avoided. The block ULV method we propose depends on a short sequence of Householder transformations to compute U_L and V_L. Hence we use the WY representation [7, 11] to represent them implicitly with $U_L = I_m + W_u Y_u^T$ and $V_L = I_n + W_v Y_v^T$, where W_u and Y_u are $m \times k$ matrices and W_v and Y_v are $n \times k$ matrices. The main advantages with these implicit representations are that matrix-vector products involving the ULV factors can be computed efficiently (e.g., $U_{Lk} x^{(i)}$ and $V_{L0}^T v^{(i-1)}$) using $\mathcal{O}(mk + nk)$ flops, and modest $\mathcal{O}(mk + nk)$ storage requirements are required for the representations of U_L and V_L. Furthermore, block algorithms generally perform faster than algorithms based on lower level operations on high-speed computers due to parallelization of block operations and reduced data movement (some empirical results are given in [6]). Therefore, this new method for computing the SVD factors needed in RINVIT and RSVD-LSI lead to a more scalable approach to the encoding and updating of term and document vectors used by the LSI model.

2.2 The first step

The computations on the first step are:

Step 1.1: Define $U^{(0)} = I$, $V^{(0)} = I$, and $E^{(0)} = (U^{(0)})^T A V^{(0)}$. Let $X^{(1)}$ represent an $m \times p_1$ orthonormal matrix such that $\mathcal{X}^{(1)}$ approximates a left singular subspace of $E^{(0)}$.

Step 1.2: Define $m_1 = m - p_1$ and $n_1 = n - p_1$. Determine an $m \times m$ orthogonal matrix $\tilde{U}$ with the partition

$$\tilde{U} = \begin{array}{c} \\ \end{array} \overset{\begin{array}{cc} p_1 & m_1 \end{array}}{\left[\begin{array}{cc} \tilde{U}_1 & \tilde{U}_2 \end{array}\right]} \; m$$

such that $\tilde{\mathcal{U}}_1 = \mathcal{X}^{(1)}$ using a QR decomposition of $X^{(1)}$ based on a sequence of Householder transformations:

$$X^{(1)} = \tilde{U} \begin{bmatrix} R_{11} \\ 0 \end{bmatrix}, \qquad \tilde{U} = H_1 \cdots H_{p_1}.$$

Step 1.3: Form the $n \times p_1$ matrix $J = (E^{(0)})^T X^{(1)}$. Determine an $n \times n$ orthogonal matrix $\tilde{V} = (\tilde{V}_1 \; \tilde{V}_2)$ with the partition

$$\tilde{V} = \overset{\begin{array}{cc} p_1 & n_1 \end{array}}{\left[\begin{array}{cc} \tilde{V}_1 & \tilde{V}_2 \end{array}\right]} \; n$$

such that $\tilde{V}_1 = \mathcal{J}$ using a QR decomposition of J based on a sequence of Householder transformations:

$$J = \tilde{V} \begin{bmatrix} (L_{11})^T \\ 0 \end{bmatrix}, \qquad \tilde{V} = \tilde{H}_1 \cdots \tilde{H}_{p_1}.$$

The matrix L_{11} is a $p_1 \times p_1$ lower triangular matrix.

Step 1.4: Given a deflation tolerance δ, we shall assume the deflation condition

$$\|H_{21}^{(1)}\| < \delta \ , \ \text{where } H_{21}^{(1)} = \tilde{U}_2^T E^{(0)} \tilde{V}_1, \tag{12}$$

is satisfied. The sequences of Householder transformations obtained in Steps 1.2-1.3 can be incorporated into the orthogonal factors $U^{(0)}$ and $V^{(0)}$ using the WY representation to obtain $U^{(1)} = I + W^{(1)}(Y^{(1)})^T$ and $V^{(1)} = I + \tilde{W}^{(1)}(\tilde{Y}^{(1)})^T$. Both $W^{(1)}$ and $Y^{(1)}$ have dimensions $m \times p_1$, whereas $\tilde{W}^{(1)}$ and $\tilde{Y}^{(1)}$ have dimensions $n \times p_1$. This completes the first step.

The middle matrix $L^{(1)} = (U^{(1)})^T A V^{(1)}$ has the 2×2 block partition

$$L^{(1)} = \begin{array}{c} \\ \left[\begin{array}{cc} L_{11} & 0 \\ H_{21}^{(1)} & E^{(1)} \end{array} \right] \end{array} \begin{array}{l} \\ p_1 \\ m_1 \end{array}, \quad \text{columns: } p_1 \ \ n_1$$

where

$$H_{21}^{(1)} = \tilde{U}_2^T E^{(0)} \tilde{V}_1 = AV^{(1)}(: ,1 : p_1) - U^{(1)}(: ,1 : p_1)\, L_{11}$$

and

$$E^{(1)} = \tilde{U}_2^T E^{(0)} \tilde{V}_2 = (U^{(1)}(: ,p_1 + 1 : m))^T A V^{(1)}(: ,p_1 + 1 : n)\ .$$

Thus the deflation criterion ensures the norm of the off-diagonal block $H_{21}^{(1)}$ is sufficiently small (with respect to δ). The problem is essentially decoupled and naturally leads to block deflation where the same procedure is then applied to

$$E^{(1)} = \tilde{U}_2^T E^{(0)} \tilde{V}_2 = (U^{(1)}(: ,p_1 + 1 : m))^T A\, V^{(1)}(: ,p_1 + 1 : n)\ .$$

2.3 The second step

Define $m_1 = m - p_1$, $n_1 = n - p_1$, $m_2 = m - (p_1 + p_2)$, and $n_2 = n - (p_1 + p_2)$, where p_2 represents a natural number such that $p_1 + p_2 < \min(m, n)$.

Step 2.1: Let $X^{(2)}$ represent an $m \times p_2$ orthonormal matrix such that $\mathcal{X}^{(2)}$ approximates a left singular subspace of $E^{(1)}$.

Step 2.2: Determine an $m_1 \times m_1$ orthogonal matrix $\tilde{U}$ with the partition

$$\tilde{U} = \begin{array}{c} \begin{array}{cc} p_2 & m_2 \end{array} \\ \left[\begin{array}{cc} \tilde{U}_1 & \tilde{U}_2 \end{array} \right] m \end{array}$$

such that $\tilde{\mathcal{U}}_1 = \mathcal{X}^{(2)}$ using a QR decomposition of $X^{(2)}$ based on a sequence of Householder transformations:

$$X^{(2)} = \tilde{U} \begin{bmatrix} R_{22} \\ 0 \end{bmatrix}, \qquad \tilde{U} = H_1 \cdots H_{p_2}.$$

Step 2.3: Form the $n \times p_2$ matrix $J = (E^{(1)})^T X^{(2)}$. Determine an $n_1 \times n_1$ orthogonal matrix $\tilde{V} = (\tilde{V}_1\ \tilde{V}_2)$ with the partition

$$\tilde{V} = \begin{array}{c} \begin{array}{cc} p_2 & n_2 \end{array} \\ \left[\begin{array}{cc} \tilde{V}_1 & \tilde{V}_2 \end{array} \right] n \end{array}$$

such that $\tilde{\mathcal{V}}_1 = \mathcal{J}$ using a QR decomposition of J based on a sequence of Householder transformations:

$$J = \tilde{V} \begin{bmatrix} (L_{22})^T \\ 0 \end{bmatrix}, \qquad \tilde{V} = \tilde{H}_1 \cdots \tilde{H}_{p_1}.$$

The matrix L_{22} is a $p_2 \times p_2$ lower triangular matrix.

Step 2.4: We shall assume the deflation condition

$$\|H_{32}^{(2)}\| < \delta \ , \ \text{where } H_{32}^{(2)} = \tilde{U}_2^T E^{(1)} \tilde{V}_1, \tag{13}$$

is satisfied. After padding $\tilde{U}$ and $\tilde{V}$ with zeros, they can be incorporated into $U^{(1)}$ and $V^{(1)}$ using the WY representation to get $U^{(2)} = I + W^{(2)}(Y^{(2)})^T$ and $V^{(2)} = I + \tilde{W}^{(2)}(\tilde{Y}^{(2)})^T$. Here, $W^{(2)}$ and $Y^{(2)}$ have dimensions $m \times (p_1 + p_2)$, and $\tilde{W}^{(2)}$ and $\tilde{Y}^{(2)}$ have dimensions $n \times (p_1 + p_2)$. This completes the second step.

The middle matrix $L^{(2)} = (U^{(2)})^T A V^{(2)}$ has the 3×3 block partition

$$L^{(2)} = \begin{array}{c} \begin{array}{ccc} p_1 & p_2 & n_2 \end{array} \\ \left[\begin{array}{ccc} L_{11} & 0 & 0 \\ H_{21}^{(2)} & L_{22} & 0 \\ H_{31}^{(2)} & H_{32}^{(2)} & E^{(2)} \end{array} \right] \begin{array}{l} p_1 \\ p_2 \\ m_2 \ , \end{array} \end{array}$$

where

$$H_{32}^{(2)} = \tilde{U}_2^T E^{(1)} \tilde{V}_1 = AV^{(2)}(:,p_1{+}1 : p_1{+}p_2) - U^{(2)}(:,p_1{+}1 : p_1{+}p_2)\, L_{22}$$

and

$$E^{(2)} = \tilde{U}_2^T E^{(1)} \tilde{V}_2 = (U^{(2)}(:,p_1+p_2+1 : m))^T AV^{(2)}(:,p_1+p_2+1 : n)\ .$$

Thus the deflation criterion in Step 2.4 ensures the norm of the lower off-diagonal block $H_{32}^{(2)}$ is sufficiently small (with respect to δ) before the third step of the block algorithm.

In general, it can be shown the subspace $\mathcal{R}(X^{(i)})$ needs to adequately approximate some left singular subspace of $E^{(i-1)}$ to satisfy the inequality $\|H_{i+1,i}^{(i)}\| < \delta$ so that block deflation can occur. One choice for δ is $\hat{\sigma}_1\epsilon/k$, where $\hat{\sigma}_1$ is an approximation to $\|A\|$ and the user wants a ULV decomposition such that $\|H\|/\|A\| < \epsilon$.

3. Exploiting structure

Since m corresponds to the size of the dictionary and n corresponds to the number of documents, the term-by-document matrix A is typically large and sparse. Hence, the matrix $E^{(i-1)} = (U^{(i-1)}(:,s_{i-1} + 1 : m))^T AV^{(i-1)}(:,s_{i-1}+1 : n)$ should never be formed explicity, especially since a Lanczos-based method can form matrix-vector or matrix-matrix products involving $E^{(i-1)}$ in product form to compute the $(m - s_{i-1}) \times p_i$ orthonormal matrix $X^{(i)}$. Similarly, the dominant SVD factors of B_k can also be computed using the block algorithm by exploiting the structure of B_k in an obvious way with a Lanczos-based algorithm.

4. Approximating the correct singular values

In this section we address the issue of isolating approximations to the dominant k singular values σ_j, for $j = 1, 2, \ldots k$. A practical test involves accepting $X^{(i)}$ provided that $\|H_{i+1,i}^{(i)}\| < \delta$ for some appropriate tolerance δ, such as $\delta = 10^{-4}/k$. $X^{(i)}$ must also be accompanied by approximations to singular values. Suppose $s_i = \sum_{j=1}^{i} p_j$ singular value approximations $\tilde{\sigma}_i$, for $i = 1, 2, \ldots, s_i$, have been accepted, where $k \le s_i$. The block algorithm terminates when $\|E^{(i)}\| < \tilde{\sigma}_k$, but one cannot be sure the singular values have converged in the correct order. Therefore, we must find representations of L_k, U_{Lk}, U_{L0}, V_{Lk}, and V_{L0}.

At this point the current orthogonal ULV factors are $U^{(i)} = I + W^{(i)}(Y^{(i)})^T$ and $V^{(i)} = I + \tilde{W}^{(i)}(\tilde{Y}^{(i)})^T$, where $W^{(i)}$ and $Y^{(i)}$ are $m \times s_i$

matrices and $\tilde{W}^{(i)}$ and $\tilde{Y}^{(i)}$ are $n \times s_i$ matrices. Then

$$A = U^{(i)} L^{(i)} (V^{(i)})^T = U^{(i)} \begin{array}{c} \begin{array}{cc} s_i & n - s_i \end{array} \\ \left[\begin{array}{cc} L'_{s_i} & 0 \\ H' & E' \end{array} \right] \end{array} \begin{array}{l} s_i \\ m - s_i \end{array} (V^{(i)})^T,$$

where $L'_{s_i} = (U^{(i)}(:,1{:}\,s_i))^T A V^{(i)}(:,1{:}\,s_i)$. Let L'_{s_i} have the SVD

$$L'_{s_i} = (\hat{U}_k \, \hat{U}_0) \begin{array}{c} \begin{array}{cc} k & s_i - k \end{array} \\ \left[\begin{array}{cc} \hat{\Sigma}_k & 0 \\ 0 & \hat{\Sigma}_0 \end{array} \right] \end{array} \begin{array}{l} k \\ s_i - k \end{array} (\hat{V}_k \, \hat{V}_0)^T .$$

Then it follows $A = U_L L V_L^T$, where

$$L = \begin{array}{c} \begin{array}{ccc} k & s_i - k & n - s_i \end{array} \\ \left[\begin{array}{ccc} \hat{\Sigma}_k & 0 & 0 \\ 0 & \hat{\Sigma}_0 & 0 \\ H'_{11} & H'_{12} & E' \end{array} \right] \end{array} \begin{array}{l} k \\ s_i - k \\ m - s_i \end{array}, \quad \text{with } H' = (H'_{11} \; H'_{12}),$$

$$L_k = \hat{\Sigma}_k , \quad H = \left[\begin{array}{c} 0 \\ H'_{11} \end{array} \right], \quad E = \left[\begin{array}{cc} \hat{\Sigma}_0 & 0 \\ H'_{12} & E' \end{array} \right],$$

$$U_L = (U_{Lk} \, U_{L0}) = U^{(i)} \begin{array}{c} \begin{array}{ccc} k & s_i - k & m - s_i \end{array} \\ \left[\begin{array}{ccc} \hat{U}_k & \hat{U}_0 & 0 \\ 0 & 0 & I \end{array} \right] \end{array} \begin{array}{l} s_i \\ m - s_i \end{array} \quad \text{and}$$

$$V_L = (V_{Lk} \, V_{L0}) = V^{(i)} \begin{array}{c} \begin{array}{ccc} k & s_i - k & n - s_i \end{array} \\ \left[\begin{array}{ccc} \hat{V}_k & \hat{V}_0 & 0 \\ 0 & 0 & I \end{array} \right] \end{array} \begin{array}{l} s_i \\ n - s_i \end{array}$$

Obviously, matrix-vector multiplications involving U_{Lk}, U_{L0}, V_{Lk}, and V_{L0} in the improved version of RINVIT as well as RSVD-LSI can now be done efficiently with $\mathcal{O}(mk + nk)$ flops per iteration along with modest $\mathcal{O}(mk + nk)$ storage requirements.

References

[1] R. Baeza-Yates and B. Ribeiro-Neto. *Modern Information Retrieval.* Addison-Wesley, 1999.

[2] M.W. Berry and R.D. Fierro. Low-rank Orthogonal Decompositions for Information Retrieval Applications. *Num. Lin. Alg. with Applics.*, 3:301–327, 1996.

[3] M.W. Berry and M. Browne. *Understanding Search Engines: Mathematical Modeling and Text Retrieval.* SIAM, Philadelphia, 1999.

[4] M.W. Berry, Z. Drmač, and E. Jessup. Matrices, Vector Spaces, and Information Retrieval. SIAM Review, 41:335–362, 1999.

[5] M.W. Berry, S.T. Dumais, and G.W. O'Brien. Using Linear Algebra for Intelligent Information Retrieval. SIAM Review, 37:573–595, 1995.

[6] Christian H. Bischof. Fundamental Linear Algebra Computations on High-Performance Computers. Preprint MCS-P150-0490, Mathematics and Computer Science Division, Argonne National Laboratory, August 1990.

[7] C.H. Bischoff and C. Van Loan. The WY Representation for Products of Householder Matrices. *SIAM J. Sci. Comp.*, 8:s2-s13, 1987.

[8] B. De Moor. Structured Total Least Squares and L_2 Approximation Problems. *Lin. Alg. Appl.*, 188/189:163–205, 1993.

[9] S. T. Dumais. Improving the Retrieval from External Sources. *Behav. Res. Meth. Intrum. Comput.* 23:229-236, 1991.

[10] S. Deerwester, S. Dumais, G. Furnas, T. Landauer and R. Harshman. Indexing by Latent Semantic Analysis. *J. Amer. Soc. Info. Sci.*, 41:391-407, 1990.

[11] R. Fierro. Lanczos, Householder Transformations, and Implicit Deflation for Fast and Reliable Dominant Singular Subspace Computation. *Num. Lin. Alg. with Applics.*, 8:245–264, 2001.

[12] R.D. Fierro. Perturbation Analysis for Two-Sided (or Complete) Orthogonal Decompositions. *SIAM J. Matrix Anal. Appl.*, 17:383–400, 1996.

[13] G. Golub and C. V. Loan. *Matrix Computations.* John Hopkins University Press, 1996.

[14] D. Harman, E.M. Voorhees. in: Proceedings of the Fifth Text Retrieval Conference (TREC-5), Department of Commerce, National Institute of Standards and Technology. Gaithersburg, MD, NISC Special Publication 500-238, 1997.

[15] E.R. Jessup and J.H. Martin. *Taking a New Look at the Latent Semantic Analysis Approach to Information Retrieval.* in Computational Information Retrieval, M. Berry (Ed.), SIAM, Philadelpha, 2001, pp. 129-152.

[16] E.P. Jiang. Information Retrieval and Filtering Using the Riemannian SVD, Ph.D. Thesis, Department of Computer Science, The University of Tennessee, Knoxville, TN, 1998.

[17] E.P. Jiang and M. W. Berry. Solving Total Least Squares Problems in Information Retrieval. *Lin. Alg. Appl.* , 316:137–156, 2000.

[18] H. Park, M. Jeon, and J.B. Rosen, *Lower Dimensional Representation of Text Data in Vector Space Based Information Retrieval.* in Computational Information Retrieval, M. Berry (Ed.), SIAM, Philadelpha, 2001, pp. 3-23.

[19] G. Salton. *The SMART Retrieval System.* Prentice Hall, 1971.

[20] G. Salton and M.J. McGill. *Introduction to Modern Information Retrieval.* McGraw-Hill, New York, 1983.

CONSTRAINED TOTAL LEAST SQUARES FOR COLOR IMAGE RECONSTRUCTION

Michael K. Ng
Department of Mathematics, The University of Hong Kong, Pokfulam, Hong Kong
mng@maths.hku.hk

Nirmal K. Bose and Jaehoon Koo
Department of Electrical Engineering
Signal Processing Center, The Pennsylvania State University
University Park, PA 16802, U.S.A.
nkb@stspbkn.ee.psu.edu

Abstract Multiple undersampled color images of a scene are often obtained by using a charge-coupled device detector array of sensors which are shifted relative to each other by subpixel displacements. This geometry of sensors, where each color sensor has a subarray of sensing elements of suitable size has recently been popular in the task of attaining spatial resolution enhancement from the acquired low-resolution degraded color images that comprise the set of observations. With the objective of improving the performance of the signal-processing algorithms in the presence of the ubiquitous perturbation errors of displacements around the ideal subpixel locations (because of imperfections in fabrication) in addition to noisy observations, the regularized constrained total least squares (RCTLS) method is deployed here. The expected superiority of this RCTLS approach over the conventional least squares theory based algorithm is demonstrated by example.

Keywords: constrained total least squares, Hankel/Toeplitz matrix, image reconstruction, multisensors, regularization.

1. Introduction

Single image processing methods often provide unacceptable results because of the ill-conditioned nature of associated inverse problems. Therefore, multiple image processing has developed into an active research area because multiple deconvolution operators can be used to make the prob-

S. Van Huffel and P. Lemmerling (eds.), Total Least Squares and Errors-in-Variables Modeling, 365–374.

lem well-posed. Rapid progress in computer and semiconductor technologies is making it possible to implement multiple image processing tasks reasonably quickly, but the need for processing in real time requires attention to design of efficient and robust algorithms for implementation on current and future generations of computational architectures.

The spatial resolution of an image is often determined by imaging sensors. In a CCD camera, the image resolution is determined by the size of its photo-detector. An ensemble of several shifted images could be collected by a prefabricated planar array of CCD sensors and one may reconstruct an image with higher resolution which is equivalent to an effective increase of the sampling rate by interpolation. Fabrication limitations are known to cause subpixel displacement errors, which coupled with observation noise limit the use of least squares techniques in this scenario. It is well-known that TLS is an effective technique for solving a set of error contaminated equations [6] and therefore is an appropriate method for consideration in our high-resolution image reconstruction applications. However, a possible drawback of using the conventional TLS approach is that the formulation is not constrained in point spread functions obtained from multisensors. In this paper, an image processing technique that leads to the use of *constrained* total least squares (CTLS) theory is described. Then a computational procedure is advanced and the role of regularization is outlined.

Our paper is outlined as follows. In the next section, the mathematical model of color image reconstruction with multisensors is given. The regularized constrained total least squares (RCTLS) formulation is considered in Section 3. An alternating minimization algorithm for regularized constrained TLS-based high-resolution color image reconstruction is developed in Section 4. Finally, numerical results are presented in Section 5.

2. The mathematical model

We begin with a brief introduction of the mathematical model in high-resolution color image reconstruction. Details can be found in [1, 2, 3].

Consider a sensor array with $L_1 \times L_2$ sensors, each sensor has $N_1 \times N_2$ sensing elements (pixels) and the size of each sensing element is $T_1 \times T_2$. Our aim is to reconstruct an image of resolution $M_1 \times M_2$, where $M_1 = L_1 \times N_1$ and $M_2 = L_2 \times N_2$. To maintain the aspect ratio of the reconstructed color image, we consider the case where $L_1 = L_2 = L$ only. For simplicity, we assume that L is an even number in the following discussion. In order to have enough information to resolve the high resolution color image, there are subpixel displacements between the sensors. In the ideal case, the sensors are shifted from each other by a value proportional to $T_1/L \times T_2/L$. However, in practice there can be small perturbations around these ideal subpixel locations due to imperfection of the mechanical imaging system. Thus, for $l_1, l_2 = 0, 1, \cdots, L-1$ with $(l_1, l_2) \neq (0,0)$, the horizontal and vertical displacements $d^x_{l_1 l_2}$ and $d^y_{l_1 l_2}$ of the $[l_1, l_2]$-th sensor array with respect to the $[0,0]$-th reference sensor array are given by

$$d^x_{l_1 l_2} = \frac{T_1}{L}(l_1 + \epsilon^x_{l_1 l_2}) \quad \text{and} \quad d^y_{l_1 l_2} = \frac{T_2}{L}(l_2 + \epsilon^y_{l_1 l_2}).$$

Here $\epsilon^x_{l_1 l_2}$ and $\epsilon^y_{l_1 l_2}$ denote, respectively, the normalized horizontal and vertical displacement errors. We remark that the parameters $\epsilon^x_{l_1 l_2}$ and $\epsilon^y_{l_1 l_2}$ can be obtained by manufacturers during camera calibration. We assume that $|\epsilon^x_{l_1 l_2}| < \frac{1}{2}$ and $|\epsilon^y_{l_1 l_2}| < \frac{1}{2}$ for $0 \leq l_1, l_2 \leq L-1$. For if not, the low resolution color images observed from two different sensor arrays will be overlapped so much that the reconstruction of the high resolution image is rendered impossible.

Let $\mathbf{f}^{(r)}$, $\mathbf{f}^{(g)}$ and $\mathbf{f}^{(b)}$ be the original scene in red, green and blue channels respectively. Then the observed low resolution image in the i-th ($i \in \{r, g, b\}$) channel $\bar{\mathbf{g}}^{(i)}_{l_1 l_2}$ for the (l_1, l_2)-th sensor is modeled by:

$$\bar{\mathbf{g}}^{(i)}_{l_1 l_2}[n_1, n_2] = \eta^{(i)}_{l_1 l_2} +$$
$$+ \textstyle\sum_{j \in \{r,g,b\}} w_{ij} \left\{ \int_{T_2(n_2 - \frac{1}{2}) + d^y_{l_1 l_2}}^{T_2(n_2 + \frac{1}{2}) + d^y_{l_1 l_2}} \int_{T_1(n_1 - \frac{1}{2}) + d^x_{l_1 l_2}}^{T_1(n_1 + \frac{1}{2}) + d^x_{l_1 l_2}} \mathbf{f}^{(j)}(x_1, x_2) dx_1 dx_2 \right\}.$$

for $n_1 = 1, \ldots, N_1$ and $n_2 = 1, \ldots, N_2$. Here $\eta^{(i)}_{l_1 l_2}$ is the noise corresponding to the (l_1, l_2)-th sensor in the i-th channel, and w_{ii} and w_{ij} ($i \neq j$) are the within-channel and the cross-channel degradation parameters. We note that $w_{ij} \geq 0$ for $i, j \in \{r, g, b\}$ and $\sum_{j=r,g,b} w_{ij} = 1$, for $i \in \{r, g, b\}$. We intersperse the low resolution images $\bar{\mathbf{g}}^{(i)}_{l_1 l_2}[n_1, n_2]$ to form an $M_1 \times M_2$ image by assigning

$$\bar{\mathbf{g}}^{(i)}[L(n_1 - 1) + l_1, L(n_2 - 1) + l_2] = \bar{\mathbf{g}}^{(i)}_{l_1 l_2}[n_1, n_2], \quad i \in \{r, g, b\}.$$

The image $\bar{\mathbf{g}}^{(i)}$ so formed is called the *observed high resolution image from the i-th channel.*

For $i \in \{r, g, b\}$, let $\bar{g}^{(i)}$ and $f^{(i)}$ be respectively the discretization of $\bar{\mathbf{g}}^{(i)}$ and $\mathbf{f}^{(i)}$ using a column by column ordering. Let

$$\bar{g} = [\bar{g}^{(r)}\ \bar{g}^{(g)}\ \bar{g}^{(b)}]^t \quad \text{and} \quad f = [f^{(r)}\ f^{(g)}\ f^{(b)}]^t.$$

Under the Neumann boundary condition assumption [4], the degradation matrices in each channel are banded matrices with bandwidth $L+1$, but there are entries added to the upper left part and the lower right part of the matrices (see the second matrix in (1)). The resulting matrices, denoted by $H^x_{l_1l_2}(\bar{\epsilon}^x_{l_1l_2})$ and $H^y_{l_1l_2}(\bar{\epsilon}^y_{l_1l_2})$, have a Toeplitz-plus-Hankel structure:

$$H^x_{l_1l_2}(\bar{\epsilon}^x_{l_1l_2}) = \frac{1}{L}\begin{bmatrix} 1 & \cdots & 1 & \frac{1}{2}-\bar{\epsilon}^x_{l_1l_2} & & 0 \\ \vdots & \ddots & \ddots & \ddots & \ddots & \\ 1 & \ddots & \ddots & \ddots & \ddots & \frac{1}{2}-\bar{\epsilon}^x_{l_1l_2} \\ \frac{1}{2}+\bar{\epsilon}^x_{l_1l_2} & \ddots & \ddots & \ddots & \ddots & 1 \\ & \ddots & \ddots & \ddots & \ddots & \vdots \\ 0 & & \frac{1}{2}+\bar{\epsilon}^x_{l_1l_2} & 1 & \cdots & 1 \end{bmatrix} + \frac{1}{L}\begin{bmatrix} 1 & \cdots & 1 & \frac{1}{2}+\bar{\epsilon}^x_{l_1l_2} & & 0 \\ \vdots & ⋰ & ⋰ & & & \\ 1 & ⋰ & & & & \frac{1}{2}-\bar{\epsilon}^x_{l_1l_2} \\ \frac{1}{2}+\bar{\epsilon}^x_{l_1l_2} & & & & ⋰ & 1 \\ & & & ⋰ & ⋰ & \vdots \\ 0 & & \frac{1}{2}-\bar{\epsilon}^x_{l_1l_2} & 1 & \cdots & 1 \end{bmatrix}, \quad (1)$$

and $H^y_{l_1l_2}(\bar{\epsilon}^y_{l_1l_2})$ is defined similarly. The degradation matrix corresponding to the (l_1, l_2)-th sensor under the Neumann boundary condition is given by

$$H_{l_1l_2}(\bar{\epsilon}_{l_1l_2}) = H^x_{l_1l_2}(\bar{\epsilon}^x_{l_1l_2}) \otimes H^y_{l_1l_2}(\bar{\epsilon}^y_{l_1l_2}).$$

The degradation matrix for the whole sensor array is made up of degradation matrices from each sensor:

$$H_L(\bar{\epsilon}) = \sum_{l_1=0}^{L-1}\sum_{l_2=0}^{L-1} D_{l_1l_2}H_{l_1l_2}(\bar{\epsilon}_{l_1l_2}), \quad i,j \in \{r,b,g\}. \quad (2)$$

Here $D_{l_1l_2}$ are diagonal matrices with diagonal elements equal to 1 if the corresponding component of the observed low resolution image comes from the (l_1, l_2)-th sensor and zero otherwise. We note that we have the same matrix $H_L(\epsilon)$ within the channels and across the channel, therefore the overall degradation matrix is given by

$$A_L(\bar{\epsilon}) = \begin{bmatrix} w_{rr} & w_{rg} & w_{rb} \\ w_{gr} & w_{gg} & w_{gb} \\ w_{br} & w_{bg} & w_{bb} \end{bmatrix} \otimes H_L(\bar{\epsilon}) \equiv W \otimes H_L(\bar{\epsilon}). \quad (3)$$

3. Image reconstruction formulation

In this paper, the displacement errors are not known exactly. The spatial invariance of the blurring function translates into the spatial invariance of the displacement error in the blurring matrix. The "true" blur function is represented as follows. For each $l_1, l_2 \in \{0, 1, \cdots, L-1\}$

$$\begin{aligned} \bar{h}^z_{l_1 l_2} &= \frac{1}{L}\overbrace{[\frac{1}{2} - \bar{\epsilon}^z_{l_1 l_2}, 1, \cdots, 1, \cdots, 1, \frac{1}{2} + \bar{\epsilon}^z_{l_1 l_2}]}^{\text{there are } L+1 \text{ entries}} \\ &= h^z_{l_1 l_2} + \delta h^z_{l_1 l_2}, \quad \forall z \in \{x, y\}, \end{aligned} \tag{4}$$

where

$$h^z_{l_1 l_2} = \frac{1}{L}[\frac{1}{2} - \epsilon^z_{l_1 l_2}, 1, \cdots, 1, \cdots, 1, \frac{1}{2} + \epsilon^z_{l_1 l_2}], \quad z \in \{x, y\}$$

and

$$\delta h^z_{l_1 l_2} = \frac{1}{L}[-\delta\epsilon^z_{l_1 l_2}, 0, \cdots, 0, \cdots, 0, \delta\epsilon^z_{l_1 l_2}], \quad z \in \{x, y\}.$$

Here $h^z_{l_1 l_2}$ is the estimated (or measured) point spread function and $\delta h^z_{l_1 l_2}$ is the error component of the point spread function at the (l_1, l_2)th sensor in the z-direction ($z \in \{x, y\}$).

The observed signal is also subject to errors. It is assumed that the observed signal $g = [g_1, \ldots, g_{M_1 M_2}]^t$ can be represented by

$$\bar{g} = g + \delta g, \tag{5}$$

where

$$\delta g = [\delta g_1, \delta g_2, \cdots, \delta g_{M_1 M_2}]^t$$

and δg_i is independent uniformly distributed noise with zero-mean and variance σ_g^2. Then the image reconstruction problem requires the recovery of the vector f from the given inexact point spread function $h_{l_1 l_2}$ ($l_1 = 0, 1, \cdots, L_1 - 1$, $l_2 = 0, 1, \cdots, L_2 - 1$) and the observed noisy signal vector g.

In the following, the constrained TLS formulation for multisensors is considered. Using

$$W \otimes H_L(\bar{\epsilon}) f = W \otimes \sum_{l_1=0}^{L-1} \sum_{l_2=0}^{L-1} D_{l_1 l_2} H_{l_1 l_2}(\bar{\epsilon}_{l_1, l_2}) f = \bar{g} = g + \delta g,$$

(4) and (5), the preceding equation can be reformulated as follows:

$$W \otimes \sum_{l_1=0}^{L-1}\sum_{l_2=0}^{L-1} D_{l_1 l_2} H^x_{l_1 l_2}(\epsilon_{l_1,l_2}) \otimes H^y_{l_1 l_2}(\epsilon_{l_1,l_2}) f - g + \sum_{l_1=0}^{L-1}\sum_{l_2=0}^{L-1} \delta\epsilon^x_{l_1 l_2} \cdot f^y_{l_1 l_2}$$
$$\sum_{l_1=0}^{L-1}\sum_{l_2=0}^{L-1} \delta\epsilon^y_{l_1 l_2} \cdot f^x_{l_1 l_2} + \sum_{l_1=0}^{L-1}\sum_{l_2=0}^{L-1} \delta\epsilon^x_{l_1 l_2} \cdot \delta\epsilon^y_{l_1 l_2} \cdot W \otimes D_{l_1 l_2} E \otimes E f - \delta g = 0, \tag{6}$$

where E has the Toeplitz-Hankel structure below,

$$\frac{1}{L}\begin{bmatrix} 0 & \cdots & 0 & -1 & & 0 \\ \vdots & \ddots & \ddots & \ddots & \ddots & \\ 0 & \ddots & \ddots & \ddots & \ddots & -1 \\ 1 & \ddots & \ddots & \ddots & \ddots & 0 \\ & \ddots & \ddots & \ddots & \ddots & \vdots \\ 0 & & 1 & 0 & \cdots & 0 \end{bmatrix} + \frac{1}{L}\begin{bmatrix} 0 & \cdots & 0 & 1 & & 0 \\ \vdots & \cdot^{\cdot^{\cdot}} & \cdot^{\cdot^{\cdot}} & & & \\ 0 & \cdot^{\cdot^{\cdot}} & & & & -1 \\ 1 & & & & \cdot^{\cdot^{\cdot}} & 0 \\ & & & \cdot^{\cdot^{\cdot}} & \cdot^{\cdot^{\cdot}} & \vdots \\ 0 & & -1 & 0 & \cdots & 0 \end{bmatrix}$$

and

$$f^x_{l_1 l_2} = W \otimes D_{l_1 l_2} H^x_{l_1 l_2}(\epsilon_{l_1,l_2}) \otimes E f \quad \text{and} \quad f^y_{l_1 l_2} = W \otimes D_{l_1 l_2} E \otimes H^y_{l_1 l_2}(\epsilon_{l_1,l_2}) f.$$

The constrained TLS formulation amounts to determining the necessary "minimum" quantities $\{\delta\epsilon^x_{l_1 l_2}, \delta\epsilon^y_{l_1 l_2}\}_{l_1,l_2=0}^{L-1}$ and δg such that (6) is satisfied. For simplicity, denote

$$\Delta = [\delta\epsilon^x_{00}, \delta\epsilon^y_{00}, \delta\epsilon^x_{01}, \delta\epsilon^y_{01}, \cdots, \delta\epsilon^x_{L-1L-2}, \delta\epsilon^y_{L-1L-2}, \delta\epsilon^x_{L-1L-1}, \delta\epsilon^y_{L-1L-1}]^t.$$

Mathematically, the constrained TLS formulation can be expressed as

$$\min_f \left\| \begin{bmatrix} \Delta \\ \delta g \end{bmatrix} \right\|_2^2 \qquad \text{subject to (6).}$$

We remark that color image reconstruction problems are ill-conditioned inverse problems [4]. Regularization can be used to achieve stability. By introducing a regularization operator P and a regularization parameter μ to restrict the set of admissible solutions, the regularized solution f is computed as the solution to

$$\min_f \left\| \begin{bmatrix} \Delta \\ \delta g \end{bmatrix} \right\|_2^2 + \mu \|Pf\|_2^2 \qquad \text{subject to (6).} \tag{7}$$

The term $\|Pf\|_2^2$ is added in order to regularize the solution. The regularization parameter μ controls the degree of regularity of the solution.

In [4], we have considered to use the weighted discrete Laplacian matrix as the regularization matrix. The cross-channel weights of this regularization matrix not only enforce the within-channel spatial smoothness, but also capture the changes of reflectivity across the channels.

4. Alternating minimization algorithm

In this section, we introduce an approach to minimizing the objective function arising from the constrained total least squares computation. By combining (6) and (7), we minimize $J(f, \triangle)$ where

$$J(f,\triangle) \equiv \| \triangle \|_2^2 + \mu \|Pf\|_2^2 + \left\| W \otimes H_L(\epsilon) f - g + \sum_{l_1=0}^{L-1}\sum_{l_2=0}^{L-1} \delta\epsilon^x_{l_1 l_2} \cdot f^y_{l_1 l_2} \right.$$
$$\left. + \sum_{l_1=0}^{L-1}\sum_{l_2=0}^{L-1} \delta\epsilon^y_{l_1 l_2} \cdot f^x_{l_1 l_2} + \sum_{l_1=0}^{L-1}\sum_{l_2=0}^{L-1} \delta\epsilon^x_{l_1 l_2} \cdot \delta\epsilon^y_{l_1 l_2} \cdot W \otimes D_{l_1 l_2} E \otimes E f \right\|_2^2 \quad (8)$$

Below we develop an alternating minimization algorithm for solving the regularized constrained total least problem (8). The algorithm is stated as follows. Assume we have $\triangle_{n-1}$:

- Determine f_n by solving the following least squares problem:

$$\min_{f_n} \left\{ \|W \otimes H_L(\epsilon + \triangle_{n-1}) f_n - g\|_2^2 + \mu \|P f_n\|_2 \right\}. \quad (9)$$

- Given f_n, we want to compute $\triangle_n$ from (8). However, we first note that (6) is a nonlinear system involving the terms $\{\delta\epsilon^x_{l_1 l_2}, \delta\epsilon^y_{l_1 l_2}\}_{l_1,l_2=0}^{L-1}$. Because $\delta\epsilon^z_{l_1 l_2}$ $(l_1, l_2 = 0, 1, \cdots, L-1)$ should be very small, the quantity $|\delta\epsilon^x_{l_1 l_2} \cdot \delta\epsilon^y_{l_1 l_2}|$ can be negligible in (6), and therefore the nonlinear term

$$\sum_{l_1=0}^{L-1}\sum_{l_2=0}^{L-1} \delta\epsilon^x_{l_1 l_2} \cdot \delta\epsilon^y_{l_1 l_2} \cdot D_{l_1 l_2} E \otimes E f$$

can be ignored in (6). In this case, we can determine $\triangle_n$ by solving the following least squares problem:

$$\min_{\triangle_n} \left\{ \| \triangle_n \|_2^2 + \left\| W \otimes H_L(\epsilon) f_n - g + \sum_{l_1,l_1=0}^{L-1} \delta\epsilon^x_{l_1 l_2} f^y_{n\ l_1 l_2} + \sum_{l_1,l_2=0}^{L-1} \delta\epsilon^y_{l_1 l_2} f^x_{n\ l_1 l_2} \right\|_2^2 \right\}. \quad (10)$$

The minimum 2-norm of the above least squares solution $\triangle_n$ can be computed efficiently.

With our proposed iterative scheme, one must solve a symmetric positive definite linear least squares problem in (9) at each iteration of the alternating minimization algorithm. It is obvious that the size of this system is large in general (the number $M_1 M_2$ of image pixels is large), but has structures that can be utilized. We apply a preconditioned conjugate gradient (PCG) method with cosine transform based preconditioner to solving this linear least squares problem [1, 5]. In [4], it has been shown that the singular values of the preconditioned matrices are clustered around 1 for sufficiently small subpixel displacement errors. Hence when the conjugate gradient method is applied to solving the preconditioned system, we expect fast convergence. Numerical results in [4] have shown that the cosine transform preconditioners can indeed speed up the convergence of the method.

5. Numerical example

In this section, we illustrate the effectiveness of using constrained total least squares for solving high resolution color image reconstruction problems. The 128 × 128 original image is shown in Figure 1. We tried the following degradation matrix to degrade the original color image

$$\begin{bmatrix} 0.8 & 0.1 & 0.1 \\ 0.1 & 0.8 & 0.1 \\ 0.1 & 0.1 & 0.8 \end{bmatrix} \otimes H_L(\bar{\epsilon}).$$

Gaussian white noises with signal-to-noise ratios of 30 and 40 dBs were added to each degraded image plane (cf. Figure 2). The parameters $\bar{\epsilon}^x_{l_1 l_2}$ and $\bar{\epsilon}^y_{l_1 l_2}$ are random values chosen between $\frac{1}{2}$ and $-\frac{1}{2}$.

To show the advantage of the proposed RCTLS algorithm over the conventional RLS (regularized least squares) algorithm, we compare two methods. The reconstructed image by RLS and the reconstructed image by RCTLS are shown in Figures 3 and 4, respectively. The reconstructed high-resolution color image using RCTLS shows slight improvement both in image quality and relative error. We remark that the optimal regularization parameters are used in both RLS and RCTLS methods. (The color images in this paper can be found in http://hkumath.hku.hk/~mng/tls.)

We conclude that the main contribution of this paper is the regularized constrained total least squares formulation and solution of the high resolution color image reconstruction problem with multisensors. The numerical algorithm developed is iterative and involves a two-step minimization strategy at each iteration. Each step, in turn, involves the least-squares solution of a convex optimization problem, though the overall problem is nonconvex. Some preliminary numerical examples are presented to illustrate the usefulness of the algorithm for high-resolution color image reconstruction from multisensor.

Acknowledgments

This work is supported by Research Grants Council grant nos. HKU 7147/99P and 7132/00P, HKU CRCG grant nos. 10202720, 10203501, and Army Research Office Grant DAAD 19-00-1-0539.

References

[1] N. Bose and K. Boo, High-resolution image reconstruction with multisensors, *International Journal of Imaging Systems and Technology*, 9:294–304, 1998.

[2] N. Galatsanos, A. Katsaggelosm, R. Chin, and A. Hillery, Least Squares Restoration of Multichannel Images, *IEEE Trans. Signal Processing*, 39:2222–2236, 1991.

[3] B. Hunt and O. Kubler, Karhunen-Loeve Multispectral Image Restoration, Part I: Theory, *IEEE Trans. Acoust., Speech, Signal Processing*, 32:592–600, 1984.

[4] M. Ng and W. Kwan, High-Resolution Color Image Reconstruction with Neumann Boundary Conditions, Annals of Operations Research, to appear.

[5] M. Ng and A. Yip, *A Fast MAP Algorithm for High-Resolution Image Reconstruction with Multisensors*, Multidimensional Systems and Signal Processing, 12(2):143–164, 2001.

[6] S. Van Huffel and J. Vandewalle, *The Total Least Squares Problem: Computational Aspects and Analysis*, SIAM Publications, Philadelphia, 1991.

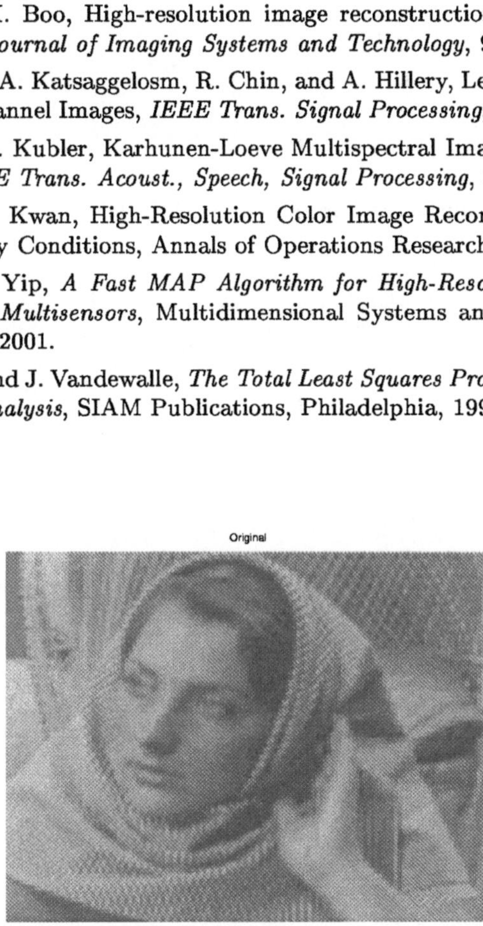

Figure 1. Original image.

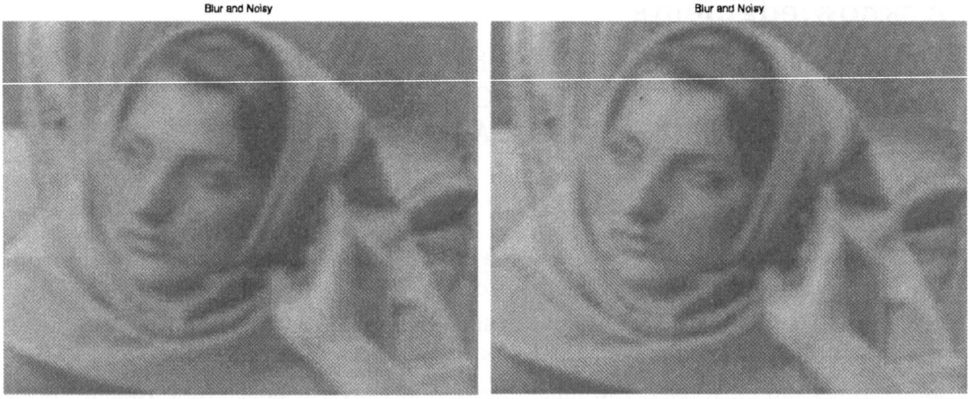

Figure 2. Blurred and noisy images 30dB (left) and 40dB (right).

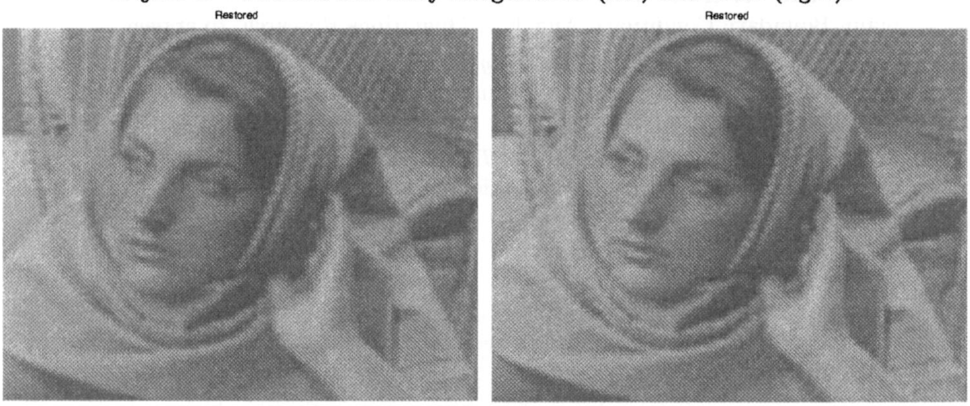

Figure 3. Reconstructed images using RLS [rel. err. = 6.6e-2](left) and RCTLS [rel. err. = 6.4e-2](right) from Figure 2 (left).

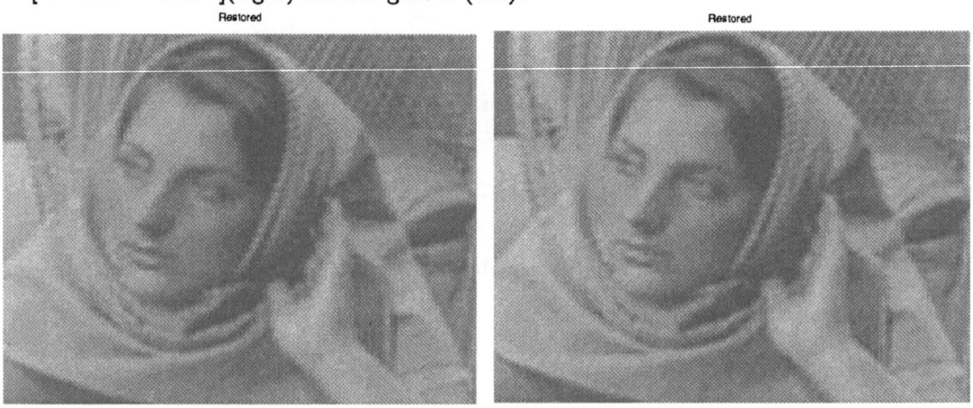

Figure 4. Reconstructed images using RLS [rel. err. = 6.1e-2](left) and RCTLS [rel. err. = 5.9e-2](right) from Figure 2 (right).

TOTAL LEAST SQUARES IN ASTRONOMY

Richard L. Branham, Jr.
Regional Center for Scientific and
Technogicial Research (CRICYT)
C.C. 131, 5500 Mendoza, Argentina
email: rlb@lanet.com.ar

Abstract Although astronomers have been involved with the development and use of least squares, they have made insufficient use of total least squares. Astronomers, however, have examined alternatives that also permit error in the equations of condition. There exist, nevertheless, problems of astronomical data reduction for which total least squares represents the ideal mathematical tool. Among these problems are the differential correction of an orbit and the determination of parameters of Galactic kinematics. Total least squares, although more computationally demanding than ordinary least squares, can be used in any situation where the latter is applicable. But care must be paid to the proper scaling of the data matrix. The method merits greater attention by the astronomical community.

Keywords: astronomy, Galactic kinematics, linear errors-in-variables model, total least squares.

1. Introduction

Astronomers have been involved with methods for solving overdetermined systems ever since Tobias Mayer published the method of averages in 1755 to reduce observations of the moon. Laplace, in 1799 in his monumental *Celestial Mechanics* [8, Book 3, Ch. 5, Sec. 39], advocated use of the min-max criterion. In 1805 Legendre [9, Appendix], in his *New Methods for Determining Orbits of Comets*, also advocated the min-max criterion, but found the method of solution so laborious that he recommended least squares as easier. Gauss, in 1809 in his *Theoria Motus Corporum Coelestium* , expounded on least squares, mentioning that he had been using the method since 1795 [4, Secs. 175–186].

S. Van Huffel and P. Lemmerling (eds.), Total Least Squares and Errors-in-Variables Modeling, 375–384.

During the 19th century least squares became the dominant data reduction technique. Standard treatises, such as Chauvenet's *Spherical and Practical Astronomy* published in 1863 [1, Appendix] and Newcomb's *Compendium of Spherical Astronomy* published in 1906 [10, Ch. 3], contained extensive summaries of how to apply the method in astronomy. Astronomers also studied techniques for dealing with discordant observations, largely based on the probability arguments used to derive the normal equations.

But some astronomers have also realized the limitations of ordinary least squares. In 1940 Strömberg proposed a reduced major axis regression, which treats two variables—call them x and y-as equal [19]. This reduced major axis regression minimizes the product of the x and y deviations. In 1953 Trumpler and Weaver, in their treatise *Statistical Astronomy* [20, p. 184], wrote that many problems of astronomical data reduction are carried out by the precepts of OLS even though error occurs in more than one variable. To use their own words, "...the results of such investigations are open to question and should be rediscussed using more adequate mathematical techniques."

This lacuna was addressed when Jefferys [6], [7] and Eichhorn [2] developed what astronomers call "general least squares" or "generalized least squares", although these terms should not be confused with what numerical analysts refer to as generalized least squares; the methods are different. Although the Jefferys-Eichhorn general least squares can calculate the same solution as TLS (Eichhorn, personal communication), it is a more cumbersome approach relying on iteration from an assumed first approximation to the solution, with the usual disadvantages of an iterative versus the direct solution that TLS provides, and an assumed covariance matrix. Astronomers have also used methods similar to that given by Press *et al.* in their *Numerical Recipes in C* [12, Sec. 15.3]: perform univariate regression, although with an intercept term, by a nonlinear iterative algorithm that minimizes the dispersions in the two coordinates. Pont, Mayor, and Burkes studied stellar radial velocities (a star's velocity along the line of sight) with an algorithm similar to this [11]. But this approach compares unfavorably with the TLS found in the numerical analysis literature. It permits only univariate, rather than also multivariate, regression, replaces a direct linear solution by an iterated nonlinear solution, calculates no covariance matrix, and requires ancillary input data, the initial dispersions of the coordinates, that may not be readily available. But astronomers have been slow to adopt genuine TLS (in the sense used by numerical analysts).

2. Astronomical terminology

To understand the sections that follow I should explain some astronomical terms. The previous section already mentioned *radial velocity,* usually measured in $km\,\sec^{-1}$. To specify a star's angular position in space various coordinate systems are used. For work with Galactic structure and kinematics it is customary to use *Galactic longitude*, an angle measured parallel to the Galactic plane and denoted by l, and *Galactic latitude*, an angle measured perpendicular to the plane and denoted by b. A star's *proper motion* is its angular displacement with respect to the line of sight caused by the star's motion in space. The proper motion is usually decomposed into two components. One decomposition expresses the proper motion parallel to the plane of the Galaxy, the *proper motion in Galactic longitude* (denoted by μ_l), and the motion perpendicular to the plane, *proper motion in Galactic latitude* (denoted by μ_b). A star's *parallax* is its angular displacement caused by the Earth's motion around the Sun. The reciprocal of the parallax gives the star's distance. The distance is usually measured in parsecs (*pc*) or kiloparsecs (*kpc*). 1 pc= $3.0857{\cdot}10^{13}\,km$. Given a star's parallax and proper motion we can calculate its *transverse velocity.* The transverse velocity and radial velocity provide the three components of the star's velocity in space with respect to the Sun.

The atomic lines and molecular bands present in the atmosphere of a star and their intensity determine a star's *spectral* and *luminosity class.* Star's of spectral classes O and B are hot, surface temperatures of 20,000–35,000 $^\circ K$. The surface temperature decreases to about 6,000 $^\circ K$ for stars of class G, like the Sun, and to about 2,000 $^\circ K$ for stars of class M. The size of a star determines its luminosity class. Stars of class I, with radii up to 200 times the Sun's radius, are called supergiants. The classes pass through bright giants (II), normal giants (III), sub-giants (IV), down to stars of class V, like the Sun, called dwarfs.

Problems of Galactic kinematics involve the search for certain parameters of interest to astronomers. The motion of the Sun with respect to the local stars, the local solar motion, is one of these. The Sun and other local stars orbit around the center of the Galaxy, 8.5 *kpc* away. The Oort A and B constants are associated with the shearing motion of the volume element containing the local stars caused by the differential rotation of the stars about the center (the Galaxy does not rotate like a wheel, rather each star has its individual orbit about the center). Hot stars of classes O and B also exhibit a K term in their radial velocities, a systematic offset in the velocities showing that they are not normally distributed about 0.

3. TLS and Galactic kinematics

For many problems of multivariate regression in astronomy TLS becomes an ideal adjustment method and obviates debatable assumptions that must be made when OLS is used. One of these problems is the estimation of the parameters of Galactic kinematics. Consider a model for Galactic rotation in which we desire to find: the components X, Y, Z of the reflex local solar motion; the A and B constants of Galactic rotation; a K term for the early stars; and corrections Δp and Δq for systematic errors in the proper motions. Let $\dot{r}$ be the star's radial velocity in $km\,\sec^{-1}$ and measure the proper motion in $arc - \sec yr^{-1}$. Smart [18, Sec. 8.42] gives equations of condition, valid for distances up to 1 kpc, linking the unknowns:

$$\cos l \cos bX + \sin l \cos bY + \sin bZ + K + \frac{A \sin 2l}{\pi} \cos^2 b = \dot{r}; \quad (1)$$

$$-\sin lX + \cos lY + \frac{\kappa h \Delta p + \kappa g \Delta q + (A \cos 2l + B) \cos b}{\pi} = \frac{\kappa \mu_l}{\pi} \cos b; \quad (2)$$

$$\begin{aligned} &-\cos l \sin bX - \sin l \sin bY + \cos bZ + \\ &\qquad + \frac{\kappa f \Delta p + \kappa k \Delta q - \frac{1}{2} A \sin 2l \sin 2b}{\pi} = \frac{\kappa}{\pi} \mu_b, \end{aligned} \quad (3)$$

where π is the star's parallax, κ a conversion constant with value 4.74 $km\,\sec^{-1}$, and f, g, h, k algebraic expressions whose form Smart gives.

The presence of the parallax π on *the left-hand-side*, not merely in the data on the right-hand-side, constitutes an important consideration for solving Eqns. (1)–(3), which therefore represent a situation where the assumptions of OLS become invalid. The errors of parallax measurements are substantial, typically 23% of the value itself. (This percentage comes from the median parallax and median parallax error in the Hipparcos catalog of stellar parallaxes [3].) Thus, in the equations of condition the coefficients of $A, B, \Delta p, \Delta q$ incorporate significant error whereas the coefficients of X, Y, Z, K are basically error-free. This situation is thus more complicated than that of pure TLS and should be solved by mixed TLS-LS, TLS-LS for short, a situation where some of the columns of the matrix of the equations of condition contain error whereas other columns are error-free; thus neither pure TLS, error in all of the columns, nor OLS, error in no column, is applicable. But before the equations can be solved by mixed TLS-LS they should be recast because the parallax in the denominator causes complications: although the parallaxes themselves may have a normal error distribution, at least approximately, the same will not be true for their reciprocals, the distances, where the error distribution is skewed and of infinite variance. To ameliorate these complications one should multiply Eqns. (1)–(3) by π to transfer the error from the denominator.

What assumptions must be made when we use OLS with Eqns. (1)–(3)? There exists substantial feeling that solutions should only be performed on subsets of the data, such as spectrum-luminosity groups, for the putative reason that each subgroup exhibits different kinematical behavior. But the substantive reason arises from in the past using only OLS as the reduction technique. Because OLS assumes that the equations of condition are error-free, significant bias in the solution is only avoided by dividing the stars into groups of the same spectrum-luminosity class where the parallax is about the same for all of the stars in the group. To quote Smart [18, p. 57], "Indeed, progress can be made by assuming, for example, that the parallaxes are distributed closely around a value...which we may describe as a *mean parallax.*" Stars of the same spectrum-luminosity class follow closely this assumption. When the equations of condition are accumulated, either by forming normal equations of by use of orthogonal transformations, the resulting matrix becomes what is referred to in sparse matrix terminology as "doubly-bordered block diagonal." Performing separate solutions for different spectrum-luminosity groups and subsequently statistically combining them to obtain values for the Oort constants, for example, assumes that the borders are null, a demonstrably incorrect assumption. This usually invalid assumption nevertheless ameliorates partially, but only partially, the nefarious effects of the errors in the parallaxes.

The advantages of TLS over OLS for problems of Galactic kinematics when one employs all of the data rather than subsets arise from: the greater distances for many of the proper motions assure that random motions do not dominate the genuine proper motions; spectrum-luminosity groups are an idealization resulting from histogramming continuous properties into discrete groups, whereas a global solution treats the astrophysical properties as continuous; one proceeds from a firm mathematical basis, the correct reduction model.

I have used TLS-LS [16] applied to nearly 100,000 stars taken from the Hipparcos Catalog to calculate twelve parameters of Galactic kinematics based on a more general model for Galactic kinematics, the Ogorodnikov-Milne (O-M) model, than that incorporated in Eqns. (1)–(3). Nine variables of the O-M model represent the nine components of a tensor of the deformations of the volume element containing the local stars. Quantities such as the Oort A and B parameters and the K term can be obtained from these nine components when we make certain simplifying assumptions about the nature of the deformations. The three components of the solar motion constitute the remaining three variables.

Reasonable values for all of these are obtained, showing that high quality astrometric data and an adequate reduction method can produce good results for global solutions when in the past this proved difficult or impossible. An interesting aspect of this study arises from the Gould belt O-B stars. (The brighter stars of classes O and B are not distributed about the plane of the Galaxy, but rather about another plane known as Gould's belt.) To obtain consistent parameters these stars should be suppressed because they correspond to different kinematics than the Galactic belt stars. 3,630 Gould belt O-B stars were finally eliminated, but interestingly enough the results differ little if they are excluded or left in the solution. These stars may be considered noise in the data, and TLS, by allowing for noise in the equations of condition, performs admirably to suppress noise.

Very well. TLS_LS gives good results when applied to problems of Galactic kinematics. But are the results *better* than what OLS would give? Two studies suggest that the answer is yes. I applied both TLS_LS and OLS to determining various parameters of Galactic kinematics [14]. 3,100 stars taken from the Yale Bright Star Catalog constituted the observational data base. TLS_LS gave good solutions whereas the results from OLS were not merely bad, but absolutely wretched. Another study of mine, submitted to *The Astrophysical Journal,* examines the kinematics, based on a generalized O-M model, for 1,276 O-B5 stars. The generalization consists of including second-order terms in the Taylor series expansions that define the model rather than only the first-order terms usually used. The second-order terms extend the model's validity to distances out to about 3 *kpc* from the Sun rather than the1 *kpc* of the simpler model. Because O-B5 stars are observable out to great distances, it is important to include the second-order terms. A comparison of certain variables in my solution, such as the circular motion of the local standard of rest about the Galactic center, with value 232 $km\,\sec^{-1}$, against what some other investigators have obtained, using OLS with the first-order Ogorodnikov-Milne model, for this quantity, values in the 270–280 $km\,\sec^{-1}$range, indicates that the TLS_LS solution, coupled with the inclusion of second-order terms, seems preferable; it lies within the range of 220±20 $km\,\sec^{-1}$ recommended by the International Astronomical Union.

4. The Galactic plane and the position of the sun

The determination of the Sun's position with respect to the Galactic plane, a problem associated with the determination of the Galactic pole, represents another area where TLS, or mixed TLS-LS, becomes an ideal mathematical tool. From the mathematical definition of a plane one has

$$[\ \cos l \cos b \quad \sin l \cos b \quad \sin b\ ] \begin{bmatrix} \cos l_g \cos b_g \\ \sin l_g \cos b_g \\ \sin b_g \end{bmatrix} + \pi q = 0, \qquad (4)$$

where l_g, b_g are the coordinates of the pole and q the Sun's distance from the Galactic plane. Eq. (4) has seldom been used in this form because of the paucity, until recently, a lacuna filled by the Hipparcos mission, of sufficient parallaxes to permit a strong determination. Rather, approximations have been made. Trumpler and Weaver [20, pp. 422–423], for example, use the equation in the form

$$p + (\cos l_g \cos b_g) \cos l + (\sin l_g \cos b_g) \sin l = -b, \qquad (5)$$

where p corresponds to πq of Eq. (4) and represents angular distance from the plane and one takes $\cos b \approx 1$ and $\sin b \approx b$. An assumed distance to the Galactic center converts p to q. Eq. (5) is applied to counts of areal star density. Gum, Kerr, and Westerhout [5], using exclusively radio observations of the 21-cm hydrogen line, apply the equation as

$$r \sin b + r \cos l \sin \Delta \cos l_g + r \sin l \sin \Delta \sin l_g + q = z, \qquad (6)$$

where r and z are observed distances to a hydrogen cloud and Δ is the displacement, assumed small, from a given pole. Both Eqns. (5) and (6) are open to objection. Neither one enforces the condition that the Galactic pole have unit Euclidean norm; Eq. (4) should be solved subject to the condition that $(\cos l_g \cos b_g)^2 + (\sin l_g \cos b_g)^2 + \sin^2 b_g = 1$. And Eq. (6), by incorporating observational error in both r and z and hence on both sides of the equation, should be solved by the precepts of TLS-LS rather than OLS.

Both of these exigencies can be complied with by use of nonlinear TLS-LS. Let y be the vector of our unknowns,

$$y = [\ \cos l_g \cos b_g \quad \sin l_g \cos b_g \quad \sin b_g \quad q\ ]^T.$$

Because the positions of the stars are well known, but their parallaxes incorporate substantial error, Eq. (4) shows that the first three columns of the equations of condition are basically error-free whereas the fourth column contains error. The equations of condition may be solved by minimization. If we have a total of m stars and define

$$Ay = \begin{bmatrix} \cos l_1 \cos b_1 & \sin l_1 \cos b_1 & \sin b_1 & \pi_1 \\ \vdots & \vdots & \vdots & \vdots \\ \cos l_m \cos b_m & \sin l_m \cos b_m & \sin b_m & \pi_m \end{bmatrix} \cdot \begin{bmatrix} \cos l_g \cos b_g \\ \sin l_g \cos b_g \\ \sin b_g \\ q \end{bmatrix}, \tag{7}$$

then we solve the non-linear problem

$$\| Ay \| / \sqrt{1 + y_4^2} + \lambda(y_1^2 + y_2^2 + y_3^2 - 1) = \min, \tag{8}$$

where λ is a Lagrangian multiplier. I have used 93,106 Galactic belt stars from the Hipparcos catalog and Eq. (8) to determine that the Sun lies 34.56±0.56 *pc* above the galactic plane [17]. This solution, relying on fewer assumptions than alternative studies referenced in Branham [17], should probably be superior to many others, and this is reflected in the smaller mean error compared with those of the other studies, where the mean errors are greater than 1 *pc*.

5. Differential correction of planetary orbits

Yet another area where TLS, in particular its noise suppression properties, has proved useful comes from the differential correction of orbits. I have, however, already discussed this application previously [13] and will not go into detail.

6. Scaling

Scaling the equations of condition becomes important for obtaining a reliable TLS solution. I have found that in general column scaling by the Euclidean norm works well. Sometimes first scaling the rows by the Euclidean norm and then the columns seems indicated. No scaling at all performs poorly. Column scaling, perhaps surprisingly, works even for problems of Galactic kinematics, where the error in the distance increases with increasing distance: although the error for a parallax is approximately constant, the relative error in the parallax, and hence in its reciprocal, the distance, increases as the parallax decreases. I have found that assigning weights on the basis of the relative errors actually performs *worse* for Galactic kinematics than simple column scaling by the Euclidean norm, most likely because the higher weighting of the nearby stars results in irregular local motions dominating over the systematic effects induced by Galactic rotation that are of interest.

7. A covariance matrix

For a complete least squares adjustment one must calculate not only the solution, but also the covariance matrix, which permits finding the mean errors of the unknowns and the correlations among them. For OLS the unscaled covariance matrix is merely the inverse of the matrix of the normal equations, or the seminormal equations should one employ orthogonal transformations. But the covariance matrix for TLS is more complicated. I have analyzed in detail a covariance matrix for TLS, which embodies all possibilities from pure TLS with homo- and heteroscedastic data, mixed TLS_LS with homo- and heteroscedastic data, to, as a limiting case, OLS with homo- and heteroscedastic data [15]. That paper gives detailed expressions for the partial derivatives that enter into the covariance matrix.

8. Conclusions

TLS and TLS-LS, therefore, represent reduction techniques of genuine utility to astronomy and deserving of greater attention by astronomers. Part of this lack of interest arises not from inertia, but from TLS's being discussed almost entirely in the numerical analysis literature. Standard software packages, moreover, contain little that permits easy use of TLS: there is not a single routine in Matlab or the IMSL library, for example, for straightforward implementation of TLS. There are, however, papers on TLS appearing in the astronomical literature. My personal intuition tells me that with the demonstration of TLS's utility the method will soon become, if not standard, at least one of the mathematical tools that astronomers keep in their arsenal.

References

[1] W. Chauvenet. *A Manual of Spherical and Practical Astronomy.* Dover, New York, 1960.

[2] H. Eichhorn. Generalized least-squares adjustment, a timely but much neglected tool. *Celest. Mech.*, 56:337–351, 1993.

[3] ESA. The hipparcos and tycho catalogues. Technical Report SP-1200, European Space Agency, 1997.

[4] K. F. Gauss. *Theory of the Motion of the Heavenly Bodies Moving about the Sun in Conic Sections.* Dover, New York, 1963.

[5] C. S. Gum, F. J. Kerr, and G. A. Westerhout. 21-cm determination of the principal plane of the galaxy. *Not. R. Astron. Soc.*, 121:132–149, 1960.

[6] W. H. Jeffreys. On the method of least squares. *Astron. Jour.*, 85:177–181, 1980.

[7] W. H. Jeffreys. On the method of least squares ii. *Astron. Jour.*, 86:149–155, 1981.

[8] P. S. Laplace. *Celestial Mechanics.* Celsea, New York, 1966.

[9] A. M. Legendre. *Nouvelles Méthodes pour la Détermination des Orbites des Comètes.* Courcier, Paris, 1805.

[10] S. Newcomb. *A Compendium of Spherical Astronomy.* Dover, New York, 1960.

[11] F. Pont, M. Mayor, and G. Burki. New radial velocities for classical cepheids. local galactic rotation revisited. *Astron. and Astrophys.*, 285:415–439, 1994.

[12] W. H. Press, S. A. Teukolsky, W. T. Vettering, and B. P. Flannery. *Numerical Recipes in C: the Art of Scientific Computing, 2nd ed.* Cambridge U. Press, Cambridge, 1992.

[13] R. L. Branham, Jr. Total least squares in astronomy. In S. Van Huffel and J. Wandewalle, editors, *Total Least Squares and Errors-in-Variables Modeling*, pages 371–377. SIAM, Philadelphia, 1996.

[14] R. L. Branham, Jr. Total least squares: an ideal mathematical tool for galactic kinematics. *Astrophys. and Space Sc.*, 257:107–121, 1998.

[15] R. L. Branham, Jr. A covariance matrix for total least squares with heteroscedastic data. *Astron. Jour.*, 117:1942–1948, 1999.

[16] R. L. Branham, Jr. The parameters of galactic kinematics determined from total least squares. *Rev. Mex. Astron. and. Astrophys.*, 36:97–112, 2000.

[17] R. L. Branham, Jr. The sun's distance from the galactic plane. *Astrophys. and Space. Sc.*, 2001. submitted.

[18] W.M. Smart. *Stellar Kinematics.* Wiley, New York, 1967.

[19] G. Strömberg. Accidental and systematic errors in spectroscopic absolute magnitudes for dwarf g0-k2 stars. *Astrophys. Jour.*, 92:156–169, 1940.

[20] R. J. Trumpler and H. Weaver. *Statistical Astronomy.* Dover, New York, 1962.

TLS AND CONSTRAINED TLS NEURAL NETWORKS FOR COMPUTER VISION

Giansalvo Cirrincione
University of Picardie, Department of Electrical Engineering,
33, rue St. Leu, 80039 Amiens, France
exin@dag.it

Abstract This paper deals with the exploitation of the epipolar constraint information for the construction of the essential matrix (fundamental matrix for uncalibrated images), which, once decomposed, solves the structure from motion problem. This technique has been longly considered inferior to the techniques which use the optical flow information, because of its high sensitivity to noise. The approach used here, which is particularly robust to noise, both demonstrates the validity and presents an extension of the essential matrix approach. Once established the fact that the problem at hand is Total Least Squares (TLS) with a certain structure, a statistical analysis of the problem suggests the use of the Constrained TLS in order to take in account the linear dependences of the noise components in the coefficients. This leads to the CTLS and CTLSn EXIN neurons which are simple variants (change of metrics) of the TLS EXIN linear neuron which solves the basic TLS problem. These neurons are able to yield good results even in the presence of an outlier contamination of half the point matches.

Keywords: computer vision, essential matrix, fundamental matrix, constrained TLS, minor component analysis.

1. Introduction

In the computer vision literature, the problem of noise and outliers is often underestimated. In fact, a lot of existing techniques work well only for noiseless images. As a consequence, many researchers claim that the fact that noisy images imply bad results is sufficient to reject the whole methodology. In [6], instead, it is pointed out that it is enough to reconsider the classical techniques in order to manage noise for having very good results. This reassessment often implies simple changes, as a preprocessing of the input data. This paper deals with the essential

S. Van Huffel and P. Lemmerling (eds.), Total Least Squares and Errors-in-Variables Modeling, 385–395.

matrix estimation. Given a sequence of images, captured e.g. by one mobile camera (egomotion), the first step is the extraction of the feature image points. A second step is tracking of the feature points throughout the sequence, i.e., for each image pair, the matching of two corresponding points. These matches are then used for the essential matrix (E) estimation if the camera is calibrated. In the uncalibrated case, by using the same techniques, the fundamental matrix (F) can be recovered. The essential matrix, after decomposition, yields the motion parameters. Solving for these matrices requires the same approach. In absence of noise, the eight-point algorithm solves the set of equations:

$$q_i' G q_i = 0 \qquad i = 1, \dots, N \tag{1}$$

$N = 8$, $G = (G_{ij})$, G_{ij} is defined as $g_{3(i-1)+j}$ and $q_i = [x_i, y_i, \zeta_i]$ in the first image and $q_i' = [x_i', y_i', \zeta_i']$ in the second image represent either the homogeneous pixel coordinates if $G = F$ or the homogeneous normalized coordinates if $G = E$. This set can be solved exactly only in absence of noise. For noisy images, more matches are needed and the algorithm must solve an *orthogonal* L_2 *approximation problem* (Total Least Squares, TLS, problem [11]):

$$\min_{g \in \mathbb{R}^9} \|Zg\|_2 = \min \sum_{i=1}^{N} r_i^2 \qquad \text{s.t.} \quad g^T g = 1 \tag{2}$$

being $g = [g_1, \dots, g_9]^T = [\hat{g}^T, g_9]^T$ and $Z = [Z_1, \dots, Z_9] = [z_1, \dots, z_N]^T \in \mathbb{R}^{N\text{x}9}$ where

$$z_i = [x_i' x_i, x_i' y_i, x_i' \zeta_i, y_i' x_i, y_i' y_i, y_i' \zeta_i, \zeta_i' x_i, \zeta_i' y_i, \zeta_i' \zeta_i]^T \tag{3}$$

and $r_i = z_i^T g$ is the i-th residual. This problem is solved by the eigenvector of $Z^T Z$ (*moment matrix*) associated to the smallest eigenvalue. This way of solving TLS uses the Minor Component Analysis (MCA) (see [7]) and can be solved by using linear neurons as MCA EXIN and MCA EXIN+ [6], [7]. The homogeneous problem (2) can be transformed in an inhomogeneous overdetermined set of equations by imposing that one of the elements of g be equal to -1. This allows the use of the TLS neurons, e.g. the TLS EXIN neuron [8], [6], [7].

The TLS solution is suboptimal [11] because the perturbations in the z_i^T rows are not Gaussian distributed as their elements involve the product of two spatial coordinates. This analysis justifies the introduction of the Constrained Total Least Squares (CTLS) technique. Curiously, nearly all researchers who have used these algorithms have not understood they were solving a TLS problem, just giving rise to the opinion that these methods are useless because too sensitive to noise. It is interesting to notice that the first researches who have used the TLS, i.e. Chaudhuri and Chatterjee in 1991 [4], have thought it was a new method and not a correct interpretation of the existing ones. Furthermore, they have completely neglected the correlation in the coefficients of Z and have illustrated their ideas with simulations which are false as shown in [6]. Suddenly, in 1996 [5], they have admitted the existence of the problem without mentioning their old results. Only Cirrincione [6] and Mühlich and Mester [14] in 1998 have shown the exact nature of the problem. The problem of correlation in the coefficients becomes still more serious if N-vectors[1] [12] are used: interestingly, in [13] N-vectors are used in a plain TLS and very bad results are found without understanding why!

A lot of techniques have been tried in order to improve the accuracy of the eight point algorithm in presence of noise. In case of a large image, the condition number of $Z^T Z$ worsens because of the *lack of homogeneity in the image coordinates*, [10]. In order to avoid this problem, several scalings of the point coordinates have been proposed [10] with good results. One way of scaling is to use the N-vectors [12]. Chaudhuri and Chatterjee in 1991 [4] and 1996 [5] use this preprocessing before a plain TLS (this approach yields very bad results). Another preprocessing used in the literature is the statistical scaling of Hartley [10] which requires a centring and a scaling (either isotropic or non-isotropic) of the image feature points. This preprocessing has found a theoretical justification in the paper of Mühlich and Mester [14], limited to the assumption of noise confined only in the second image. These authors only justify the isotropic scaling in the second image while accepting the two scalings in the first image, and correctly propose the use of the mixed OLS-TLS algorithm . Strangely, they claim that the Hartley's method is not consistent, unlike their method, which is unbiased. Simulations in [10] show that improvements in the condition number of even 10^8 are possible and in case of low accurate matches, the advantage of normalization is dramatic.

2. Nonlinear methods

The best nonlinear technique is the gradient method (called *S2* in [15]) which has never been used for the N-vectors and assumes that the third homogeneous coordinate of the image points is constant and can be set to

[1]The N-vector $\mathbf{m}$ of a point P on the image plane can be interpreted as the unit vector starting from the view point and pointing toward (or away from) P.

1. As seen before, the TLS residual in the minimization $r_i = q_i' G q_i = z_i^T g$ is not Gaussian distributed. This expression is known as the Bookstein [3] *algebraic distance*. It has no geometrical significance and does not, for example, measure the perpendicular distance of a match point to the conic represented by S (in the $4D$ image coordinate space (x, y, x', y')). The TLS criterion for S can be considered as a generalization of this distance. The idea is to approximate the true distance of the point match $h_i \in \mathbb{R}^4$ to the conic (h_0 is the closest point of S to h_i) by its first order approximation (assuming that the gradient has the same direction at h_i and h_0):

$$d_i = \|h_i - h_0\| \cong \frac{r(h_i)}{\|\nabla r|_{h_i}\|} \tag{4}$$

The Sampson criterion requires the minimization of $\sum_i d(h_i, S)^2$. Weng et al. [16] adapted Sampson's method to the computation of G [2]. They proposed to minimize:

$$D^2 = \sum_{i=1}^{N} (d_i)^2 = \sum_{i=1}^{N} (w_{Si} r_i)^2 \qquad \text{being} \tag{5}$$

$$w_{Si} = \frac{1}{\|\nabla r|_{h_i}\|} = \frac{1}{(Gq_i)_1^2 + (Gq_i)_2^2 + (G^T q_i')_1^2 + (G^T q_i')_2^2} \tag{6}$$

where r is the generic residual and $h_i = [x_i, y_i, x_i', y_i']^T$.

Another similar, but inferior nonlinear technique is the epipolar distance method [13], which does not work for N-vectors. Another technique is the *optimal Hartley* algorithm [10], which is iterative and yields the F matrix corresponding to the most likely true placement of the matched points.

3. The constrained total least squares method

Consider an overdetermined set of equations $Ax \approx b$ where $A \in \mathbb{R}^{M \times L}$ and $b \in \mathbb{R}^M$ have complex elements and define $C = [A; b]$ and corresponding perturbation matrix $\Delta C = [\Delta C_1, \ldots, \Delta C_{L+1}]$. TLS operates on this system under the assumption that the noise components of A and b are i.i.d. and zero mean. If there is correlation among the components of A or b, a noise whitening transformation can be applied and the error norm is appropriately modified. However, if there is *linear* dependence among the noise components, then TLS must be reformulated to take

[2] A method has also been proposed [13] which performs the exact orthogonal distance, but it is computationally very expensive (several hours on powerful computers !).

into account the reduced dimensionality of the noise entries. Abatzoglou and Mendel [1] have reformulated TLS as Constrained TLS (CTLS) in order to take in account the linear algebraic relations among the noise entries of A and b. This criterion follows the principle of parsimony and so uses the minimum number of noise components when reformulating TLS, to take in account the reduced entropy in the data. Let $v = \{v_1, v_2, \ldots, v_K\}$ be the minimal algebraic set of linearly independent random variables such that, for $i = 1, \ldots, L+1$ it holds $\Delta C_i = F_i v$ where v and $F_i \in \mathbb{R}^{MxK}$. Assume v is a white random vector. Define now the vector of noise columns:

$$\text{vec}\,(\Delta C) = \begin{bmatrix} \Delta C_1 \\ \vdots \\ \Delta C_{L+1} \end{bmatrix} = \begin{bmatrix} F_1 v \\ \vdots \\ F_{L+1} v \end{bmatrix} = \tilde{F} v \tag{7}$$

where $\tilde{F} \in \mathbb{R}^{(L+1)MxK}$. Since v has dimension K, minimal by assumption, $\operatorname{rank} \tilde{F} = K$. Define a norm on vec$(\Delta C)$ by $\|\text{vec}\,(\Delta C)\|_Q^2 = (\text{vec}\,(\Delta C))^H Q\, \text{vec}(\Delta C)$ and select Q so that all components of v are equally weighted. In [2] it is proved that the choice $Q = U\left(\Sigma\Sigma^T\right) U^T$ where $U \Sigma V^T$ is the SVD of $\tilde{F}$, implies $\|\text{vec}\,(\Delta C)\|_Q^2 = \|v\|_2^2$. The CTLS method consists of determining a vector x_{CTLS} and the minimum perturbation v in order to solve the problem:

$$\min_{u,x} \|v\|_2^2 = \min_{\Delta C, x} \|\text{vec}\,(\Delta C)\|_Q^2 \tag{8}$$

$$\text{s.t.} \qquad \left[C + \left[F_1 v, \ldots, F_{L+1} v\right]\right] [x; -1]^T = 0 \tag{9}$$

This is a quadratic minimization problem that is subject to a quadratic constraint equation. A closed form solution may not exist. However, the problem can be transformed into an unconstrained one [1] over the variable x under certain assumptions. Indeed, assume the exact relation $C_0 \left[x_0; -1\right]^T = 0$ in the noiseless case. Note that the dimension of the random vector v, K, must be less than $M\,(L+1)$ which is the dimension of vec(ΔC). Also, the larger K, the smaller the number of algebraic relations the entries of ΔC satisfy. Then, the CTLS solution can be obtained [2] as the x that minimizes the function:

$$E\,(x) = \begin{bmatrix} x \\ -1 \end{bmatrix}^H C^H \left(H_x H_x^H\right)^{-1} C \begin{bmatrix} x \\ -1 \end{bmatrix} \tag{10}$$

where

$$H_x = \sum_{i=1}^{L} x_i F_i - F_{L+1} \tag{11}$$

under the hypotheses $M < K$ and $H_x \in \mathbb{R}^{M \times K}$ of full rank.

In [2] Newton's method is used for the minimization. A drawback of this method is that its region of convergence depends on the condition number of the Hessian, which is noisy. Hence, the convergence properties need investigation and no theory exists for the CTLS in this case. Simulations have shown instability of the algorithm. In [6], the Hessian is analysed and a particular minimization based on the scaled conjugate gradient technique is proposed.

4. CTLS estimation of the G matrix

As seen in eq.(3), every row z_i is function of the set of variables $s = \{x, y, x', y'\}$ even if the points are represented as N-vectors. Consider M matches. Given the matrix $Z = [Z_1, \ldots, Z_9] \in \mathbb{R}^{M \times 9}$, define the corresponding vector $\text{vec}(Z) = \left[Z_1^T, \cdots, Z_9^T\right]^T$ consisting of the column vectors of the Z matrix that are lined up. Define as $\text{vec}(Z(\varepsilon))$ the noise corrupted version of $\text{vec}(Z)$, viz. $\text{vec}(Z(\varepsilon)) = \text{vec}(Z) + \delta_Z$ where δ_Z is the corresponding first order noise vector. Define the covariance matrix of Z as $\Gamma_Z = E\left[\delta_Z \delta_Z^T\right]$ and the covariance matrix of the row z_i as $\Gamma_{z_i} = E\left[\delta_{z_i} \delta_{z_i}^T\right]$ where δ_{z_i} is the first order noise vector of z_i. This first order perturbation is given by:

$$\delta_{z_i} = \left[\nabla Z_{i1}|_{s_i} \cdot \delta_{s_i}, \cdots, \nabla Z_{i9}|_{s_i} \cdot \delta_{s_i}\right]^T = (J_{s_i} z_i)\, \delta_{s_i} \tag{12}$$

where δ_{s_i} is the perturbation vector $\left[\delta_{x_i}, \delta_{y_i}, \delta_{x'_i}, \delta_{y'_i}\right]^T$, Z_{ij} is the ij-th element of matrix Z and $J_{s_i} z_i$ is the Jacobian of z_i w.r.t. s, computed at s_i. Assume the noise components of the image points in the two images are i.i.d., zero mean with common isotropic variance σ^2 and express the perturbation vector as:

$$v = \delta_s = \left[\delta_{s_1}^T, \cdots, \delta_{s_M}^T\right]^T \in \mathbb{R}^{4M} \tag{13}$$

Hence:

$$\Gamma_{z_i} = E\left[\delta_{z_i} \delta_{z_i}^T\right] = \sigma^2 (J_{s_i} z_i)(J_{s_i} z_i)^T = \sigma^2 M_i \tag{14}$$

$$\Gamma_Z = \text{diag}\ (\Gamma_{z_1}, \Gamma_{z_2}, \ldots, \Gamma_{z_M}) \in \mathbb{R}^{9M \times 9M} \tag{15}$$

In the case of normalization of matrix G, by imposing $g_9 = -1$ and points not represented by N-vectors, i.e. $q_i = [x_i, y_i, 1]^T$ and $q'_i = [x'_i, y'_i, 1]^T$, matrix M_i is given by:

$$M_i = \begin{bmatrix} q_i q_i^T & 0 & 0 \\ 0 & q_i q_i^T & 0 \\ 0 & 0 & 0 \end{bmatrix} + \begin{bmatrix} x'_i x'_i \Sigma & x'_i y'_i \Sigma & x'_i \Sigma \\ x'_i y'_i \Sigma & y'_i y'_i \Sigma & y'_i \Sigma \\ x'_i \Sigma & y'_i \Sigma & \Sigma \end{bmatrix} \tag{16}$$

where $0 \in \mathbb{R}^{3X3}$ and $\Sigma = \text{diag}\ (1,1,0)$.

Define as $\Delta Z = [\Delta Z_1, \ldots, \Delta Z_9]$ the first order perturbation of matrix Z. Recalling the expression (13) for v, it follows ($k = 1, \ldots, 9$):

$$\Delta Z_k = \text{diag}\ (\nabla Z_{1k}\,|_{s_1}, \ldots, \nabla Z_{Mk}\,|_{s_M})\, \delta_s = F_k v \in \mathbb{R}^{Mx4M} \qquad (17)$$

Hence, using the notation of sec.3,

$$H_g = \sum_{i=1}^{9} g_i F_i = \text{diag}\ \left(H_{g_1}^T, \ldots, H_{g_M}^T\right) \in \mathbb{R}^{Mx4M} \qquad (18)$$

$$H_{g_i}^T = \beta_i = (\nabla r_i\,|_{s_i})_{const\, g} = \frac{\partial\left(z_i^T g\right)}{\partial s} = (J_{s_i} z_i)^T g \in \mathbb{R}^4 \qquad (19)$$

and $r_i = z_i^T g$ is the i-th residual. It follows:

$$H_g H_g^T = \text{diag}\ (\alpha_1, \alpha_2, \ldots, \alpha_M) \in \mathbb{R}^{M\, x\, M} \qquad (20)$$

where

$$\alpha_i = \beta_i^T \beta_i = \left\| (\nabla r_i\,|_{s_i})_{\text{const}\, g} \right\|_2^2 \qquad (21)$$

The generalization of $E(g)$ (see eq.(10)), i.e. without setting any component of g to a fixed value, is given by:

$$E(g) = g^T Z^T \left(H_g H_g^T\right)^{-1} Z g = \sum_{i=1}^{M} E^{(i)}(g) \qquad (22)$$

where

$$E^{(i)}(g) = \frac{r_i^2}{\beta_i^T \beta_i} = \frac{r_i^2}{\left\| (\nabla r_i\,|_{s_i})_{const\, g} \right\|_2^2} \equiv (d_i)^2 \qquad (23)$$

Hence, the Sampson distance d_i has been refound from a different and original point of view (CTLS) and, above all, for the general case of image point representation.

5. Neurons for the CTLS problem

The CTLS method leads to the minimization of (22) which can be used as cost function of a neural learning law, which represents a gradient flow. Several learning laws can be proposed according to the type of normalization of the coordinates of the points of the matches. All neurons are linear, i.e. are represented by $y(t) = w^T(t)x(t)$ where $w(t)$ is the weight vector, $x(t)$ is the input vector and $y(t)$ is the scalar output. The weight vector converges to the required solution. The neural energy function, computed for each match i, is given by (23) where the residual r_i and the coefficient β_i depend on the type of normalization and so change for each learning law. These neurons can work either *on-line* (at each iteration a row z_i, computed from the corresponding match, is fed to the neuron) or in *batch* mode (at each epoch, all matrix Z is fed to the neuron).

5.1 The CTLS EXIN neuron

The CTLS EXIN neurons [6] differ for the choice of the normalized parameter. Here, the case of $g_9 = -1$ is taken in consideration. For each match $q_i \leftrightarrow q_i'$ (third component equal to 1), the residual r_i and β_i are given by:

$$r_i = -Z_{i9} + \hat{z}_i^T w \tag{24}$$

$$\beta_i = \left[x_i' g_1 + y_i' g_4 + g_7, x_i' g_2 + y_i' g_5 + g_8, x_i g_1 + y_i g_2 + g_3, x_i g_4 + y_i g_5 + g_6\right]^T \tag{25}$$

where $w = [g_1, \cdots, g_8, -1]^T$ and $\hat{z}_i$ is equal to z_i except for the last component which is set to zero. In this case, the learning law is given by:

$$w(t+1) = w(t) - \alpha(t)\,\bar{\gamma}_i(t)\hat{z}_i + \left[\alpha(t)\,\bar{\gamma}_i^2(t)\right] M_i w(t) \tag{26}$$

where $\alpha(t)$ is the learning rate, M_i is given by (16) and $\bar{\gamma}_i = \frac{r_i}{\beta_i^T \beta_i}$. This learning law derives from the TLS EXIN learning law by a changing of adaptive metrics ($\bar{\gamma}_i(t)$ and M_i depend on the i-th match input).

5.2 The CTLSn EXIN neuron

The CTLSn EXIN neurons require the N-vector normalization of the match inputs (preprocessing), i.e. the N-vectors $\mathbf{m}_i = [U_i, V_i, K_i]^T = \frac{q_i}{\|q_i\|_2}$ and $\mathbf{m}_i' = [U_i', V_i', K_i']^T = \frac{q_i'}{\|q_i'\|_2}$. Unless the CTLS EXIN case, the normalization on the vector g does not influence the stability and the behavior of the neural law [6]. Hence, consider, without loss of generality, $g_9 = -1$. The only difference from the CTLS EXIN case stands in $\beta_i = [\beta_{i1}, \beta_{i2}, \beta_{i3}, \beta_{i4}]^T$ which is expressed as:

$$[\beta_{i1},\ \beta_{i3},\ \beta_{i2},\ \beta_{i4}]^T = \begin{bmatrix} J_i G^T K_i \mathbf{m}_i' \\ J_i' G K_i' \mathbf{m}_i \end{bmatrix}$$

where $J_i = J(\mathbf{m}_i) = \begin{bmatrix} K_i^2 + V_i^2 & -U_i V_i & -U_i K_i \\ -U_i V_i & K_i^2 + U_i^2 & -(V_i K_i) \end{bmatrix} \in \mathbb{R}^{2\times 3}$ and $J_i' = J(\mathbf{m}_i')$. The CTLSn EXIN learning law is given by (26), where

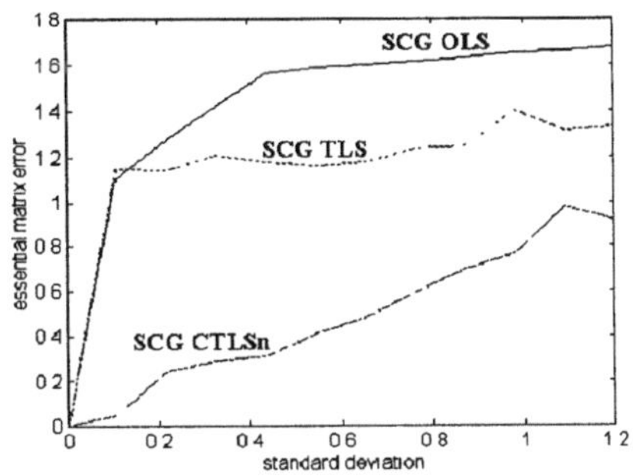

Figure 1. Effect of noise in case of N-vector normalization.

$M_i w(t) = \left(\frac{\partial \beta_i}{\partial g}\right)^T \beta_i$ and

$$\left[\frac{\partial \beta_{i1}}{\partial g}, \frac{\partial \beta_{i3}}{\partial g}, \frac{\partial \beta_{i2}}{\partial g}, \frac{\partial \beta_{i4}}{\partial g}\right]^T = \left[\begin{array}{c} K_i \mathbf{m}_i'^T \otimes J_i \\ K_i' J_i' \otimes \mathbf{m}_i^T \end{array}\right] \in \mathbb{R}^{4\times 9} \qquad (27)$$

being $\otimes$ the Kronecker product (recall that for this choice of normalization, the last column of (27) must be set to 0).

6. Simulations

The following simulations deal with synthetic images and are extracted from [6], which also presents results from real images. The camera is normalized and only egomotion is considered. The Frobenius norm of the difference between the true and the computed essential matrices, before the singularity constraint enforcement, is taken as figure of merit. The effect of image noise and mismatch is analysed. In the second case, a certain fraction of the given feature points is randomly permuted to obtain mismatch in feature correspondences. These mismatches must be considered outliers: indeed, mismatches are quite likely to occur among the neighboring points and so these errors reflect the worst case scenario. Every epoch of the learning is composed of all the extracted matches. All methods are accelerated by the Scaled Conjugate Gradient (SCG) technique [6]. Fig. 1 shows the accuracy of the estimation in the case of white Gaussian noise added to the image points. For each value of standard deviation the average over 50 experiments is plotted. The image is very large and so the image points are normalized (here N-vectors are used). SCG OLS and SCG TLS EXIN give similar results because they are not optimized for this kind of normalization. This is in contrast with the results of Chaudhuri and Chatterjee [4] who have used plain TLS using N-vectors. SCG CTLSn EXIN gives the best results.

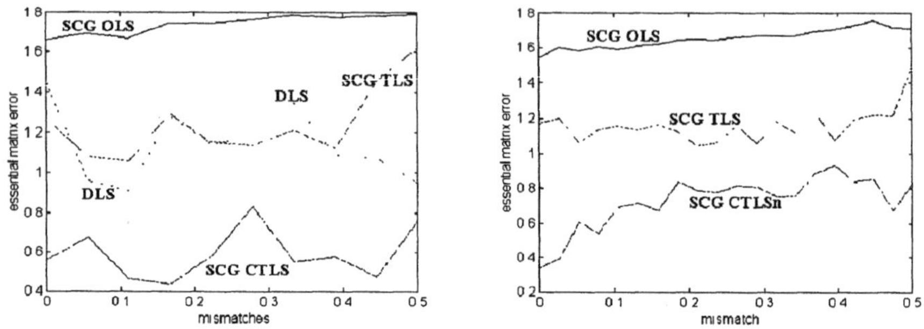

Figure 2. Left: Effect of mismatch in case of N-vector normalization; Right: Effect of mismatch without image point normalization.

Fig. 2 deals with a medium size image. Each point in the plot is the average over 50 experiments. The image point standard deviation is equal to 0.5. Notice the inaccuracy of SCG TLS EXIN and DLS, also because of the noise level and the image size, and the robustness of SCG CTLS EXIN until 50% of mismatch. Fig. 2 deals with a large size image with the same noise level as before. The average is computed over 100 experiments. Notice the better performance of SCG CTLSn EXIN, whose estimation error is 0.8 at 50% of mismatch.

7. Conclusions

In the literature about the essential matrix technique, the problem of dealing with noisy data and mismatches has been often underestimated and has given the erroneous idea this method is inappropriate. Instead, in this paper, the method is correctly classified as structured Total Least Squares (CTLS). This novel viewpoint revalues the essential matrix and extends the nonlinear gradient algorithm (justification of the N-vector preprocessing). At this proposal, according to the size of the image, the CTLS and CTLSn EXIN linear neurons have been introduced. They are very robust, even in the presence of a 50% mismatch. They can also be implemented together with other neural networks (see [6]) in order to build a completely neural SFM module which is able to recover the motion and scene parameters by exploiting the parallelism and the capability of tracking slowly varying inputs which are typical features of the neural networks. Furthermore, their robustness justifies the implementation of the case deletion diagnostics in a neural network (CASEDEL EXIN, [9]) composed of these neurons.

References

[1] T.J. Abatzoglou and J.M. Mendel. Constrained total least squares. *Proc. ICASSP*, pages 1485–1488, 1987.

[2] T.J. Abatzoglou, J.M. Mendel, and G.A. Harada. The constrained total least squares technique and its application to harmonic superresolution. *IEEE Trans. Signal Processing*, 39(5):1070–1087, May 1991.

[3] F. Bookstein. Fitting conic sections to scattered data. *Computer Vision, Graphics and Image Processing*, 9:56–71, 1979.

[4] S. Chaudhuri and S. Chatterjee. Performance analysis of total least squares methods in three-dimensional motion estimation. *IEEE Trans. On Robotics and Automation*, 7(5):707–714, October 1991.

[5] S. Chaudhuri, S. Sharma, and S. Chatterjee. Recursive estimation of motion parameters. *Computer Vision and Image Understanding*, 64(3):434–442, November 1996.

[6] G. Cirrincione. *A Neural Approach to the Structure from Motion Problem*. PhD thesis, LIS INPG Grenoble, December 1998.

[7] G. Cirrincione. Neural minor component analysis and tls. In S. Van Huffel and P. Lemmerling, editors, *Total Least Squares and Errors-in-Variables Modeling: Analysis, Algorithms and Applications*, 2002.

[8] G. Cirrincione and M. Cirrincione. Linear system identification using the tls exin neuron. In *NEURAP 98*, Marseille (France), March 1998.

[9] G. Cirrincione and M. Cirrincione. Robust neural approach for the estimation of the essential parameters in computer vision. *International Journal on Artificial Intelligence Tools*, 8(3), September 1999.

[10] R.I. Hartley. In defense of the eight-point algorithm. *IEEE Transactions on Pattern Analysis and Machine Intelligence*, 19(6):580–593, June 1997.

[11] S. Van Huffel and J. Vandewalle. *The Total Least Squares Problems: Computational Aspects and Analysis*. Frontiers in Applied Mathematics. SIAM Philadelphia, 1991.

[12] K. Kanatani. *Geometric Computation for Machine Vision*. The Oxford engineering science series. Clarendon Press Oxford, 1993.

[13] Q.T. Luong, R. Deriche, O. Faugeras, and T. Papadopoulo. On determining the fundamental matrix: Analysis of different methods and experimental results. Technical Report 1894, INRIA, April 1993.

[14] M. Muhlich and R. Mester. The role of total least squares in motion analysis. In H. Burkhardt, editor, *Proc. European Conf. Computer Vision*, pages 305–321. Springer Lecture Notes on Computer Science, June 1998.

[15] P.H.S. Torr and D.W. Murray. The development and comparison of robust methods for estimating the fundamental matrix. *International Journal of Computer Vision*, 24(3):271–300, 1997.

[16] J. Weng, T.S. Huang, and N. Ahuja. *Motion and Structure from Image Sequences*. Springer Series in Information Sciences. Springer-Verlag, 1993.

Adaptive filter, 191
Astronomy, 375
Audio coding, 331
Bayesian methods, 121
Behavioural models, 281
Bounded uncertainties
  ellipsoidal, 213
Computer vision, 251, 385
Cox model, 179
Damped sinusoids, 321, 331
Decimation, 321
Displacement rank, 79, 94
Downdating, 47
Errors-in-variables model, 121, 155
  bilinear, 155
  dynamic, 271, 281
  linear, 155, 213
  nonlinear, 121, 145, 225
  semi-linear, 165
Errors-in-variables, 179
  linear, 15
  polynomial, 131
Essential/fundamental matrix, 155, 385
Estimating function, 15, 145
Frisch scheme, 271
Functional relationship, 121, 131, 145
Galactic kinematics, 375
Generalized linear model, 121
Hankel TLS, 47, 293, 321, 331
Hankel/Toeplitz matrix, 47, 79, 94, 321, 331, 365
Image reconstruction, 365
Impulse noise, 341
Information geometry, 15
Information retrieval, 353
Instrumental variables, 271
Kalman filter, 191, 281
Kernel regression, 121
Latency error, 107
Least squares, 203
  adjusted, 131, 155
  data, 25
  extended, 107
  ordinary, 25, 35
  recursive, 341
  residual, 35
  structural, 131
Line segment fitting, 251, 261
Measurement error, See Errors-in-variables
Method of moments, 251
Minimum contrast estimation, 145
Minor component analysis, 235, 385
Orthogonal distance regression, 225
Overdetermined, 2
Parametric quadratic form, 67
Parametric uncertainty, 191
Prediction error, 271
Rank revealing decomposition, 47
Regularizaion, 203
Regularization, 57, 191, 321, 365
Robust fitting, 213
Semiparametric approach, 15
SIMEX, 121
Singular value decomposition, 25, 35, 47, 321, 353
  Riemannian, 293, 353
Sparse QR factorization, 225
Spectral estimation, 321
Splines, 121
Structural relationship, 15, 121, 131, 251
Subspace estimation, 47, 321, 331
Survival analysis, 179
System identification, 107, 271, 293, 305
Total L_p norm approximation, 145, 213, 341
Total least mean squares, 341
Total least squares, 15, 57, 131
  constrained, 365, 385
  elementwise-weighted, 67, 155, 293
  extended structured, 107
  generalized, 305
  scaled, 25, 35
  structured, 79, 94, 107, 225
Updating, 47

MIX
Papier aus verantwortungsvollen Quellen
Paper from responsible sources
FSC® C105338

If you have any concerns about our products,
you can contact us on
ProductSafety@springernature.com

In case Publisher is established outside the EU,
the EU authorized representative is:
Springer Nature Customer Service Center GmbH
Europaplatz 3, 69115 Heidelberg, Germany

Printed by Libri Plureos GmbH
in Hamburg, Germany